Sensors

Volume 4

Thermal Sensors

Sensors

A Comprehensive Survey

Edited by
W. Göpel (Universität Tübingen, FRG)
J. Hesse (Zeiss, Oberkochen, FRG)
J. N. Zemel (University of Pennsylvania, Philadelphia, PA, USA)

Published in 1989:
Vol. 1 Fundamentals and General Aspects
(Volume Editors: T. Grandke, W. H. Ko)
Vol. 5 Magnetic Sensors
(Volume Editors: R. Boll, K. J. Overshott)

Remaining volumes of this closed-end series will be published by 1992:

Vol. 2 Chemical and Biochemical Sensors (scheduled for 1991)
Vol. 3 Mechanical Sensors (scheduled for 1992)
Vol. 4 Thermal Sensors (scheduled for 1990)
Vol. 6 Optical Sensors (scheduled for 1991)

Distribution

VCH, P. O. Box 101161, D-6940 Weinheim (Federal Republic of Germany)

Switzerland: VCH, P. O. Box, CH-4020 Basel (Switzerland)

United Kingdom and Ireland: VCH (UK) Ltd., 8 Wellington Court, Wellington Street, Cambridge CB1 1HZ (England)

USA and Canada: VCH, Suite 909, 220 East 23rd Street, New York, NY 10010-4606 (USA)

ISBN 3-527-26770-0 (VCH, Weinheim) ISBN 0-89573-676-4 (VCH, New York)

Sensors

A Comprehensive Survey

Edited by
W. Göpel, J. Hesse, J. N. Zemel

Volume 4

Thermal Sensors

Edited by
T. Ricolfi, J. Scholz

Series Editors:
Prof. Dr. W. Göpel
Institut für Physikalische und
Theoretische Chemie der Universität
Auf der Morgenstelle 8
D-7400 Tübingen, FRG

Prof. Dr. J. Hesse
Carl Zeiss,
ZB „Entwicklung“
Postfach 1380
D-7082 Oberkochen, FRG

Prof. Dr. J. N. Zemel
Center for Sensor Technology
University of Pennsylvania
Philadelphia, PA 19104-6390, USA

Volume Editors:
Dr. T. Ricolfi
CNR, Istituto di Metrologia
Strada delle Cacce 73
I-10135 Torino, Italy

Dr. J. Scholz
Sensycon GmbH
Postfach 2203
D-6450 Hanau, FRG

Published jointly by
VCH Verlagsgesellschaft mbH, Weinheim (Federal Republic of Germany)
VCH Publishers Inc., New York, NY (USA)

Editorial Directors: Dipl.-Phys. W. Greulich, Dipl.-Chem. Dr. M. Weller, N. Banerjea-Schultz
Production Manager: Dipl.-Wirt.-Ing. (FH) H.-J. Schmitt

Library of Congress Card No.: applied for

British Library Cataloguing-in-Publication Data:
Sensors.
Vol. 4, Thermal sensors.
1. Sensors
I. Göpel, W. (Wolfgang) II. Hesse, J. III. Zemel, J.
N. (Jay N.) IV. Ricolfi, T. V. Scholz, J.
620.0044
ISBN 3-527-26770-0

Deutsche Bibliothek Cataloguing-in-Publication Data:
Sensors : a comprehensive survey / ed. by W. Göpel ... –
Weinheim ; Basel (Switzerland) ; Cambridge ; New York, NY :
VCH.
NE: Göpel, Wolfgang [Hrsg.]
Vol. 4. Thermal sensors / ed. by T. Ricolfi ; J. Scholz. – 1990
ISBN 3-527-26770-0 (Weinheim ...)
ISBN 0-89573-676-4 (New York)
NE: Ricolfi, Teresio [Hrsg.]

Printed on acid-free paper

Composition: Filmsatz Unger + Sommer GmbH, D-6940 Weinheim.
Printing: Diesbach Medien, D-6940 Weinheim.
Bookbinding: Großbuchbinderei J. Schäffer, D-6718 Grünstadt.
Printed in the Federal Republic of Germany

Preface to the Series

The economic realities of productivity, quality, and reliability for the industrial societies of the 21st century are placing major demands on existing manufacturing technologies. To meet both present and anticipated requirements, new and improved methods are needed. It is now recognized that these methods must be based on the powerful techniques employing computer-assisted information systems and production methods. To be effective, the measurement, electronics and control components, and sub-systems, in particular sensors and sensor systems, have to be developed in parallel as part of computer-controlled manufacturing systems. Full computer compatibility of all components and systems must be aimed for. This strategy will, however, not be easy to implement, as seen from previous experience. One major aspect of meeting future requirements will be to systematize sensor research and development.

Intensive efforts to develop sensors with computer-compatible output signals began in the mid 1970's; relatively late compared to computer and electronic measurement peripherals. The rapidity of the development in recent years has been quite remarkable but its dynamism is affected by the many positive and negative aspects of any rapidly emerging technology. The positive aspect is that the field is advancing as a result of the infusion of inventive and financial capital. The downside is that these investments are distributed over the broad field of measurement technology consisting of many individual topics, a wide range of devices, and a short period of development. As a consequence, it is not surprising that sensor science and technology still lacks systematics. For these reasons, it is not only the user who has difficulties in classifying the flood of emerging technological developments and solutions, but also the research and development scientists and engineers.

The aim of "Sensors" is to give a survey of the latest state of technology and to prepare the ground for a future systematics of sensor research and technology. For these reasons the publishers and the editors have decided that the division of the handbook into several volumes should be based on physical and technical principles.

Volume 1 (editors: T. Grandke/Siemens AG (FRG) and W. H. Ko/Case Western Reserve University (USA)) deals with general aspects and fundamentals: physical principles, basic technologies, and general applications.

Volume 2 (editors: W. Göpel/Tübingen University (FRG), L. Lundström/Linköping University (Sweden), T. A. Jones†/Health and Safety Executive (UK), M. Kleitz/LIES-ENSEEG (France) and T. Seiyama/Kyushu University (Japan)) concentrates on chemical and biochemical sensors.

Volume 3 (editors: N. F. de Rooij/Neuchâtel University (Switzerland), B. Kloeck/Hitachi Ltd. (Japan) and H. H. Bau/University of Pennsylvania (USA)) presents mechanical sensors.

Volume 4 (editors: J. Scholz/Sensycon GmbH (FRG) and T. Ricolfi/Consiglio Nazionale Delle Ricerche (Italy)) refers to thermal sensors.

Volume 5 (editors: R. Boll/Vacuumschmelze GmbH (FRG) and K. J. Overshott/Brighton Polytechnic (UK)) deals with magnetic sensors.

Volume 6 (editors: E. Wagner and K. Spenner/Fraunhofer-Gesellschaft e. V. (FRG) and R. Dändliker/Neuchâtel University (Switzerland)) treats optical sensors.

Each volume is, in general, divided into the following three parts: specific physical and technological fundamentals and relevant measuring parameters; types of sensors and their technologies; most important applications and discussion of emerging trends.

It is planned to close the series with a volume containing a cumulated index.

The series editors wish to thank their colleagues who have contributed to this important enterprise whether in editing or writing articles. Thank is also due to Dipl.-Phys. W. Greulich, Dr. M. Weller, and Mrs. N. Banerjea-Schultz of VCH for their support in bringing this series into existence.

W. Göpel, Tübingen J. Hesse, Oberkochen J. N. Zemel, Philadelphia, PA

August 1990

Preface to Volume 4 of "Sensors"

According to all recent major market surveys, sales of temperature sensors are far higher in numbers than those of other kinds. Due to the wide variety of applications and the wide range of temperatures to be measured and monitored, a multitude of varying physical principles is applied in temperature sensing. The aim of this book is to describe the constructional and applicative aspects of thermal sensors while preserving a rigorous treatment of the underlying physical principles. Emphasis was laid on those principles which are established in the fields of industrial temperature measurement. Other principles which may increase in importance in the future are outlined in the introductive chapter.

General considerations of the physics of temperature measurement are dealt with in Chapter 2. Part of this chapter is devoted to the description of the new International Temperature Scale of 1990 (ITS-90). As reference to the new scale is made wherever applicable in the successive chapters, this book is likely to be the first to have been updated in this respect. (Unfortunately, industrial standards for temperature sensors such as resistance thermometers and thermocouples are not yet adapted to the new temperature scale and therefore all references made to these standards are still on the basis of the former IPTS-68.)

Chapter 2 also includes a systematic survey of the physical principles of all important thermal sensors. Specific categories of sensors are treated in Chapters 3 to 9. The three main categories of temperature sensors. ie, resistance thermometers, thermocouples, and radiation thermometers are thoroughly discussed in Chapters 3 to 5. In the chapter on radiation thermometers greater emphasis has been placed on applications. This is due to the fact that the modalities of use are often more important than the sensor itself. Chapters 6 and 7 are devoted to noise and acoustical thermometers, respectively. These two sensor types are now the subject of renewed interest due to some recent advances and to their ability to solve special measurement problems in nuclear and very high temperature applications, for example. Heat-flow and mass-flow sensors, ie, those thermal sensors which differ from thermometers, are described in Chapters 8 and 9. The usefulness of these sensors is mainly found in the wide variety of energy-related applications where they can make important contributions.

Examples of applications are given in Chapters 1 to 9. The applications of thermal sensors in some specific fields, namely, process control, automotive technology and cryogenics are systematically analyzed in Chapters 10 to 12.

When we started our task we were conscious of the difficulties we were going to meet in coordinating the work of a number of colleagues with different backgrounds and perhaps points of view. Now that the task is completed, we realize that it has been a stimulating and gratifying experience and this is due to the enthusiasm and patience of both the authors and the editorial staff to whom we are pleased to express our sincere thanks.

T. Ricolfi and J. Scholz Torino and Hanau, August 1990

Contents

List of Contributors

Dipl.-Phys. Hans Jürgen Aulfes
Universität GH Paderborn
Fachbereich Physik
Warburger Str. 100
D-4790 Paderborn, FRG
T: (5251) 602735

Roy Barber, B. Sc.
Land Instruments International Ltd.
Dronfield, Sheffield S18 6DJ, UK
T: (246) 417691
Fax: (246) 410585

Dr. Ronald E. Bedford
Division of Physics
National Research Council
Ottawa, K1A OR6, Canada
T: (613) 9547708
Fax: (613) 9521394
Tx.: 053-4322

Dr. Heinz Brixy
Institut für Reaktorentwicklung
Postfach 1913
D-5170 Jülich, FRG
T: (2461) 616162
Fax: (2461) 615327

Prof. Luigi Crovini
Consiglio Nazionale delle Ricerche
Istituto di Metrologia „G. Colonetti"
Strada delle Cacce 73
I-10135 Torino, Italy
T: (11) 39771
Fax (11) 346761

Dr. Ingo Gessler
Degussa AG
FB Lizenzen
Postfach 110533
D-6000 Frankfurt 11, FRG
T: (69) 2183197
Fax: (69) 2182854

Ing. Frederik van der Graaf
TNO Institute of Applied Physics
P. O. Box 155
2600 AD Delft, The Netherlands
T: (15) 692117
Fax: (15) 692111
Tx: 38091 tpd dt nl

Dr. Martin Hohenstatt
Degussa AG
FB Forschung Metall
Postfach 1345
D-6450 Hanau 1, FRG
T: (6181) 593551
Fax: (6181) 593030

Dr.-Ing. Gerd Kleinert
VDO Adolf Schindling AG
Sodener Str. 9
D-6231 Schwalbach a. Ts., FRG
T: (6196) 872127
Tx: 4072441 vdo d

Franco Pavese
Consiglio Nazionale delle Ricerche
Istituto di Metrologia „G. Colonetti"
Strada delle Cacce 73
I-10135 Torino, Italy
T: (11) 39771
Fax: (11) 346761

Dipl.-Phys. Wolfgang Porth
VDO Adolf Schindling AG
Sodener Str. 9
D-6231 Schwalbach a. Ts., FRG
T: (6196) 872385
Tx: 4072441 vdo d

Dipl.-Phys. Hubert Quint
Universität GH Paderborn
Fachbereich Physik
Warburger Str. 100
D-4790 Paderborn, FRG
T: (5251) 602735

Dr. Teresio Ricolfi
Consiglio Nazionale delle Ricerche
Istituto di Metrologia „G. Colonetti"
Strada delle Cacce 73
I-10135 Torino, Italy
T: (11) 39771
Fax: (11) 346761

Dr.-Ing. Jörg Scholz
Sensycon GmbH
Bereich Sensorik
Postfach 2203
D-6450 Hanau 1, FRG
T: (6181) 369487
Fax: (6181) 369392

Dipl.-Ing. Herbert Vanvor
Haydnstr. 17
D-6450 Hanau 1, FRG
T: (6181) 84901

Prof. Dr. Horst Ziegler
Universität GH Paderborn
Fachbereich Physik
Warburger Str. 100
D-4790 Paderborn, FRG
T: (5251) 602735

1 General Aspects

Teresio Ricolfi, CNR-Istituto di Metrologia "G. Colonnetti," Torino, Italy
Joachim Scholz, Sensycon GmbH, Hanau, FRG

Contents

1.1 Introductory Remarks

The basic principles of thermal sensors are well established since several years. For this reason, spectacular advances in this area have not been made in recent years nor are to be expected in the near future. Nevertheless, a great deal of work has been and is currently being made in scientific and industrial laboratories to provide adequate solutions to modern measurement needs.

Practical needs and recent solutions are shortly reviewed in this chapter. This also offers the opportunity to outline the present trends in thermal sensors and to give some hints on modern approaches like, for example, fiber-optic techniques that are not treated elsewhere in this volume.

1.2 General Needs in Practical Measurements

The growth of process control and automation, the increasing significance of quality control, the recourse to sophisticated technologies in manufacturing processes, the increasing concern to safety problems, are some major aspects of modern manufacturing and technological processes. They entail new requirements for the quality of sensors and measurement approaches.

1.2.1 Accuracy

An increased *accuracy* is the main requirement originating from a more rigorous quality control or from the adoption of advanced technologies like, for example, semiconductor processing (silicon, germanium, gallium-indium arsenic mixtures).

The measurement accuracy is partly determined by the intrinsic features of the sensor itself (sensitivity, repeatibility, stability) and, in many cases to a larger extent, by the conditions of measurement (thermal gradients, aggressive atmospheres, electromagnetic interferences, etc.).

High sensitivity requirements have produced, for example, a renewed interest for the quartz thermometer in applications up to 300 °C and the development of special fiber-optic thermometers in the high temperature range up to 2000 °C.

Insofar as stability and repeatibility are primary concerns, the platinum resistance thermometer (PRT) appears to be the optimum choice in many applications up to about 850 °C. In the high temperature range great improvements in contactless measurements have been derived from the use of stable silicon photodetectors in radiation thermometers.

As to the effect of environmental conditions, again PRTs represent a good choice. In fact, they are less affected by thermal gradients than thermocouples and less expensive than platinum metal thermocouples that are required to withstand most aggressive atmospheres. Some of the advantages of PRTs can also be ascribed to noise thermometers that, although much less diffused, can be of interest for some applications (eg, in presence of neutron irradiation) in a wide temperature range up to 2000 °C. However, electrical thermometers may be of limited use in the presence of electromagnetic interferences (eg, in microwave heating processes). In these cases advantages are offered by the recently developed thermometers based on optical effects (see Section 1.3).

1.2.2 Reliability

In many manufacturing processes, in order to reduce downtime and maintenance requirements, more emphasis is being placed on improving instrument *reliability* than on the measurement accuracy. This is particularly true for automated processes. It is generally recognized that mechanical thermometers (liquid in glass, bimetal, gas, vapor-pressure thermometers) are reliable sensors. Moreover, they are also characterized by intrinsic safety and immunity to electromagnetic interferences. However, they can hardly fulfill the requirements of process automation. So, one of the present points of interest in industrial instrumentation is their replacement with thermometers providing an electrical output.

1.2.3 Interchangeability

Interchangeability of sensors is another important requirement. Many parameters of a control system in an automated process are determined by the characteristics of the sensor being used. So, in case of failure, it should hopefully be replaced by another sensor of identical characteristics. Otherwise, all the control parameters have to be readjusted. A full interchangeability of sensors even from different sources is ensured by internationally recognized standards for commonly used sensors. From this point of view as well, PRTs appear to be a good choice. On the contrary, other thermometers, like the quartz thermometer, generally need an individual calibration, so their interchangeability is low and no standards are available.

1.2.4 Cost

The enhanced performance demands to sensors and their massive use in process control require a *cost reduction* in their fabrication in order to keep the overall cost of a control system within reasonable limits. Mechanical sensors are generally more expensive than electrical and electronic ones, so their replacement also contributes to the objective of cost reduction. As for PRTs, which have been shown to provide the best solution to many requirements, cost reduction is obtained by using thick and thin-film techniques in their fabrication (see Chapter 3). A noticeable cost reduction has also been obtained in infrared thermometers for fixed installation with the use of relatively cheap silicon and pyroelectric detectors and thin-film thermopiles.

1.2.5 Safety

Safety in industrial plants is another requirement of major concern. Examples of measures dictated by safety requirements are the replacement of mercury thermometers in the food industry and of electrical thermometers without special safety features where explosion hazards are prevalent. A valuable alternative in the latter case is offered by fiber-optic thermometers.

1.2.6 Technical diagnostics

The efficiency and safety in industrial plants are often determined by an accurate *technical diagnostic.* The measurement of thermal contours in tool machines, the monitoring of motor winding temperatures, or the location of defective components in electronic circuits are examples of technical diagnostics concerning thermal measurements. One particular aspect of technical diagnostic refers to the capability of a measuring systems of checking its own integrity, including calibration data. However, these studies are in their early stage insofar as thermal sensors are concerned.

1.3 Developments in Sensors and Measurement Techniques

Some recent achievements refer to the development of new sensors that, although based on well known principles of measurement, nevertheless are better suited to overcome some typical limitations of other sensors of the same type. New thermocouples and infrared thermometers may be included in this category.

A second category of achievements has been obtained with the application of modern technologies or devices to the fabrication of sensors or to the setting up of measuring systems. Examples are the use of thick- and thin-film techniques, micromachining, fiber optics and microprocessors.

1.3.1 New Thermocouples and Infrared Thermometers

Good results have been obtained with Nicrosil/Nisil, Pt/Au and Pt/Pd thermocouples. The Nicrosil/Nisil thermocouple is characterized by a better stability between 300 °C and 600 °C and above 1000 °C as compared with the popular type K thermocouple. The Pt/Au thermocouple is particularly suited for precision measurements due to its high reproducibility that is better than the ones of the various platinum-rhodium/platinum thermocouples. In fact, reproducibilities of the order of ± 0.02 °C have been found between 0 °C and 1000 °C [1]. Further benefits may stem from the high homogeneity of the wire materials of pure-metal thermocouples. As compared with alloy thermocouples, they should be much less affected by thermal gradients along their legs.

A strong limitation in infrared thermometers is the dependence of their readings on the emissivity of the target materials. To overcome this problem, many solutions have been realized but none of them is universally applicable. So, the predominant philosophy in recent years has been to design thermometers individually for each material or category of materials. Representative examples are described in Chapter 5.

1.3.2 Film Techniques

As previously said, cost reduction is one reason for using film techniques in the fabrication of noble-metal thermometers. This is especially true for PRTs that are progressively replacing thermocouples in many applications. Film PRTs are now commercially available and they are characterized by good interchangeability and stability up to 600 °C. The stability is such that after 1000 hours at 600 °C a thin-film PRT can be still within ±0.05 °C of its initial reading at 0 °C.

A second reason for adopting film more than wire configurations refers to contact measurements of surface temperatures where it is important to minimize the disturbance of heat transfer on the surface to be measured. A frequent requirement in surface temperature measurements is also a high speed of measurement. This is the case, for example, for laser or electron beam machining where fast temperature transients occur. Thin-film thermocouples can meet both the requirements of low disturbance and high speed. The problems of contamination from the atmosphere or contacting surfaces and that of variable stresses on the conductors, which are typical in thin-film thermocouples for high temperature applications are going to find satisfactory solutions [2], [3]. As to the achievable measuring speed, a response time of 60 ns has been recently found for a Pt/Ir thermocouple up to 790 °C [4].

1.3.3 Micromachining

Micromachining techniques utilizing the anisotropic etching of silicon are promising possibilities to manufacture miniaturized sensors in large numbers at low cost.

Although the main emphasis in this technology is put on pressure-, acceleration-, and frequency sensors, some prototypes of thermal mass-flow sensors have been developed. One possible application is the airflow-measurement with ignition control systems of automobiles [23].

1.3.4 Fiber-optic Sensors

The use of fiber-optic components is one outstanding feature of the current R & D activity on thermal sensors. Fiber-optic sensors offer immunity to electrical interference and inherent safety. Furthermore, optical fibers are both electrical insulators and poor thermal conductors and these properties can be utilized to insulate the sensor from the monitoring unit or to minimize the heat loss from the object of measurement.

An increasing use of fiber optics is being made in radiation thermometers (see Chapter 5). Current studies are aimed at finding alternative materials to glass and silica (eg, chalcogenide glasses [5]) that can transmit at longer wavelengths up to 10 μm.

A number of physical effects other than thermal radiation have been utilized to make fiber-optic temperature sensors. A comprehensive review on this subject has been recently prepared by *Grattan* [6]. The most promising devices rely upon the following temperature-related physical effects:

1) *Light scattering by cholesteric liquid crystals.* Sensors utilizing this effect can operate over a 10 °C temperature span within the range 10 °C–50 °C depending on the nature of the liquid crystals used [7].

2) *Color change.* This property is exhibited by the so-called thermochromic materials. A device utilizing a cobalt salt solution in isopropyl alcohol and water, a substance showing a marked color change between 25 °C and 75 °C, has been described by *Scheggi* et al. [8].

3) *Change of refractive index.* Different schemes have been tried utilizing this principle. The sensing substance can be either a liquid of high index range with temperature [9], or the plastic cladding of silica fibers [10].

4) *Change of birefringence.* The temperature variation of birefringence in quartz has been utilized in a prototype sensor [11].

5) *Change of light transmission.* Optical absorption techniques have been described utilizing this effect in filters [12] and in GaAs [13] and ruby glass [14] samples.

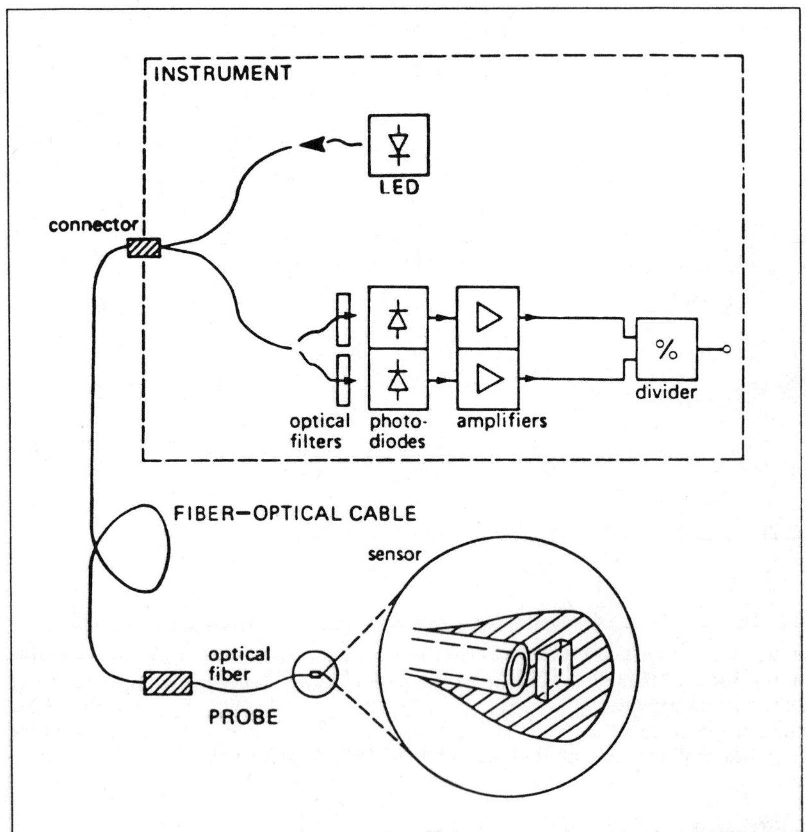

Figure 1-1. Schematic of ASEA type 1010 fiber-optic sensor. A GaAs crystal bound at the end of the fiber-optic probe is irradiated by directly modulating light from a LED. The light emitted due to the fluorescent properties of the crystal is conveyed through the fiber and a beam divider to two photodiodes. The ratio of the signals of the photodiodes depends on the wavelength shift of the emitted light and hence on the temperature of the crystal. This sensor can be used from 0 °C to 200 °C. Resolution and accuracy are 0.1 °C and ±0.5 °C, respectively.

6) *Change of light intensity.* The intensity of light emitted by rare earth phosphors is temperature dependent. This principle has been utilized in the Luxtron type 1000/2000 system [15].

7) *Wavelength shift of fluorescence.* This effect in GaAs crystals has been utilized in the ASEA type 1010 system [16] (Figure 1-1).

8) *Fluorescence decay time.* Many materials, like neodymium or chromium doped glasses and crystals, magnesium fluorogermanate, alexandrite, and others, show a high rate of change with temperature of the fluorescence decay time. This effect has been utilized in commercial instruments by *Luxtron* [17] and *Degussa* (now Sensycon) [18] (Figure 1-2). A major advantage in this technique is that an accurate measurement of the intensity of the excitation light is not needed.

Current problems in the development and use of fiber-optic thermometers refer to inaccuracies due to transmission losses in the fibers, long-term instabilities and cost, with the latter being generally higher than that of conventional measuring systems.

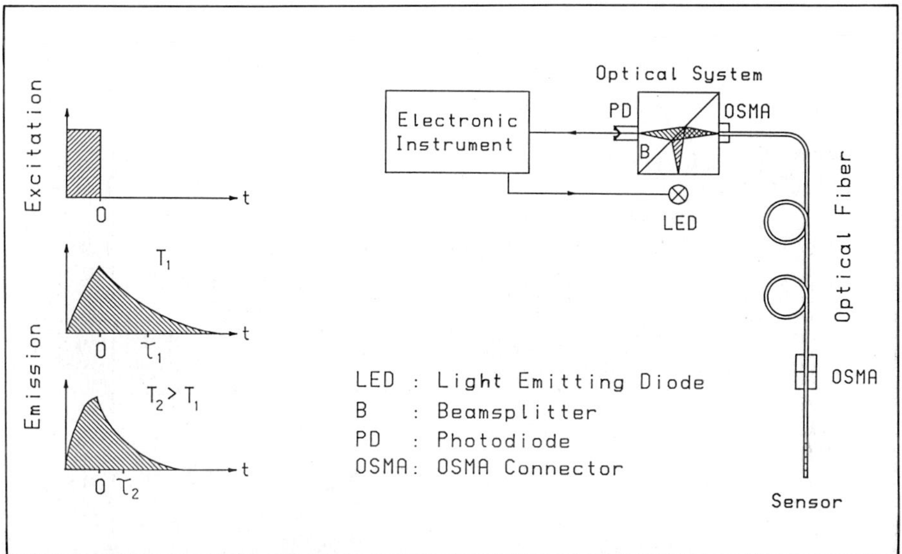

Figure 1-2. Schematic of a sensor based on the decay time of fluorescence (after [18]). The excitation radiation is turned off at time $t = 0$. The emission of fluorescence radiation is shown for two different temperatures T_1 and T_2 of the sensor element. With T_2 being the higher temperature, the corresponding decay constant τ_2 is smaller than τ_1. LED, light emitting diode; *B*, beamsplitter; *PD*, photodiode; OSMA, OSMA connector.

1.3.5 Use of Microprocessors

Microprocessors are other important tools for improving and widening the measurement capabilities [19]. The benefits that can be derived from their use in data acquisition and processing are practically unlimited.

Apart from the general convenience in the automatic acquisition of measurement data, in some cases the acquisition would be impossible without microprocessors. This occurs, for example, in the measurement of very fast temperature transients with thin-film contact thermometers or high-speed radiation thermometers having a response time of the order of 1 μs [20].

As for data processing, the applications of microprocessors include signal linearization, correction for dynamic response, cold junction compensation, implementation of time functions in radiation thermometers, real-time processing of thermal images, and others.

Also the realization of intelligent instrumentation is an important achievement. For example, "cybernetic" radiation thermometers have been realized that can recognize the conditions of measurement (eg, the degree of oxidation) on the basis of some preliminary information on the emissivity of the target material [5]. Another example is that of intelligent temperature transmitters for thermocouples or resistance thermometers [21]. Their capabilities include that of accepting imputs from sensors of different types and of being customer configurable (Figure 1-3).

Figure 1-3. Microprocessor-based temperature transmitter with remote transmitter interface (after [21]).

1.4 Needs for Future Developments

Although adequate solutions have been found for most practical measurement problems, there are applications where the available solutions are not fully satisfactory. In some cases, better solutions have already been envisaged, but they are still needing further developments.

In manufacturing processes, problems still exist in using contact thermometers under chemically aggressive atmospheres (eg, sulphuric atmospheres), in recording temperature profiles inside ducts, in measuring the temperature of molten glass and of metal surfaces during rolling and extrusion processes.

In advanced technological processes, examples of difficulties in temperature measurement are found in laser machining processes, in the adaptive control of tool machines in high-precision machining of large pieces, in cryogenic applications below 10 K under intense magnetic fields (>10 T).

In medicine, the problem of noninvasive internal temperature monitoring has not yet been given a satisfactory solution, although microwave radiometry and ultrasound thermography appears to be the most promising techniques [22].

In instrumentation improvement, there is a need for developing multi-channel systems for fiber-optic sensors and advantages are expected from the replacement of mechanical scanners with sensors arrays in thermal-mapping systems.

In calibration approaches, a growing use of sealed cells for fixed-point temperature calibration is to be expected, since they provide a simple and accurate way for the user to check its own instrumentation for calibration drifts (Figure 1-4). A calibration problem that is still needing a satisfactory solution is the evaluation of the convective heat transfer in radiant flux meters for fire-testing applications.

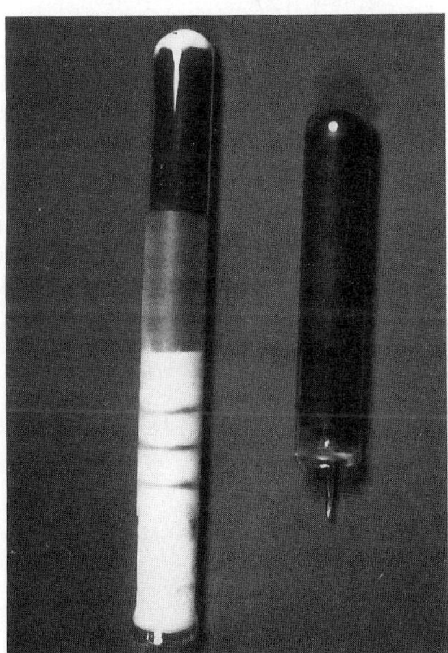

Figure 1-4.
Sealed cells for metal freezing points (courtesy IMGC).

Finally, some urgent needs stem from the adoption of the new International Temperature Scale of 1990 (ITS-90). They are the need for the standardizing organizations of preparing new industrial standards for temperature sensors and the need for the calibration laboratories at any level of taking appopriate measurements to adequate their standards and procedures to the ITS-90.

1.5 References

[1] McLaren, E. H., Murdock, E. G., *The Pt/Au Thermocouple:* Part 1 & Part 2, NRCC/27703; Ottawa: National Research Council, 1987.

[2] Kreider, K. G., *J. Vac. Sci. Techn.* **A4-6** (1986) 2618.

[3] Kreider, K. G., Yust, M., *Proc. of Sensors Expo 88;* Chicago, 1988.

[4] Tong, H. M., Arjavalingam, G., Haynes, R. D., Hyer, G. N., Ritsko, J. J., *Rev. Sci. Instrum.* **58** (1987) 875-877.

[5] Svet, D. Ya., et al., *Proc. Symp. on Major Problems on Present-day Radiation Thermometry;* IMEKO TC 12, Moscow, 1986, pp. 105-112.

[6] Grattan, K. T. V., *Measurement and Control,* **20** (1987) 32-39.

[7] Harmer, L. A., *Measurement and Control,* **15** (1982) 143-151.

[8] Scheggi, A. M., et al., in: *Fiber Optic Sensor,* Arditty, H. J., Jeunhomme, L. B. (eds); Washington: SPIE 1985, Proc. 586, pp. 110-113.

[9] Ramakrishnan, S., Kersten, R. T., *Proc. 2nd Opt. Fibre Sensors Conf.;* IEE, Stuttgart (FRG), 1984, pp. 105-110.

[10] Pinchbeck, D., Kitchen, C. A., *Proc. Conf. Electronics in Oil and Gas Industries;* London, 1985.

[11] Rogers, A. J., *Appl. Opt.* **21** (1982) 882-885.

[12] Christiansen, C., *Ann. Phys. Chem.* **23** (1984) 298.

[13] Kyuma, K., et al., *IEEE J. Quant. Electron.* **QE-18,** (1982) 676-679.

[14] Theocharous, E., *Proc. 1st Opt. Fibre Sensors Conf.;* IEE, London, 1983, p. 10.

[15] Wickensheim, K. A., Alves, R. V., in: *Biomedical Technology;* New York: Alan Lirs, 1982, pp. 547-554.

[16] Ovrén, C., Adolfsson, M., Hok, B., *Proc. Int. Conf. Opt. Tech. in Process Control, The Hague (The Netherlands),* 1983, pp. 67-81.

[17] Wickensheim, K. A. et al., *Proc. Digitech '85;* ISA, Boston, 1985, pp. 87-94.

[18] Fehrenbach, G., *Proc. Sensor 88;* Nuremberg (FRG), 1988, pp. 49-61.

[19] Various contributions in *Proc. Symp. on Microprocessors in Temperature and Thermal Measurement;* IMEKO TC 12, Lodz (Poland), 1989; 175 pages.

[20] Righini, F., Rosso, A., Cibrario, A., *High Temp.-High Press.* **17** (1985) 153-160.

[21] Vincent, P., *Measurement and Control* **20** (1987) 29-31.

[22] Pavese, F., in: *Temperature Measurement. Proc. Intern. Symp. on Temperature Meas. in Industry and Science;* Beijing: China Academic Publ., 1986, pp. 206-211.

[23] Binder, S., „Anforderungen an Mikromechanik-Sensoren im Kraftfahrzeug“, Lecture held during AMA-seminary on "Mikromechanik", Heidelberg, March 14/15, 1989.

2 Physical Principles

RONALD E. BEDFORD, National Research Council of Canada, Ottawa, Canada

Contents

2.1 Introduction

Thermal sensors, as the name implies, are those that sense changes in temperature. Their chief use, then, is for the measurement or control of temperature, and when so-used they are called thermometers. Thermometers may be of two general classes: purely thermal sensors that respond to a temperature change caused by absorbed thermal energy, or photo-detectors that depend upon photon-electron interactions instigated by absorbed photons. Thermal sensors are also used to measure heat fluxes as, for example, energy losses through walls, solar radiation (pyrheliometry) and, more broadly, radiometric quantities in general. They can also be used to measure mass flow. Thus, most thermal sensors are thermometers and even when they are not, they are based upon temperature sensing devices. This chapter is concerned chiefly with temperature measurement and here the term thermometer and thermal sensor become almost synonymous. The physical principles underlying thermal sensors require discussion of temperature itself – its concept, measurement methods, and scales – in addition to elucidation of the principles of the various sensors themselves.

Temperature is one of the most important of physical quantities. It appears explicitly in some of the most fundamental laws of physics and pervades almost every part of physical measurements. Its unit, the kelvin, *K*, is one of the seven base units of le Système International d'Unités (SI). Measurement of temperature, i.e., the assignment of numerical values to temperature according to some reasonably ordered system that has physical significance, is more complex than for the other base quantities for reasons connected both with temperature being intensive (ie, non-additive) and with the lack of any intuitive quantitative understanding of it. There are few physical properties that are not temperature-dependent. It follows that there is a very large number of different physical properties that can serve, at least in principle, as the basis for thermal sensors with which to measure temperature and temperature changes.

It is the intent in this chapter to describe the thermodynamic basis of temperature, the general principles required for its measurements, and the advantages to be derived from temperature scales that are universal and are related to thermodynamic temperatures in some known way. The International Temperature Scale of 1990 and the reasons for its introduction are outlined. This is followed by a detailed discussion of the physical principles that underlie a large number of thermal sensors (or thermometers) – those that are important for the establishment of thermodynamic temperature and others that are in widespread practical use. There are also included some comments on the measurement of heat flow and heat flux, and some generalities on associated sensors.

2.2 Concept of Temperature

2.2.1 Historical Resumé

Although the ancients could measure length, time, and mass moderately precisely, they had no notion at all of the fundamental nature of temperature nor is there any record of their attempting to measure it. The invention of the thermometer is relatively recent, being ascribed to Galileo in 1592. His instrument was based upon the thermal expansion of air. Very soon thereafter liquid-expansion thermometers were introduced in much the same form as they exist today. The development of thermometers was accompanied by the concurrent development of temperature scales with which to compare and record measurements of temperature. Most of these scales were based upon one or other of two principles: (a) calibration of the thermometer at two temperatures to establish two fiduciary marks, division of the intervening interval into equal parts, and linear extrapolation beyond the fiduciary marks; (b) calibration of the thermometer at one temperature with subsequent scale divisions based upon a calculated expansion of the fluid. The first of these is the obvious forerunner of modern practical scales and was the most precise and widest used; the second suggested some fundamental, rather than strictly empirical, basis and hints at present thermodynamic temperatures.

2.2.2 Thermodynamic Basis

All of the early temperature scales were simply empirical. However, the basis for a fundamental scale had been laid in experiments on the properties of gases by a succession of workers. Boyle had deduced ($\sim$1661) that the product of the pressure (P) and volume (V) of a fixed quantity of air (number of moles, N) at constant temperature (T) is constant, at least over a moderately wide range of pressure. This may be written

$$PV = \text{constant} \tag{2-1}$$

when N, T are constant. More than a century later, from the results of independent experiments, Charles and Gay-Lussac deduced the relationship

$$V = V_0(1 + \alpha\tau), \tag{2-2}$$

where V is the volume occupied by N moles of any gas at temperature τ on the arbitrary scale used, V_0 is the volume at the zero of the τ scale, and α is the volume coefficient of thermal expansion. Within the limits of their experiments, Charles and Gay-Lussac found the same value of α for most gases.

If the temperature T of Equation (2-1) is related to τ by

$$T = \tau + \alpha^{-1} \tag{2-3}$$

then Equation (2-2) has the form

$$V/T = \text{constant}\,. \tag{2-4}$$

Combining Equations (2-1) and (2-4), we obtain the general gas law

$$PV/T = \text{constant}\,. \tag{2-5}$$

The value for the constant in Equation (2-5) was deduced to be NR, where R is the gas constant (joules per mole kelvin) and N is the number of moles, so that the gas law is

$$PV = NRT\,. \tag{2-6}$$

From Equation (2-3), when $\tau = -\alpha^{-1}$, $T = 0$, which corresponds in Equation (2-2) to a contraction of the gas to zero volume, and so no lower temperature would be possible. These ideas are, of course, very simplistic, but nevertheless they introduce the notion of a temperature scale (T) with an *absolute zero* at which both pressure and volume approach zero. Equation (2-5) also indicates the basis of a gas thermometer for measuring T by maintaining V constant and measuring P, or vice versa. Detailed experiments by Regnault in the early 19th century suggested that Equation (2-5) is only approximately true and that different gases lead to different values of α^{-1}. We now know that the equation holds for all gases in the limit of very low pressures and high dilution (N small) and call such gases *ideal* or *perfect*. The corresponding scale for defining T is called the *absolute temperature scale* or the *ideal gas temperature scale*. Historically, a centigrade scale that assigned 0 °C to the freezing point of water and 100 °C to the boiling point of water had comc into use. For conformity in size of the degree, Equation (2-3) leads to $T = \alpha^{-1} = 273.15$ °C when $\tau = 0$ °C (for the current best value for α of 0.003661 (°C)$^{-1}$) or, conversely, the zero of the ideal gas scale is at -273.15 °C.

The need had become evident for a truly fundamental temperature scale independent of the properties of any particular substance. William Thomson (Lord Kelvin), building upon earlier studies by Sadi Carnot on reversible heat engines, provided the solution. A heat engine operates in such a way that during one part of its cycle its working substance absorbs heat from a reservoir at a high temperature and during another part, after using some of this heat to do external work, it rejects a lesser quantity of heat to another reservoir at a lower temperature. Carnot (in 1824) had shown that his idealistic heat engine, in which all of the operations of the cycle are reversible, worked at the maximum possible efficiency and that this efficiency was independent of the nature of the working substance and dependent only upon the temperatures of the two reservoirs. Thomson (in 1848) proposed to define temperature in terms of the efficiency of an ideal Carnot engine. Consider Carnot engines working between successive reservoirs at temperatures T_n and T_{n-1}, extracting heat Q_n from the hotter reservoir, and rejecting heat Q_{n-1} to the cooler one, as shown schematically in Figure 2-1. By analyzing the efficiencies of successive engines, one can show that

$$\frac{Q_2}{Q_1} = \frac{f(T_2)}{f(T_1)}, \tag{2-7}$$

where f is some unknown function of T that depends only upon the temperatures T_2 and T_1, and not on the nature of the working substance.

Kelvin's proposal for defining *thermodynamic temperatures* was to use for the ratio on the right hand side Equation (2-7) just the ratio of the temperatures themselves, ie,

$$\frac{Q_2}{Q_1} = \frac{T_2}{T_1}. \tag{2-8}$$

Temperature defined in this way can be shown to be identical to those defined by the ideal gas laws or, indeed, identical to those defined by all the other fundamental equations of physics such as Planck's Law or Nyquist's formula for the mean square electrical noise voltage across an unloaded resistor.

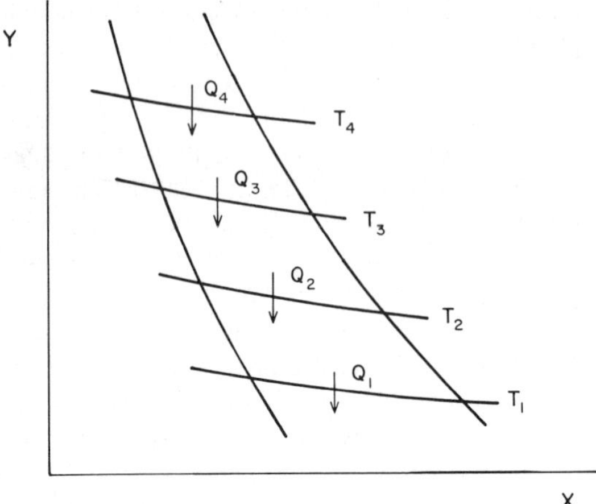

Figure 2-1.
Carnot cycles represented schematically in the (X, Y) plane (which could be, for example, the (V, P) plane). Each cycle is taken clockwise around a closed loop bounded by isotherms T_m and connecting adiabatics.

It remained to assign numerical values to thermodynamic temperatures. Thomson realized that by Equation (2-8) a single reference temperature would suffice; nevertheless, to retain numerical correspondence with the (now equivalent) ideal gas scale, he accepted the two reference point formulation.

In parallel with a thermodynamic basis for temperature obtained from consideration of macroscopic systems, a statistical basis for temperature in terms of microscopic systems was being developed. This had its origin in the studies of Maxwell (1859) and Boltzmann (1869) on the kinetic theory of gases. They derived a relation for the velocity (energy) distribution of the molecules in a fixed volume of gas in thermal equilibrium in which the probability of any particular molelcule having a translational kinetic energy in an infinitesimal range about E is proportional to exp $(-E/k_B T)$, where k_B is the Boltzmann constant. Here, $k_B T$ is a characteristic energy and T (it turns out) is just the thermodynamic temperature of Equation (2-8), so that T is linked to the kinetic energy of internal molecular motion. More generally, statistical mechanics develops for any system in thermal equilibrium the relationship

$$n_1/n_2 = \exp(-\Delta E/k_B T) \tag{2-9}$$

for the number of particles n_1 and n_2 in two internal states differing in energy by ΔE. This concept is useful for interpreting the meaning of temperature in systems for which the ordinary notions of thermodynamics are not applicable. One can deduce from kinetic theory that the average mean square velocity of a particle of mass m is $3k_B T/m$ and from classical thermodynamics that it is $3RT/M$, where M is the molar mass. Correspondence gives $R = k_B M/m = k_B N_a$ where N_a is Avogadro's number (the number of molecules per mole).

2.2.3 Temperature as a Physical Quantity

Temperature is a physical quantity that is more difficult than most others to understand and measure. One reason is connected with its being an intensive quantity with the consequent inability to multiply and subdivide its unit in order to measure over a wide range. However the unit of temperature is defined, some other means of scaling must be devised. We agree with Quinn [1] that the lack of an intuitive comprehension of temperature is not simply because it is intensive, but because it has no close association with any easily-perceived extensive quantity, such as density with mass for example. The notion that one substance is twice as dense as another is easily grasped from holding equal-sized pieces of each in the hand, but how does one sense, or what indeed does it mean, that one object is twice as hot as another? The conception of temperature is little improved by such formal definitions as: temperature is that *thermodynamic* quantity that takes the same value in two systems that are brought into *thermal contact* and are allowed to reach *thermal equilibrium.* While this definition may not help our understanding of temperature, it does nevertheless bring in several key ideas: that temperature is a thermodynamic quantity; that its measurement will involve thermal contact and thermal equilibrium; and that temperature is inseparably linked with heat flow. Heat always flows between two systems in contact but not in thermal equilibrium (ie, not at the same temperature), and by convention we choose to say that it flows from higher to lower temperatures. Thus, temperature may be defined also as the potential that governs the flow of heat.

The non-additivity of temperature leads directly to the need for a scale. To establish any temperature scale we require both a reference temperature to determine the size of the unit, and an interpolation rule to allow us to proceed from one temperature to another, ie, to measure temperature differences. In principle we may select any convenient physical system whose state can be specified by two variables X, Y to act as the thermometer, so long as the variables X, Y both remain constant when external conditions (such as temperature and pressure) are unchanged. For the hypothetical system shown in Figure 2-2, we can select any of the isotherms (say T_s) as a reference, and choose any convenient path in the X-Y plane (eg, the line $Y = Y_c$) for the interpolation rule. We can assign numerical values of temperature to

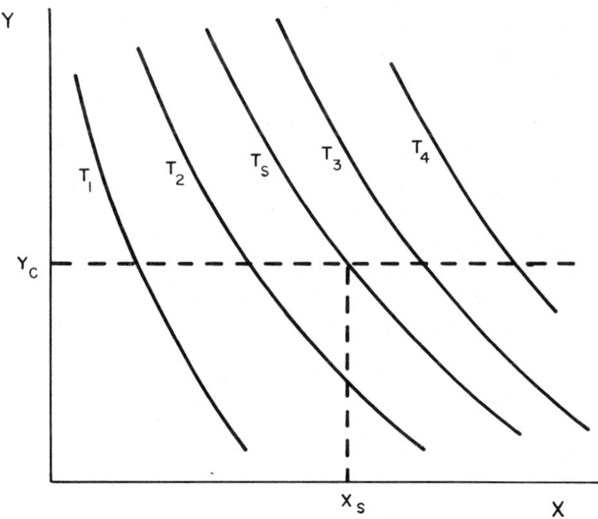

Figure 2-2. Isotherms T_i plotted in the (X, Y) plane for a hypothetical physical system with general variables X and Y.

all of the isotherms by choosing an appropriate functional relationship between the coordinate X and the isotherm temperatures T at the intersections of the isotherms with, for example, the line $Y = Y_c$. For the ideal gas temperature scale of Section 2.2.2, Y corresponds to volume and X to pressure, ie, temperature is measured as a function of pressure at constant volume.

If the relationship between T and X (and Y) can be derived theoretically from first principles (for example, Equation (2-6) or (2-8)) we have a primary or thermodynamic thermometer and a foundation for defining absolute or thermodynamic temperature; if the relationship is simply empirical, we have a secondary or practical thermometer and the basis for a practical temperature scale. For a thermodynamic thermometer the extension of the unit defined by the reference temperature over the whole temperature regime is imbedded in the fundamental equation; that is, the fundamental equation automatically extrapolates thermodynamic temperature directly in both directions from the fundamental reference temperature in terms of the chosen unit. It is taken as axiomatic that any practical scale should approximate thermodynamic temperature as closely as possible. With this proviso it is not so clear how to ensure, for practical temperatures, both constancy of size of unit (at various temperatures) and at the same time agreement with thermodynamic temperatures. In general, the measure of any physical quantity ψ is expressed as a product of a unit U and a numerical value v, ie,

$$\psi = v\mathrm{U}. \tag{2-10}$$

For a practical temperature T_p one can obviously define its unit U to be invariable (equal to the unit for thermodynamic temperature T for example), but this really begs the question because we must then find a way for expressing v so that $T_p \approx T$ everywhere. A way to do this is to extend the method for the hypothetical system described above and assign values of thermodynamic temperature to a large number of reference isotherms, one of which will define the unit. The instrument(s) and equations for interpolation between these are so chosen that the values of all interpolated practical temperatures are also very close to their thermodynamic counterparts. This is just the method used in the SI, which recognizes thermodynamic temperature (T) as the fundamental standard, but for practical use introduces in effect another physical quantity T_{90} with the same unit and so-defined as to approximate T as closely as possible. The degree of success of this method for the International Practical Temperature Scale of 1990 will be evident in Section 2.3.3. The unit for T from Kelvin's time until 1954 (then called the degree Kelvin) was defined as 1/100 of the interval between the freezing and boiling points of water. This had the undesirable feature that the location of the absolute zero was left as a value to be determined by experiment. Consequently, in 1954 the unit (now called the kelvin, K) was redefined as 1/273.16 of the interval between absolute zero and the temperature of the triple point of water. This accorded with Kelvin's original proposal that a single reference temperature sufficed. The numerical value assigned to the triple point is, of course, quite arbitrary; the value chosen was to ensure numerical compatability with earlier practice.

2.3 Temperature Scales

2.3.1 Need for a Practical Temperature Scale

In the early development of thermometry, innumerable practical scales were introduced long before any understanding of the thermodynamic basis for temperature evolved. In principle, since Kelvin's time, there has been no need for a practical scale since there exists a variety of ways of measuring thermodynamic temperature directly. The difficulty has been, however, that it cannot be measured sufficiently reproducibly for use as a metrological standard. Furthermore, thermodynamic thermometers are usually bulky, cumbersome, and slow and so unsuited for practical applications. It is possible to measure temperatures much more easily and reproducibly with practical thermometers (for example, platinum resistance thermometers) than with thermodynamic thermometers. For a great many industrial users, reproducibility of measurements on some practical scale is of higher importance than agreement with thermodynamic temperatures. At the same time, for worldwide uniformity in temperature measurement, it is obvious that there must be a single practical scale, rather than a multiplicity of them as in early times.

2.3.2 International Temperature Scales

The responsibility for the introduction of internationally-accepted temperature scales lies with the International Committee of Weights and Measures (CIPM), the organization under the authority of the General Conference of Weights and Measures (CGPM) that has the mandate for perfecting and unifying international standards. The first truly international scale was introduced in 1927 as the International Temperature Scale of 1927 (ITS-27). The ITS-27 was defined in such a way that temperatures measured on it were both highly reproducible and approximated thermodynamic temperatures as closely as they were then known. Furthermore, the ITS-27 embodied the important concept that it did not rely upon a particular artifact for its realization, but could be set up at any time directly from its definition. A scale defined in terms of an artifact can be just as, or even more, precise but has the drawback of undergoing changes (perhaps indeterminable) if the artifact changes, or irretrievable loss if the artifact is lost, damaged, or destroyed. The definition assigned values of temperature to a selected number of fixed points judiciously distributed and interpolated between them with specified types of thermometers according to assigned formulae. Phase equilibrium temperatures were chosen for the defining fixed points because they (in particular, triple, freezing, melting, and boiling points of pure substances) have better reproducibility than the thermometers used to measure them and can be reproduced more precisely than their thermodynamic temperature values can be measured. This same principle of construction has been followed with all successors of the ITS-27.

The international scale in current use is the International Temperature Scale of 1990 (ITS-90) [2], which superseded the International Practical Temperature Scale of 1968 IPTS-68 (amended edition of 1975) [3], as of January 1, 1990. In the following Sections 2.3.3 and 2.3.4 we discuss the reasons why the ITS-90 was introduced and outline its definition.

2.3.3 Shortcomings of the International Practical Temperature Scale of 1968

In the period following the introduction of the IPTS-68 there was a great deal of activity in thermometry research. A variety of ways of measuring T were exploited, leading to a manyfold improvement in the accuracy with which T is known from very low temperatures to 1064 °C. Similarly there was a considerable advance in the ability to measure thermal and electrical quantities, due in large part to the continual improvements in digital measuring equipment and computer capabilities, the latter leading to very precise automation and control of experiments. One result of all of this was the uncovering of various shortcomings of the IPTS-68.

2.3.3.1 Disagreement with Thermodynamic Temperatures

In 1968 temperatures defined by the IPTS-68 agreed with thermodynamic temperatures within the experimental uncertainty of existing measurements, almost all of these the results of gas thermometry. That the degree of agreement was highly optimistic is evident from Figure 2-3 which shows the results of current knowledge of $T - T_{68}$. The curves in Figure 2-3 are derived from a large number of experiments that incorporate about ten fun-

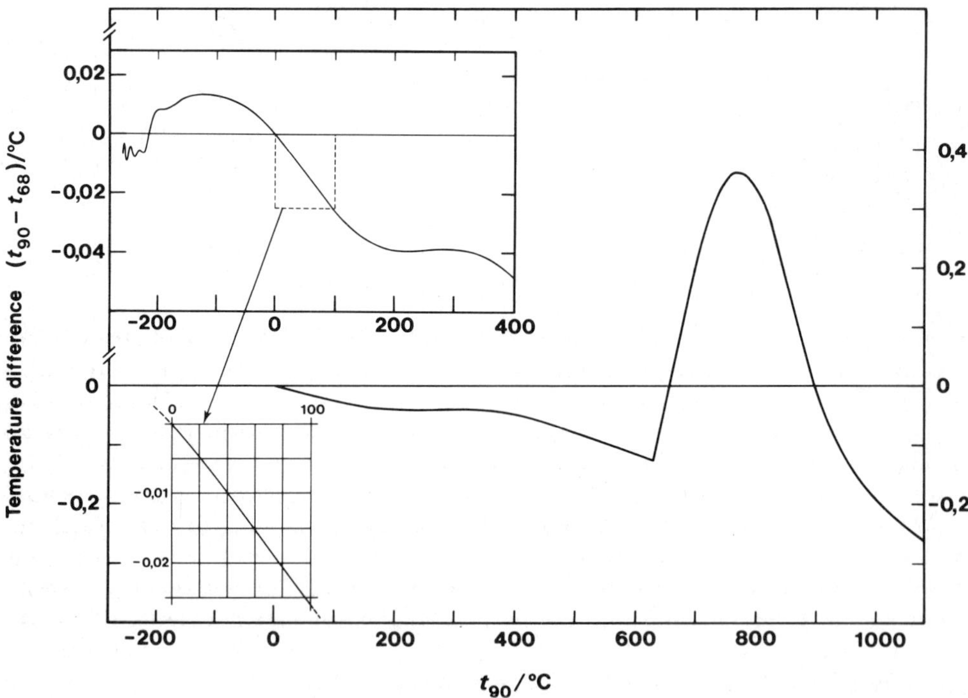

Figure 2-3. Differences of t_{68} from thermodynamic temperature t (after Preston-Thomas [2]). The figures shows $t_{90} - t_{68}$; since t_{90} is the same as t to within the uncertainty of measurement of t, this is equivalent to $t - t_{68}$.

damentally different methods for measuring thermodynamic temperature. In each temperature range (except perhaps above 650 °C) there are overlapping results and generally good agreement from at least two such methods, so the accuracy of $T - T_{68}$ is now judged to be very high. The uncertainties of the curves shown in Figure 2-3 are approximately as follows: a few tenths of a millikelvin from 14 K to 30 K; ± 2 mK from 30 K to 100 °C, improving to ± 1 mK near 80 K and 300 K; ± 2 mK at 100 °C rising to ± 20 mK at 630 °C; ± 20 mK (relatively) but ± 100 mK (absolutely) from 630 °C to 1100 °C. There is a discontinuity in the slope at 630 °C that is inherent in the IPTS-68 definition. Bedford et al. [4] have measured it directly to be 0.5%, in excellent agreement with spectral radiation measurements above and below.

The unacceptably large values of $T - T_{68}$ above 630 °C are a direct consequence of the imprecision of the platinum 10% rhodium/platinum thermocouple and the inability of its IPTS-68 specified interpolation equation to reproduce T. Above 1064 °C, T_{68} would be accurately thermodynamic if T_{68} (Au) had been accurately assigned (the uncertainty in the second radiation constant c_2 is small enough that it has negligible effect on T_{68}); since it was not, $T - T_{68}$ is calculable from the magnitude of the mis-assignment.

It is obvious that these differences between T and T_{68} are so large and vary so irregularly with temperature that the ITS had to be redefined to remove them.

2.3.3.2 *Lack of Thermodynamic Smoothness*

Not only does T_{68} differ from T, but the differences do not vary smoothly with T (Figure 2-3). When physical quantities that depend upon T are measured, it is evident that tests of theoretical models will be unreliable, and that numerical predictions from valid theories will be inaccurate, unless $T_{68} = T$. In practice, in only a few cases is the absolute magnitude of $T - T_{68}$ of importance. However, when the first derivative of a physical quantity with respect to T is required (as, for example, in measurements of specific heat or thermoelectric power), highly significant errors can occur when $\mathrm{d}(T - T_{68})/\mathrm{d}T$ changes rapidly, and this is just what does occur according to Figure 2-3, especially below 60 K and near 630 °C. In these circumstances it is impossible to distinguish between real physical effects in the measured data and spurious effects due to lack of thermodynamic smoothness of the IPTS-68.

2.3.3.3 *Range*

Within a short time after the introduction of the IPTS-68 the need for an extension below its lower limit of 13.8 K became urgent. Furthermore, evidence had been increasingly available that both the then-current helium vapor pressure scales [5, 6] and the lower end of the IPTS-68 departed substantially from thermodynamic temperatures and, moreover, were inconsistent with one another. Because of this the *1976 Provisional 0.5 K to 30 K Temperature Scale* (EPT-76) [7] was put forward for international use between 0.5 K and 30 K. It was thermodynamically smooth, continuous with the IPTS-68 at 27.1 K, and closely agreed with thermodynamic temperatures. The EPT-76 was meant to be a temporary, patchwork solution to the need for a formal scale below 13.8 K until such time as the ITS itself was revised.

2.3.3.4 Platinum 10% Rhodium/Platinum Thermocouple

The large differences of T_{68} from T between 630 °C and 1064 °C result from a combination of the limited precision of the platinum 10% rhodium/platinum thermocouple and the inability of its specified equation to properly interpolate T. The thermocouple, with a precision within only ± 0.2 K (and some consider even this an optimistic estimate), lies between the platinum resistance thermometer (PRT) with an imprecision of 1 or 2 mK at 630 °C and the optical pyrometer with an imprecision of about 10 mK at 1064 °C. This situation is highly anomalous for a scale-defining instrument; it is obvious that the thermocouple needed to be replaced as a standard instrument of the ITS. This would, of course, in no way affect its wide application as a highly important secondary temperature sensor. It has been known for some time that platinum resistance thermometers specially designed for high temperature use can operate at least to the freezing point of silver (962 °C) to within a precision of better than ± 10 mK. Logically, then, the range of definition by the platinum resistance thermometer in the ITS should be extended.

2.3.3.5 Faults in Definition Below 273 K

The IPTS-68 subdivides the range from 13.8 K to 273 K into four parts. To measure a temperature in any part requires calibration of the thermometer in all higher-temperature parts, which is inconvenient. Additionally, the interpolating equations below 273 K are unsatisfactory. Identically-calibrated platinum resistance thermometers can indicate temperatures differing by up to 3 mK (between fixed points). This non-uniqueness is quite distinct from general thermometer imprecision; it results from the inadequacy of the interpolating equations, probably because of the way they are forced to join smoothly at junction temperatures between parts of the range. It was necessary to substantially diminish the magnitude of this non-uniqueness.

2.3.4 International Temperature Scale of 1990

The ITS-90 [2] is defined along the same lines as all of its predecessors. Values of temperature are assigned to seventeen phase equilibrium states of pure substances, the details of which are given in Table 2-1. Interpolation between these fixed points is with one of four defining instruments (depending on the temperature range) using assigned formulae. The ITS-90 recognizes that thermodynamic temperature (T) with its unit the kelvin (K) is the fundamental quantity. It also allows for thermodynamic temperatures to be expressed as Celsius temperatures (t), defined by

$$t/°\mathrm{C} = T/\mathrm{K} - 273.15 \qquad (2\text{-}11)$$

where the unit for Celsius temperature (°C) is by definition equal in magnitude to the kelvin. The ITS-90 introduces corresponding practical temperatures T_{90} and t_{90}, related as in Equation (2-11), with the same units (Kelvin and degree Celsius) as T and t, respectively.

Table 2-1. Defining fixed points of the ITS-90.

Number	Temperature		Substance[(a)]	State[(b)]
	T_{90}/K	$t_{90}/{}^{\circ}\mathrm{C}$		
1	3 to 5	−270.15 to −268.15	He	V
2	13.8033	−259.3467	$e-H_2$	T
3	≈17	≈ −256.15	$e-H_2$ (or He)	V (or G)
4	≈20.3	≈ −252.85	$e-H_2$ (or He)	V (or G)
5	24.5561	−248.5939	Ne	T
6	54.3584	−218.7916	O_2	T
7	83.8058	−189.3442	Ar	T
8	234.3156	−38.8344	Hg	T
9	273.16	0.01	H_2O	T
10	302.9146	29.7646	Ga	M
11	429.7485	156.5985	In	F
12	505.078	231.928	Sn	F
13	692.677	419.527	Zn	F
14	933.473	660.323	Al	F
15	1234.93	961.78	Ag	F
16	1337.33	1064.18	Au	F
17	1357.77	1084.62	Cu	F

(a) All substances except ^{3}He are of natural isotopic composition; $e-H_2$ is hydrogen at the equilibrium concentration of the ortho- and para-molecular forms

(b) For complete definitions and advice on the realization of these various states, see "Supplementary Information for the ITS-90". The symbols have the following meanings:

V: vapor pressure point

T: triple point (temperature at which the solid, liquid, and vapor phases are in equilibrium)

G: gas thermometer point

M, F: melting point, freezing point (temperature, at a pressure of 101 325 Pa, at which the solid and liquid phases are in equilibrium)

The definition of the kelvin is unchanged as the fraction 1/273.16 of the interval between absolute zero and the temperature of the triple point of water. Temperatures T_{90} agree with T as closely as the mathematical formulation has permitted and within the experimental uncertainty of measurements of T. Hence, for the first time, temperatures measured throughout the ITS are consistent with the definition of the kelvin.

In contrast to previous practical scales, in certain temperature ranges the ITS-90 specifies alternative definitions of T_{90} of equal status. For any particular thermometer the difference in values of T_{90} between two definitions (termed subrange inconsistency) is at, or near to, the level of measurement uncertainty [8, 9]. Although non-uniqueness is substantially smaller than in the IPTS-68, it is unfortunately still somewhat larger than measurement uncertainty in some temperature ranges.

The text of the ITS-90 is restricted to the definition of the scale. All of the supporting documentation for realizing the scale appears in a separate monograph entitled *Supplementary Information for the ITS-90* [10]. Methods for approximating the ITS-90 at lower levels of accuracy are described in a second monograph entitled *Techniques for Approximating the ITS-90* [11].

2.3.4.1 Outline of the ITS-90

The ITS-90 has a lower limit of 0.65 K and an indefinite upper limit that is determined by the practicability of measurements with a monochromatic radiation thermometer. This wide span is divided into essentially five ranges, within some of which several sub-ranges are also defined. The definition is outlined below; reference should be made to the text of the ITS-90 [2] for the complete details and for the numerical values of many of the required quantities.

a) 0.65 K to 5 K

In this range T_{90} is defined in terms of the vapor pressure p of ^3He and ^4He using equations of the form:

$$T_{90}/\text{K} = A_0 + \sum_{i=1}^{9} A_i\,[\ln(p/\text{Pa}) - B)/C]^i . \tag{2-12}$$

The values of the constants A_0, A_i, B, and C are given in Table 2-2 for ^3He in the range 0.65 K to 3.2 K, and for ^4He in the ranges 1.25 K to 2.1768 K (the λ point) and 2.1768 K to 5.0 K.

b) 3 K to 24.6 K

In this range T_{90} is defined in terms of a ^3He or a ^4He gas thermometer of the constant-volume type using the equation

$$T_{90} = \frac{a + bp + cp^2}{1 + B_x\,(T_{90})\,N/V}, \tag{2-13}$$

where p is the pressure in the gas thermometer, N/V is the gas density in moles per cubic meter in the gas thermometer bulb, and $B_x\,(T_{90})$ are the values of the second virial coefficients for

Table 2-2. Values of the constants for the helium vapor pressure Equations (2-12), and the temperature range for which each equation, identified by its set of constants, is valid.

	^3He 0.65 K to 3.2 K	^4He 1.25 K to 2.1768 K	^4He 2.1768 K to 5.0 K
A_0	1.053447	1.392408	3.146631
A_1	0.980106	0.527153	1.357655
A_2	0.676380	0.166756	0.413923
A_3	0.372692	0.050988	0.091159
A_4	0.151656	0.026514	0.016349
A_5	−0.002263	0.001975	0.001826
A_6	0.006596	−0.017976	−0.004325
A_7	0.088966	0.005409	−0.004973
A_8	−0.004770	0.013259	0
A_9	−0.054943	0	0
B	7.3	5.6	10.3
C	4.3	2.9	1.9

^{3}He or ^{4}He depending upon which gas is used. The values of the coefficients a, b, and c are obtained by calibrations at the triple point of neon (24.5661 K), the triple point of equilibrium hydrogen (13.8033 K), and at a temperature between 3.0 K and 5.0 K that is determined using a ^{3}He or a ^{4}He vapor pressure thermometer as specified in a). With ^{4}He above 4.2 K the denominator in Equation (2-13) may be ignored so long as the lowest calibration point is chosen to be above 4.2 K.

c) 13.8 K to 273.16 K

In this range T_{90} is defined in terms of the resistance ratios of platinum resistance thermometers calibrated at specified sets of defining fixed points, and using specified reference and deviation functions for interpolation at intervening temperatures. The resistance ratio $W(T)$ is defined as

$$W(T_{90}) = R(T_{90})/R(273.16\ \text{K}), \tag{2-14}$$

where $R(T_{90})$ is the electrical resistance. The text of the scale give specifications that determine the quality of the platinum that may be used.

The temperature T_{90} is defined by the relation

$$W(T_{90}) = W_r(T_{90}) + \Delta W(T_{90}), \tag{2-15}$$

where $W_r(T_{90})$ is a resistance ratio that is related to T_{90} by the reference function

$$\ln[W_r(T_{90})] = A_0 + \sum_{i=1}^{12} A_i \left[\frac{\ln(T_{90}/273.16\ \text{K}) + 1.5}{1.5}\right]^i \tag{2-16}$$

or by the equivalent inverse function

$$T_{90}/273.16\ \text{K} = B_0 + \sum_{i=1}^{15} B_i \left[\frac{W_r(T_{90})^{1/6} - 0.65}{0.35}\right]^i. \tag{2-17}$$

The values of the coefficients A_i and B_i are given in Table 2-3. The deviations $\Delta W(T_{90}) = W(T_{90}) - W_r(T_{90})$ of individual platinum resistance thermometers from $W_r(T_{90})$ are measured at specified fixed points and T_{90} interpolated between them by deviation functions that relate $\Delta W(T_{90})$ to T_{90}. Four different (overlapping) sub-ranges with equal status are defined in the range 13.8 K to 273.16 K. The upper limit for each is 273.16 K; the lower limits are 13.8 K, 24.6 K, 54.4 K, and 83.8 K respectively. Within the first three of these, $\Delta W(T_{90})$ is given by

$$\Delta W(T_{90}) = a[W(T_{90}) - 1] + b[W(T_{90}) - 1]^2 + \sum_{i=1}^{5} c_i [\ln W(T_{90})]^{i+n}. \tag{2-18}$$

c.1) 13.8 K to 273.16 K

In this range $n = 2$ in Equation (2-18), and the values of the coefficients a, b, c_i are obtained by calibration at the fixed points 2 to 9 of Table 2-1.

Table 2-3. Coefficients A_i, B_i, C_i and D_i of the ITS-90 reference functions (Equations (2-16), (2-17), (2-20) and (2-21), respectively).

i	A_i	B_i	C_i	D_i
0	−2.13534729	0.183324722	2.78157254	439.932854
1	3.18324720	0.240975303	1.64650916	472.418020
2	−1.80143597	0.209108771	−0.13714390	37.684494
3	0.71727204	0.190439972	−0.00649767	7.472018
4	0.50344027	0.142648498	−0.00234444	2.920828
5	−0.61899395	0.077993465	0.00511868	0.005184
6	−0.05332322	0.012475611	0.00187982	−0.963864
7	0.28021362	−0.032267127	−0.00204472	−0.188732
8	0.10715224	−0.075291522	−0.00046122	0.191203
9	−0.29302865	−0.056470670	0.00045724	0.049025
10	0.04459872	0.076201285		
11	0.11868632	0.123893204		
12	−0.05248134	−0.029201193		
13		−0.091173542		
14		0.001317696		
15		0.026025526		

c.2) 24.6 K to 273.16 K

In this sub-range, $c_4 = c_5 = n = 0$ in Equation (2-18), and the values of the coefficients a, b, c_1, c_2, and c_3 are obtained by calibration at the fixed points 2, and 5 to 9 of Table 2-1.

c.3) 54.4 K to 273.16 K

In this sub-range $n = 1$ and $c_2 = c_3 = c_4 = c_5 = 0$ in Equation (2-18), and the values of the coefficients a, b, c_1, are obtained by calibration at the fixed points 6 to 9 of Table 2-1.

c.4) 83.8 K to 273.16 K

In this sub-range $\Delta W(T_{90})$ is given by

$$\Delta W(T_{90}) = a\,[W(T_{90}) - 1] + b\,[W(T_{90}) - 1] \ln W(T_{90}) \,. \tag{2-19}$$

The values of a and b are obtained by calibration at the fixed points 7 to 9 of Table 2-1.

d) 0 °C to 961.8 °C

In this range T_{90} is defined in terms of the resistance ratios of platinum resistance thermometers using Equation (2-15) as in c). Here, however, the reference function that relates $W_r(T_{90})$ to T_{90} is

$$W_r(T_{90}) = C_0 + \sum_{i=1}^{9} C_i \left[\frac{T_{90}/\mathrm{K} - 754.15}{481}\right]^i . \tag{2-20}$$

The equivalent inverse function is

$$T_{90}/\mathrm{K} - 273.15 = D_0 + \sum_{i=1}^{9} D_i \left[\frac{W_r(T_{90}) - 2.64}{1.64} \right]^i . \qquad (2\text{-}21)$$

The values of the coefficients C_i and D_i are given in Table 2-3. Six (overlapping) sub-ranges of equal status are defined between 0 °C and 961.8 °C, all with a lower limit of 0 °C and with upper limits of 961.8 °C, 660.3 °C, 419.5 °C, 231.9 °C, 156.6 °C, and 29.8 °C respectively. The deviation function for each of these is

$$\Delta W(T_{90}) = a[W(T_{90}) - 1] + b[W(T_{90}) - 1]^2 + c[W(T_{90}) - 1]^3 + + d[W(T_{90}) - W(660.323\ °\mathrm{C})]^2 . \qquad (2\text{-}22)$$

d.1) 0 °C to 961.8 °C

In this range the values of *a, b, c* in Equation (2-22) are obtained by calibration at the fixed points 9, 12, 13, 14 in Table 2-1. For measurements below 660.3 °C, $d = 0$; for measurements above 660.3 °C, the values of *a, b,* and *c* are retained and the value of *d* obtained from calibration at the freezing point of silver.

d.2) 0 °C to 660.3 °C

In this sub-range the values of *a, b, c* of Equation (2-22) are obtained by calibration at the fixed points 9, 12, 13, 14 in Table 2-1.

d.3) 0 °C to 419.5 °C

In this sub-range $c = d = 0$ and the values of *a, b* are obtained by calibration at the fixed points 9, 12, 13 in Table 2-1.

d.4) 0 °C to 231.9 °C

In this sub-range $c = d = 0$ and the values of *a, b* are obtained by calibration at the fixed points 9, 11, 12 in Table 2-1.

d.5) 0 °C to 156.6 °C

In this sub-range $b = c = d = 0$ and the value of *a* is obtained by calibration at the fixed points 9, 11 in Table 2-1.

d.6) 0 °C to 29.8 °C

In this sub-range $b = c = d = 0$ and the value of *a* is obtained by calibration at the fixed points 9, 10 in Table 2-1.

d.7) −38.8 °C to 29.8 °C

The ITS-90 specifies one further sub-range, within the range of the platinum resistance thermometer, that is particularly convenient for measurements near to but both above and below ambient. In it, the deviation function is given by Equation (2-22) with $c = d = 0$ and the values of *a, b* are obtained by calibration at the fixed points 8, 9, 10 of Table 2-1. The required values of $W_r(T_{90})$ are obtained from Equations (2-16) or (2-20).

e) above 961.8 °C

Above the freezing point of silver T_{90} is defined by the equation

$$\frac{L_\lambda (T_{90})}{L_\lambda (T_{90}(X))} = \frac{\exp (c_2/\lambda T_{90}(X)) - 1}{\exp (c_2/\lambda T_{90}) - 1} \tag{2-23}$$

where $T_{90}(X)$ refers to any one of the silver, gold, or copper freezing points, and in which $L_\lambda (T_{90})$ and $L_\lambda (T_{90}(X))$ are the spectral radiances of a blackbody at the wavelength (in vacuum) λ at T_{90} and at $T_{90}(X)$ respectively, and $c_2 = 0.014388$ m · K.

2.4 Measurement of Temperature

Temperature differs from other basic physical quantities in that it can be measured only indirectly. One cannot, as with mass or length for example, directly compare a particular artifact with its corresponding unit. Temperature must be determined by measuring some other temperature-dependent physical quantity and from that deducing its value. The object whose temperature-dependent parameter is measured is a *thermometer* or *thermal sensor.* For this measurement procedure to be adequate, in keeping with one way of defining temperature in Section 2.2.3, it is absolutely essential that the sensor be in thermal equilibrium with the system whose temperature is required. Thermal equilibrium can be established either through intimate physical contact (contact thermometers) or through radiative contact (non-contact thermometers). Radiation thermometers do not of course attain the temperature of the system; they act in essence as comparators between the system and some previously calibrated system.

Dynamic measurements (where, by dynamic, we mean changing temperature, not changing position) have similar requirements; for a meaningful result, the temperature of system and sensor must be changing together. In some cases the system can serve as its own thermometer, but this is more often the case in the research laboratory than in the field.

A contact thermometer can measure a system temperature accurately only if extraneous avenues of heat flow are small enough not to prevent the sensor from attaining thermal equilibrium with the system. This condition requires that the thermal contact resistance between the sensor and the system be small compared to that between the sensor and the environment, that the thermal mass of the sensor be sufficiently small relative to that of the system, and that the introduction of the sensor not unduly perturb the system temperature. Some of the common mechanisms that cause the temperatures of the sensor and system to differ are:

a) Stem or lead conduction

Heat is transferred by conduction from the system to the environment through the body of the incompletely-immersed sensor or, especially for electrical sensors, along high conductivity metal leads. The solution is to ensure that the sensor is immersed sufficiently far that this heat flow has negligible effect. If the system is liquid this condition is usually easily attained; if it is solid the thermal contact is likely to be poorer so that more care is required; if it is gaseous, difficulties can be severe. In many cases the difference in temperature between the sensor and the system decreases exponentially with increasing depth of immersion. If numerical values of various physical and thermal parameters of the sensor and system can be estimated, then the temperature error is estimable by calculation [12]. Associated errors can frequently be reduced by proper thermal anchoring of lead wires or stem [13]. This type of problem is especially severe in measurement of surface temperature. Proper installation of the sensor on the surface is mandatory for the surface temperature not to be affected and for the sensor temperature to be near that of the surface. Enormous errors can result from improper installation. Jakob [14] treats the problem in detail, taking into account both conduction and free convection transfer. Baker et al. [15] discuss methods for installing the sensor.

b) Joule heating

With electrical resistance sensors through which a current is passed to measure the resistance (R) variation, joule heating ($i^2 R$) of the sensor by the measuring current (i) can introduce an appreciable quantity of heat, especially in non-metallic sensors, that raises the sensor temperature above that of the system. Readings are correctable through measurements at two or more values of current and extrapolation to zero measurement power [16]. The magnitude of the heating for a particular sensor is also dependent upon the system in which it is immersed and the heat transfer mechanisms between sensor and system. It is much lower, for example, in moving air than in still air. It is also changed if a sheath material surrounds the sensor.

c) Radiative transfer

A sensor can be prevented from attaining thermal equilibrium with the system through radiative transfer of heat either to it or from it. If any other system at a different temperature is visible to the sensor, the possibility for an error arises. The problem most often occurs in measuring the temperatures of gases, transparent liquids or solids, or surfaces of solids. A commonly-encountered example is in measurement of outdoor temperatures in the presence of sunlight. A more exotic example is in connection with the use of metal freezing points in defining the ITS. McLaren and Murdock [17] have shown that visible and near-infrared radiation transmitted out of the thermometer by total internal reflection within the silica sheath of a standard platinum resistance thermometer can cause the sensor temperature to be as much as 85 mK lower than freezing antimony temperature (630 °C) when the sensor is immersed 13 cm below the top of the antimony ingot and more than 40 cm below the top of the furnace; similar transmission of room illumination into the thermometer can raise the sensor temperature 0.5 mK above that of the triple point of water when the sensor is immersed more than 25 cm in the triple point cell [18]. Such radiative errors can be eliminated or at least reduced by suitable shielding of the sensor.

Any attempt to measure a temperature must take into account the nature of the thermal response of the sensor. When a sensor at a temperature T_0 is suddenly immersed into a system at a different (constant) temperature T_1, a finite interval of time elapses before the sensor registers the system temperature T_1; the length of this time interval is determined by the *time constant* of the sensor. This is a heat transfer problem for which, in principle, a fairly rigorous solution can be obtained from moderately complicated differential equations. Although the theory is well developed, it is usually of little quantitative help. The sensor response is governed by physical properties of the sensor, the system, and any surrounding material such as a sensor sheath or even the environment; among these properties are dimensions, thermal conductivity, surface heat transfer coefficient, thermal capacity, mass, thermal contact resistance, and the temperature coefficients of some of these properties. The response depends, for example, on whether the sensor is in contact with the system directly, or through a well, in which latter case the properties of the well are also involved. Usually the values of many of these parameters are not known well enough for the mathematical solution to be accurate. Where a time constant must be known, it is better to measure it than to calculate it.

A simple analysis is sufficient to illustrate the general nature of the sensor response. It assumes that the heat capacity of the system is sufficiently large that the introduction of the sensor does not change the system temperature. With no loss of generality, suppose $T_1 > T_0$. On insertion of the sensor, the flow of heat into it causes its temperature to rise at a rate, governed by its physical properties, given by

$$q = V c_P \rho \frac{\mathrm{d}T}{\mathrm{d}t} \tag{2-24}$$

where q is the thermal power input, V, c_P, ρ are the volume, specific heat (at constant pressure), and density respectively of the sensor, and t is the time. The heat flow is sustained through surface convection/conduction according to Newton's law

$$q = hA\,(T_1 - T) \tag{2-25}$$

where h is the coefficient of heat transfer, A is the surface area of the sensor, and T is the sensor temperature at time t. Equating q in Equations (2-24) and (2-25) and integrating we obtain

$$\int_{T_0}^{T} \frac{\mathrm{d}T}{T_1 - T} = \int_0^t \frac{hA}{V c_P \rho}\,\mathrm{d}t \tag{2-26}$$

where the physical properties in the integral on the right hand side of Equation (2-26) are assumed to be temperature independent. The solution is

$$T = T_1 - (T_1 - T_0)\,\mathrm{e}^{-t/\tau} \tag{2-27}$$

where

$$\tau = V c_P \,\rho / hA \tag{2-28}$$

is the thermometer time constant. In such a case the sensor temperature approaches the system temperature exponentially, the rate of approach being governed by τ. In a time after immersion equal to one time constant the sensor temperature has increased by 63% of the interval between the system temperature and the original sensor temperature.

A similar simplistic analysis shows that for a sudden step change in system temperature T_1, the temperature of a sensor initially in thermal equilibrium with the system will at first lag T_1 and then approach it exponentially. In reality the picture is much more complex, not the least because the system itself cannot respond instantaneously to changes in input power. More importantly, if T_1 is changing at a constant rate, the sensor temperature will always lag by an amount determined by τ and the rate of change of T_1. If T_1 is changing in some more complicated fashion, it may not be possible to deduce its value at all with any degree of precision from the corresponding sensor indications unless the time constant is small. If, for example, T_1 undergoes a regular periodic oscillation, the sensor temperature will exhibit a reduction in amplitude and displacement in phase relative to T_1, both of which are largely determined by τ.

In reality it is seldom possible to describe the sensor response in terms of a single time constant as in Equation (2-27). Physically, different portions of a sensor respond at different times and rates to system temperature change; the surface may respond first, then the interior, and so on. A more rigorous mathematical description [19] leads to a response of the form

$$T = A_0 + A_1 \, e^{-t/\tau_1} + A_2 \, e^{-t/\tau_2} + \ldots \qquad (2\text{-}29)$$

where the overall time constant τ is closely approximated by $\tau = \sum_i \tau_i$. Time constants of sensors can be as small as microseconds (eg, photon detectors) and as large as minutes (eg, clinical thermometers).

2.5 Measurement of Heat Flux

Heat flux is difficult to measure. The number of types of sensors devised for this purpose is almost as large as the number of applications. Detailed description of the methods is beyond the scope of this chapter (see Chapter 8). Distinction should be made between heat flow measurements (ie, W) and heat flux measurements (ie, W/m^2) even though, in this brief introduction, the distinction is small. Measurement of heat flux is closely related to, and indeed sometimes not clearly distinguishable from, measurement of surface temperature. Similar instrumentation can be applied to both tasks. Some infrared radiation thermometers, for example, can be supplied with alternative graduations that read either surface temperature or surface heat flux. One of the major difficulties with contact heat flux sensors is that the presence of the sensor on a surface alters the heat flow through the surface; consequently much of the design criteria is to eliminate, minimize, or correct for this deleterious effect.

The measurement of heat flow through a wall in steady-state conditions accepts Fourier's principle that the temperature difference between two planes in the wall parallel to the surface is proportional to the rate of heat flow through them. If the temperature at these planes and the thermal conductivity of the material between them were known, the heat flow could be calculated. These quantities are seldom known, however, so instead a heat flow meter for which the conductivity is known and for which the temperature of two planes (usually front and back surfaces) can be measured is attached to the surface. Ideally, the heat flow through the meter is the same as through the wall; in practice one must overcome or correct for the difficulties mentioned above. One of the first heat flow meters was a rubber band with multiple thermocouple junctions (connected as a thermopile) imbedded in such a way that alternate junctions were on either side of the band. The pre-calibrated band could be wound around cylindrical pipes or mounted on extended surfaces to measure the desired heat flow. In its original form it was of little use with high conductance materials (metals) because it markedly changed the heat flow. Many sophisticated versions of this type of meter have subsequently been described (see Chapter 8). They commonly consist of fine wire or deposited thin film thermopile sensors with alternate junctions on each side of a nonconducting wafer. The measured temperature gradient across the wafer is proportional to the heat flow through it. Or a thin electric resistance strip is attached to a thin insulating substrate, with the change in resistance proportional to the surface temperature change. Or a single differential thermocouple may measure the temperature difference created between the center and edge of a thin metal foil exposed to the heat flux with, again, this difference being proportional to the rate of heat transfer to the foil. Or the rate of temperature rise of a relatively massive slug or disk of known mass is measured, the rate being proportional to the rate of heat transfer to the slug. Such sensors can be mounted in or on a wall, in a fluid stream, or used radiometrically. Either the physical and thermal properties of the sensor must be known, or the sensor must be calibrated, to deduce the heat flux. Accuracies to better than a few percent are exceptional. Another technique uses a null method – heat is supplied electrically to a foil or disk until its temperature is the same as that of the wall in which it is mounted. The wall and sensor heat fluxes are then equal and can be deduced from the measured electrical input.

Following Baba et al. [20] the error in heat flux measurement with a thin sensor affixed to a wall can be estimated from consideration of heat transfer relationships with and without the sensor present. For the wall itself heat is transferred from the surface by convection and radiation. The local surface heat flux q'_c due to convection is $q'_c = h_w (T_w - T_e)$ and due to radiation is $q'_r = \varepsilon_w \gamma_w (T_w - T_e)$ where $\gamma_w = \sigma (T_w^3 + T_w^2 T_e + T_w T_e^2 + T_e^3)$, σ is the Stefan-Boltzmann constant, ε_w is the emissivity of the wall surface, and subscript w refers to the wall in the absence of the sensor. Thus the total heat flux is

$$q'_w = K_w (T_w - T_e) \tag{2-30}$$

where

$$K_w = h_w + \varepsilon_w \gamma_w . \tag{2-31}$$

If the sensor is thin enough that the lateral components of heat flow are negligible and if the temperature variation over the interface between the sensor and the wall is negligible, then the local heat flux from the wall through the sensor has the form

$$q' = K (T - T_e) \tag{2-32}$$

where T is the interface temperature and, corresponding to Equation (2-31),

$$K^{-1} = H^{-1} + (h_s + \varepsilon_s \gamma_s)^{-1} . \tag{2-33}$$

In Equation (2-33), γ_s has the form of γ_w with the sensor surface temperature T_s replacing T_w and H is the thermal conductance per unit area of the sensor. It is desired to measure q'_w but the sensor actually measures q'. Clearly the two will be the same if $K = K_w$, which in turn means $H \to \infty$, $h_w = h_s$, and $\varepsilon_w = \varepsilon_s$, i.e., if the sensor has the same surface properties as the wall and no thermal resistance. For a real sensor this is never so.

In a real case neither the temperature T nor the heat flux q' is constant within the interfacial area. Baba et al. calculate the temperature distribution analytically from solution of an integral equation. The result for the average interfacial temperature, $\bar{T}$, (which is the one measured by, say, a thermopile meter) when T_∞ is taken to be 0 is

$$\bar{T} = \frac{T_w}{1 + \left(\frac{K}{K_w} - 1\right) N} \tag{2-34}$$

where N is a non-dimensional parameter involving the physical properties of the wall material and the geometry of the sensor. The value of N ranges from 0 to 1. From Equations (2-30) and (2-32)

$$\frac{q'}{q'_w} = \frac{KT}{K_w T_w}, \tag{2-35}$$

so from Equations (2-34) and (2-35) the average interfacial flux, $\bar{q}'$, is

$$\frac{\bar{q}'}{q'_w} = \frac{K/K_w}{1 + \left(\frac{K}{K_w} - 1\right) N}. \tag{2-36}$$

There will be no error in the measured heat flux if $K = K_w$ (as noted above) or if $N = 1$; the error is a maximum when $N = 0$. In a typical measurement we might have $K_w \sim 8\ \mathrm{Wm^{-2}\,K^{-1}}$, $K \sim 14\ \mathrm{Wm^{-2}\,K^{-1}}$, and $N \sim 0.5$, leading to an error of about 30%. To incur smaller errors it is important to reduce as far as possible the difference between K and K_w. In this regard a mismatch of emissivities ε_w and ε_s can be especially troublesome.

2.6 Physical Principles for Thermal Sensors

We come now to a description of the physical principles that govern the operation of many of the most important and commonly-used thermal sensors. The characteristics and performance of actual sensors that utilize many of these principles are discussed in other chapters of this volume. As already noted, there are few physical properties that are not temperature dependent. Those that have been used as the basis for thermometers satisfy such criteria as simplicity, reproducibility, ease of measurability, adequate sensitivity, and monotonic variation with temperature. We speak of a primary or thermodynamic thermometer when the relationship between the physical quantity ψ and T can be derived from first principles, and of a secondary or practical thermometer when the relationship is simply empirical.

2.6.1 Primary (or Thermodynamic) Sensors

The principles described in this section include most of those that have been used in fundamental measurements of thermodynamic temperature. In each case the basic principle is outlined together with some indication of the major differences that occur between a real system that involves the principle and what an ideal system might do. These differences must be removed or corrected for a sensor to be used to high accuracy, but usually not if it is only required to measure changes in temperatures. Particular sensor or thermometer designs are not discussed.

2.6.1.1 Gas Thermometer

The theoretical law governing a gas thermometer is Equation (2-6),

$$PV = NRT.$$

It is rigorously true for a hypothetical entity – an ideal gas, but is also approximately true for a real gas at sufficiently low pressures and densities. The relationship is simple and its application straightforward; changes in T are measured in terms of changes in P (or V) when V (or P) and N are held constant, constant volume being far the most common. An equation of state which gives Equation (2-6) as a limiting case can be derived for a real gas in the form

$$PV = NRT\,[1 + B(T)\,N/V + C(T)\,(N/V)^2 + \ldots] \tag{2-37}$$

where $B(T)$, $C(T)$, ... are the second, third, ... virial coefficients. The virial coefficients are temperature dependent (as indicated) and are different for different gases. Their contribution to Equation (2-37) becomes larger at lower temperatures and high pressures where real gases depart more from ideality. In almost all gas thermometry the range of pressures used is such that terms in (N/V) to higher than the first power can be ignored, but knowledge of the second virial coefficient is required. Since constant volume gas thermometry is much the commonest, it is more convenient to express PV as a series in P, as follows

$$PV = NRT\,[1 + B(T)\,P/RT + (C(T) - B(T)^2)\,(P/RT)^2 + \ldots.] \tag{2-38}$$

where again, except for work of the highest accuracy, terms in P^2 and higher can be ignored. It is possible to calculate values of $B(T)$ [21] but usually it is simpler and more accurate to measure them. For ^4He (the most commonly-used gas) over a wide temperature range $B(T)$ varies with T approximately as

$$B(T) = a + b/T \tag{2-39}$$

where a, b are constants. If $B(T)$ is measured at a few temperatures, its values at other temperatures can be obtained from Equation (2-39). As normally practiced, the amount of gas in the sensor bulb is kept constant and temperatures calculated from measured pressures (called constant volume gas thermometry), which from Equation (2-38) gives

$$T = \frac{T_r P}{P_r} \frac{1 + B(T_r) P_r / R T_r}{1 + B(T) P / R T}, \tag{2-40}$$

ignoring the third and higher virial coefficients. This requires that some reference temperature and its associated pressure (T_r and P_r in Equation (2-40)) and the values of the second virial coefficient be known.

By using a somewhat different technique, called isotherm thermometry, one can directly measure the virial coefficients as well as the temperature. For this, one measures the quantity PV/RN (which from Equation (2-6) is essentially T) as a function of gas density N/V (or, which is equivalent from Equation (2-38), as a function of pressure) when the temperature is kept constant. From Equation (2-37), when higher order terms are neglected, the extrapolated intercept as $N/V \to 0$ gives T and the slope of the isotherm is $T B(T)$. For this, the number of moles of gas in the thermometer bulb must be known. In practice, then, the pressure is measured when the bulb is at temperature T and again when it is at a reference temperature T_r, the latter measurement essentially determining RN. The quantity of gas in the bulb is reduced and the pressure measurements repeated, and so on until a sufficient number of points on the isotherm are obtained to give T and $B(T)$. This technique has been dubbed relative isotherm thermometry as opposed to absolute isotherm thermometry; the latter term applies when the reference temperature T_r is the triple point of water (273.16 K) – the fundamental temperature of the thermodynamic scale. In this case, if T is far removed from 273.16 K, it is impractical to alternate the bulb temperature between T and T_r. Instead, two bulbs can be used, one at T and one at T_r, and an amount of gas transferred from the one at T_r to the one at T. From measurement of P_r before and after the transfer, and of P, the ratio of the two volumes can be obtained through suitable manipulation of Equations (2-37) and (2-38). The procedure is repeated for different amounts of gas initially in the reference volume to establish the isotherm.

For all of the above methods of gas thermometry, some important corrections must be applied to obtain an accurate temperature measurement. The first is for small changes in volume of the bulb due to thermal expansion when it is used at different temperatures; the correction is straightforward when the expansion coefficient of the material of the bulb is known. Secondly, in any gas thermometer, it is not possible to contain the gas completely in the bulb. A small portion is inevitably to be found in the "dead space" comprised of various capillary tubes connecting the filling and pressure measuring systems to the bulb. This dead space volume must be measured and corrected for, which requires measurement of the temperature distribution along the capillary that traverses the gradient between the bulb and room temperature. Similarly, the measured bulb pressure must be corrected for the weight of the gas in the connecting lines between the bulb and the pressure balance, which requires integration of the gas density along the length and temperature gradient of the tube. Fourth, if the capillary bore is smaller than the mean free path of the gas, a pressure difference will exist between its two ends due to thermomolecular flow. The ratio of the two end pressures is approximately equal to the square root of the ratio of the corresponding temperatures, so allowing correction. Errors can also arise from impurities in the gas (which is usually ^{4}He). The impurities can contribute in several ways, depending on the operation temperature: they can change the non-ideality of the gas through a different virial coefficient; they may be present at one temperature and freeze out at lower ones, effectively changing the number of moles of gas in the bulb; they may adsorb and desorb from bulb and tube surfaces to again affect the number of moles in the bulb. To reduce the latter error as far as possible, the bulb and associated tubing are baked for a considerable time at a temperature substantially higher than the highest to be measured while simultaneously being pumped to remove the impurities that were adsorbed on the surface. For a fuller discussion of gas thermometers, see [1, 22, 23].

The above discussion of the gas thermometer has emphasized its use for measuring thermodynamic temperatures. It can be used much more simply as a practical thermal sensor. For this it needs only be calibrated at a suitably-small number of temperatures, depending on the range of use, with linear or quadratic interpolation between them. It is used in this way with a 3-point calibration as a defining instrument of the ITS-90 from 3 K to 24.6 K (Section 2.3.4.1. b). In this mode the aerostatic head and thermomolecular pressure corrections are unnecessary as is (usually) the virial coefficient correction; they are in essence included in the calibration. They may, however, affect the form of the interpolation. In principle the interpolation should be strictly linear (Equation (2-6)), but the effect of ignoring these corrections is to require nonlinear interpolation.

2.6.1.2 *Acoustic Thermometer*

One can deduce from thermodynamic arguments that a longitudinal wave is propagated in an ideal gas of unbounded extent with a speed c_0 given by

$$c_0^2 = \frac{\gamma RT}{M} \tag{2-41}$$

where γ is the specific heat ratio c_p/c_v and M is the molecular weight of the gas. For a monatomic gas, γ is identically 5/3. Hence measurement of c_0 provides a measure of temperature. As with the gas thermometer, however, Equation (2-41) is a limiting case; several corrections are necessary to allow for non-ideality of the gas and measurement apparatus. In parallel with Equation (2-38), the speed of sound (c) in a real gas can be written in a virial expansion as

$$c^2 = c_0^2 + \sum_{i=1}^{k} A_i(T)\, P^i . \tag{2-42}$$

The first acoustic virial coefficient, $A_1(T)$, can be expressed in terms of the second virial coefficient $B(T)$ and its temperature derivatives by

$$A_1(T) = \frac{2\gamma}{M}\left[B(T) + \frac{2}{3}T\frac{\mathrm{d}B(T)}{\mathrm{d}T} + \frac{2}{15}T^2\frac{\mathrm{d}^2B(T)}{\mathrm{d}T^2}\right]. \tag{2-43}$$

Except for the very highest accuracy, terms in Equation (2-42) beyond the first are negligibly small.

The principle of acoustic thermometry is, with T kept constant, to measure the speed c at successively lower pressures and extrapolate towards $P \rightarrow 0$. Such acoustic isotherms are usually closely linear. Most applications are, in effect, sophisticated versions of the Kundt's tube experiment. A transducer operating at a known frequency is used to set up resonant standing waves in a cylindrical gas column. The corresponding wavelength is determined from interferometric measurements of the displacement of a piston over a counted number of internodal distances, the distance between adjacent nodes being one-half a wavelength. Thus c is measured as the product of wavelength and frequency and c_0 is determined from Equation (2-42) in the limit of $P \rightarrow 0$. A comprehensive treatment must also account for corrections to c arising from acoustic absorption by the gas, reflection losses at the end faces, boundary layer effects, and so on. Various gases have been used in acoustic thermometers – ^4He at low temperatures, Ar near room temperature, N_2 at higher temperatures.

There are obvious viable options for the operation of such an acoustic thermometer. One can choose between a fixed frequency and a variable path length (as outlined above) or vice versa. The former seems in general the better because the latter, although offering some mechanical simplification, requires a measure of absolute length as opposed to changes in length and demands much more complex data analysis due to the frequency dependences of various cavity parameters that enter into the corrections. Another choice is in the magnitude of the frequency. In the experimental arrangement outlined, plane wave propagation is implicitly assumed. In any such resonating cavity there are cut-off frequencies for each mode of propagation below which the mode cannot propagate. The cut-off frequencies depend upon, for example, the cavity dimensions, the gas, pressure, and temperature. For a plane wave the cut-off frequency is 0 Hz. The next lowest order mode typically has a cut-off frequency in the range 1 to 30 kHz. Plane wave propagation is then assured with appropriate choice of an audio-frequency. With an ultrasonic frequency, as with a crystal quartz oscillator, there is the possibility of the presence of higher modes whose effect on the result is not easily calculated. Plumb and Cataland [24] used a frequency near 1 MHz and were able to show that higher modes, if present, did not significantly affect their results. Colclough [25], on the other hand, used frequencies between 3 and 7 kHz.

In the audio-frequency range a boundary layer correction is necessary which is negligible at higher frequencies. Equation (2-42) applies to an unbounded gas; when the gas is confined in a tube the speed of sound is lowered because of heat conduction and viscous effects at the walls. This boundary layer correction is accurately calculable whereas the effect of multimodes is not, so that audio-frequency acoustic thermometry has a higher ultimate accuracy than ultrasonic.

In contrast to plane waves in a cylindrical gas column, Moldover and Trusler [26] have used radially-symmetric waves in a spherical resonator. Apart from using different sized spheres, this technique is of the fixed path – variable frequency type. The internal volume of the sphere must be known, but the frequencies are relatively unaffected by small geometrical imperfections so accurate measurement of internal sphere dimensions is unnecessary. Moldover and Trusler used the five lowest radially-symmetric modes in the frequency range 2 to 10 kHz. The measured resonant frequencies must be corrected for boundary layer heat conduction, attenuation in the gas, elastic changes in sphere volume, and departures from spherical symmetry. These in turn require values of the thermal expansion coefficient of the resonator and the virial coefficients and thermal parameters of the gas (argon). The speed c is obtained from the measured frequencies and volume, and T from Equation (2-42).

2.6.1.3 Dielectric Constant Thermometer

The Clausius-Mossotti equation relates the dielectric constant (ε) of a perfect gas to the molar polarizability (A_ε):

$$(\varepsilon - 1)/(\varepsilon + 2) = A_\varepsilon N/V. \tag{2-44}$$

Substitution for V from Equation (2-6) gives

$$P = \frac{(\varepsilon - 1)R}{(\varepsilon + 2)A_\varepsilon} T. \tag{2-45}$$

Thus T is directly measurable in terms of ε and P if A_ε is known either from calculation or measurement (as it is for ^{4}He). The dielectric constant is best measured in terms of capacitance, to which it is related by

$$C(P)/C(0) = \varepsilon(1 + KP) \tag{2-46}$$

where $C(P)$ is the capacitance at gas pressure P and K is the capacitor compressibility which is required to account for capacitor dimensional change when the gas pressure changes from 0 to P. For a real gas we turn again to a virial expansion for the dielectric constant which, written in terms of capacitance, is

$$P = A_1 \mu (1 + A_2 \mu + A_3 \mu^2 + \ldots) \tag{2-47}$$

where

$$\mu = \left(\frac{C(P)}{C(0)} - 1\right) \Bigg/ \left(\frac{C(P)}{C(0)} + 2\right). \tag{2-48}$$

The coefficients A_i in Equation (2-47) are related to the virial coefficients of Equation (2-37) as follows:

$$A_1 = \left(\frac{A_\varepsilon}{RT} + \frac{K}{3}\right)^{-1}$$

$$A_2 = (B - b)/A_\varepsilon$$

$$A_3 = C/A_\varepsilon^2 \qquad \text{(for the case of } ^4\text{He)}$$

where b is the second dielectric virial coefficient in the expansion of $(\varepsilon - 1)/(\varepsilon + 2)$ in powers of $(1/V)$. As before, T is obtained from the intercept of an isotherm (Equation (2-47)) determined by a series of measurements of capacitance at successively lower pressures. It is required to know both A_ε and K to find T. They either will be calculated, measured, or obtained from isotherms at known temperatures. Gugan and Michel [27] have shown that dielectric constant thermometry below 30 K has accuracy comparable to gas or acoustic thermometry.

2.6.1.4 Refractive Index Thermometer

The refractive index of a gas (n) is related to its dielectric constant by $n^2 = \varepsilon$ so that, corresponding to Equation (2-44), we have the Lorentz-Lorenz relation

$$\frac{n^2 - 1}{n^2 + 2} = A_\varepsilon N/V. \tag{2-49}$$

In principle one can determine T through measurement of n in a manner that parallels that of Section 2.6.1.3. Colclough [28] envisages an optical interferometric measurement of $(n - 1)$ by passage of a laser beam through a Michelson interferometer, one arm of which contains the gas. The optical path length (l) increases with the pressure and the relative change is proportional to the gas density, and so to P/T, through

$$\frac{\delta l}{l} = n - 1 \propto \varrho \propto \frac{P}{T}. \tag{2-50}$$

To allow for non-ideality of the gas, a refractive index virial expansion of $(n^2 - 1)/(n^2 + 2)$ in terms of pressure is used with T obtained from extrapolation to $P \to 0$. The coefficients in the expansion can be expressed in terms of the virial coefficients. The technique is likely best applied to temperature ratios through calibration at, say, the triple point of water. An accurate measurement of T involves corrections for thermal expansion of the gas cell, hydrostatic changes in cell length, and gas impurities. No successful application of this method to primary thermometry has yet been reported.

2.6.1.5 Noise Thermometer

Random thermal motion of the conduction electrons in a resistor generates a noise voltage, the mean square value ($\overline{V^2}$) of which is related to the temperature of the resistor by the Nyquist equation

$$\overline{V^2} = \int_0^\infty 4\,h\,\nu\,Z(\nu)\left[(\mathrm{e}^{h\nu/k_\mathrm{B}T} - 1)^{-1} + \frac{1}{2}\right]\mathrm{d}\nu \tag{2-51}$$

where h is Planck's constant, ν is the frequency, and $Z(\nu)$ is the resistor impedance. Except at extremely low temperatures, $h\nu \approx k_\mathrm{B}T$, and it is not difficult to ensure that the resistor impedance is frequency-independent, so Equation (2-51) can be written with no consequential error in the form

$$\overline{V^2} = 4\,k_\mathrm{B}\,TR\,\Delta\nu \tag{2-52}$$

where R is the resistance. In contrast to the foregoing types of sensors, there is no question here of non-ideality of the system and so of an approximate relationship. The voltage generated is independent of the particular composition of the resistor and, indeed, of the mass and nature of the charge carriers. For these reasons Equation (2-52) might be considered to be "more fundamental" than the earlier ones. Its exploitation to date has been rather less accurate, however, as a result of limitations of the measurement systems. Amplification of the low noise voltages is necessary, requiring that amplifier noise be distinguished from the resistor's thermal noise. Furthermore, amplifier gain and bandwidth must be known. The noise voltage of Equation (2-52) is a statistically-fluctuating quantity. If the fluctuations can be assumed to be Gaussian, then it can be shown [29] that the relative uncertainty in $\overline{V^2}$ is related to the measurement time t and bandwidth by

$$\frac{\Delta \overline{V^2}}{\overline{V^2}} \propto (t \Delta \nu)^{-1/2} . \tag{2-53}$$

The bandwidth must be chosen so that measurement times are not excessively long.

Several approaches to the measurement of $\overline{V^2}$ to obtain T have been reported. Garrison and Lawson [30] proposed to use two resistors R_0 and R (variable) with the former kept at a known reference temperature T_0 and the latter at T to be determined. The mean square voltages across each resistor are successively measured through the same amplifier and R varied until the voltages are equal. This effectively eliminates the amplifier characteristics and at balance $T = R_0 \, T_0/R$. Crovini and Actis [31] used instead two equal resistances at two temperatures. The noise voltage from the the one at the higher temperature was reduced with a precision attenuator until both were equal. From this null ratio, together with the known attenuation ratio, reference temperature, and an independent measure of the amplifier noise, T could be determined. For measurements at low temperatures, Klein et al. [32] used a single resistor with the noise voltage read simultaneously through two amplifiers and a correlator used to eliminate the uncorrelated amplifier noises. They found T from measurement of noise voltage with the resistor at T and then at T_0. At extremely low temperatures, the noise voltages become exceedingly small and normal amplification methods are insufficient. Kamper and Zimmerman [33] and later Soulen and Van Vechten [34] overcame this difficulty using a novel voltage-to-frequency conversion to obtain voltage fluctuations from more precisely-measurable frequency fluctuations. They connected a Josephson junction in parallel with the resistor, passed a small direct current through R to provide a small constant bias voltage on the junction causing it to oscillate in the audio-frequency range, and amplified the audio signal by rf parametric upconversion. Suitable analysis allows a measure of T with a relative accuracy ($\sim$0.1%) lower than the previous methods but adequate in the millikelvin range. For very high temperatures Borkowski and Blalock [35] introduced a method of noise thermometry that does not require the ohmic value of the sensing resistor. In essence, they determine T in terms of the thermal noise power, obtained from the product of the open-circuit thermal noise voltage and the short-circuit thermal noise current.

2.6.1.6 ^{3}He Melting Curve Thermometer

^{3}He has the unique property of being able to remain in either the liquid or solid state down to absolute zero. A specialty thermodynamic thermometer for below about 0.5 K makes use of the equilibrium liquid-solid coexistance curve of these two phases of ^{3}He. The ^{3}He is contained in a constant volume chamber, one wall of which is the diaphragm (usually Be-Cu) of a capacitance strain gauge for measurement of the pressure. Temperatures are calculated from measured pressures according to published tables of the $P - T$ equilibrium melting curve data. Perhaps the preferred table is that of Greywall [36] which presents $(P - P_A)$ as a polynomial in T, where P_A is a reference or calibration pressure measured at the transition fixed point between the normal liquid and the superfluid A phase at 2.7 mK. The $P - T$ values along the equilibrium melting curve have been obtained by some such procedure as the following. The slope of the melting curve is determined by the Clausius-Clapeyron equation which may be written in the form

$$\frac{\mathrm{d}P}{\mathrm{d}T} = \frac{S_L - S_S}{V_L - V_S} \tag{2-54}$$

where S_L, S_S are the molar entropies of the liquid and solid states respectively, V_L, V_S are molar volumes of the liquid and solid respectively at melting, and P, T refer to the melting pressure and temperature. Integration of Equation (2-54) would give the melting curve relationship. To determine this, with pressure and temperature constant, a measured amount of heat ΔQ is injected into the sample and the change in volume ΔV measured. In the process Δn moles of liquid are converted to solid, and $\Delta n = \Delta V/(V_S - V_L)$. Furthermore, ΔQ is an amount of latent heat given by $\Delta Q = \Delta n\, T(S_S - S_L)$. Substitution of these into Equation (2-54) gives $T\,\mathrm{d}P/\mathrm{d}T = \Delta Q/\Delta V$. ΔQ and ΔV are measured for various amounts of injected heat and integration of the data leads to the equation for the melting curve. It is roughly parabolic in shape with a minimum near 0.32 K. This minimum can serve as another useful calibration point.

2.6.1.7 Nuclear Orientation Thermometer

A nuclear orientation thermometer, another speciality primary thermometer for use at very low temperatures and especially below 100 mK, is based upon the principle that β or γ radiation from a radioactive nucleus is emitted directionally with a distribution that is determined by the degree of ordering of a system of nuclear spins. The degree of ordering, in turn, is directly related to the temperature through the Boltzmann distribution which governs the substate populations of the nuclear spin system. The relative populations a_m are given by

$$a_m = \mathrm{e}^{-E_m/k_B T} / \sum_n \mathrm{e}^{-E_n/k_B T} \tag{2-55}$$

where E_n are the energies of the nuclear hyperfine states and n take the values $I, I - 1, \ldots, -I$, where I is the nuclear spin quantum number. In Equation (2-55), T is the temperature of the spin system which will only be the same as that of the lattice if the two are in thermodynamic equilibrium. When $T \gg E_m/k_B$, all of the a_m are approximately equal (at $T = \infty$ they are exactly equal) and the γ-radiation is emitted isotropically; when $T \ll E_m/k_B$, the various a_m take on different values, resulting in nuclear orientation and anisotropic emission about the axis of quantization of the nuclear spin system with a distributional pattern rather like that from a radar antenna. At $T = \infty$ or any temperature down to about 100 mK, the distribution of intensity versus angle of emission is spherically symmetric. At $T = 0$ there are symmetric, roughly spherical lobes on either side of the orientation axis. As T increases from 0, there is a gradual transition between these two extreme cases.

The normalized directional distribution $W(\theta)$ of γ-radiation (or in essence the counting rate) from axially symmetric oriented nuclei is given by

$$W(\theta) = \sum_{\lambda=0}^{\lambda_{max}} B_\lambda(T)\, U_\lambda\, A_\lambda\, Q_\lambda\, P_\lambda(\cos\theta) \tag{2-56}$$

where θ is the angle between the direction of emission and the orientation axis; $B_\lambda(T)$ are statistical tensors (called orientation parameters) which contain a_m and so all of the temperature dependence; U_λ are angular momentum deorientation coefficients; A_λ are angular correlation coefficients that depend upon the particular radioactive decay scheme; $P_\lambda(\cos\theta)$ are Legendre polynomials; and Q_λ are solid angle correction factors to allow for non-point collection by the detector. Equation (2-56) is strictly true for point source emission, and this can usually be sufficiently well approximated. The specific forms of the various coefficients and an extended discussion of the principles are given by Marshak [37].

To use this thermometer, one incorporates suitable radioactive nuclei into substitutional sites in a host lattice that is frequently, but not necessarily, ferromagnetic; in the large hyperfine magnetic field of the host, the nuclei undergo spontaneous ordering when the temperature is in the millikelvin region. Commonly-used nuclei have been ^{54}Mn in Fe, Ni, Al, Cu, or Zn and ^{60}Co in Fe, Co, or Ni. In any case the decay scheme should be completely known to allow exact calculation of U_λ and A_λ. Since the greatest sensitivity of $W(\theta)$ to T occurs at $\theta = 0\,°$ and $90\,°$, measurement of $W(\theta)$ can be confined to one or both of these two angles. In most cases the relation between $W(0\,°C)$ and T is unique; if not, the additional measurement of $W(90\,°C)$ will be definitive. Rather than use an absolute measure of $W(0\,°C)$, $W(0\,°C)$ is determined from the ratio of counts for system at T (cold counts) to counts for system at a higher temperature where the distribution is isotropic (warm counts). The calculation of T from Equation (2-56) is a straightforward but non-trivial task.

The most desirable orientation thermometers are operable both with and without an applied magnetic field, although usually there will be an applied field that acts to polarize the host lattice. It is necessary for all magnetic fields operating on the nucleus to be colinear; otherwise corrections are necessary. Counting times are typically several minutes, so T must be constant for at least this long, but it is desirable to repeat several cycles of counts with T constant to aid with the counting statistics. The effect of background counts must be corrected for. Another uncertainty arises from radioactive heating: self heating due to β emission can be calculated easily; heating due to γ activity is more difficult to calculate; both should be estimated for any measurement.

2.6.1.8 *Total Radiation Thermometer*

Perhaps the most important primary thermometers above about 300 K are those based upon measurement of thermal radiation, either total or spectral. They differ fundamentally from all of those previously described in requiring no physical contact with the radiating system. The principles of the thermometers stem from Planck's law which relates the spectral radiance $L_\lambda(T)$ of a blackbody to wavelength λ (in vacuum) and temperature T as follows:

$$L_\lambda(T)\,d\lambda = c_1\,\lambda^{-5}\,n^2\,(e^{c_2/\lambda T} - 1)^{-1}\,d\lambda \tag{2-57}$$

where c_1 and c_2 are the first and second radiation constants whose values in terms of the fundamental constants are $c_1 = 2\,h\,c^2$ and $c_2 = hc/k_B$, and n is the refractive index of the medium into which the blackbody radiates (usually air or vacuum). Equation (2-57) is frequently written in terms of spectral radiant exitance $M_\lambda\,(T)$ instead of spectral radiance (see Chapter 5). In this case the form of the equation is unchanged but, since $M_\lambda\,(T) = \pi\,L_\lambda\,(T)$, the constant c_1 takes the value $2\,\pi\,hc^2$. A real surface is never black and can emit only some fraction $\varepsilon_\lambda\,(T)$ of the blackbody spectral radiance, where $\varepsilon_\lambda\,(T)$ is the spectral emissivity and may vary with λ and T.

The total radiance ($L\,(T)$) of the blackbody is obtained by integration over λ to be

$$L\,(T) = \int_0^\infty L_\lambda\,(T)\,\mathrm{d}\lambda$$

$$L\,(T) = \pi^{-1}\,n^2\,\sigma\,T^4 \qquad (2\text{-}58)$$

or, in terms of total radiant exitance $M\,(T) = \pi L\,(T)$,

$$M\,(T) = n^2\,\sigma\,T^4 \qquad (2\text{-}59)$$

where σ is the Stefan-Boltzmann constant ($= 2\,\pi^5\,k_B^4/\,15\,h^3\,c^2$). Equation (2-59), known as the Stefan-Boltzmann law, expresses the particularly simple principle that the total exitance from a blackbody is dependent only upon the 4th power of its temperature. Only for the most exacting measurements is the difference of n^2 from unity of consequence. The total exitance from a real surface is $\varepsilon\,(T)\;n^2\;\sigma\,T^4$, where $\varepsilon\,(T)$ is the total emissivity in the direction of view.

A blackbody may be approximated by an isothermal cavity with an aperture in the wall through which the exitance is emitted. The goodness of the approximation is governed roughly by the ratio of the area of the aperture to the total internal surface area, by the shape of the cavity, and by the emissivity of the material of the surface. Bedford [38] has described methods for calculating the effective emissivities of cavities in terms of $\varepsilon_\lambda\,(T)$ with a high degree of exactitude. If the cavity is isothermal the effective emissivity is independent of λ and T. When cavities are used to simulate blackbodies, it must be remembered that Equations (2-57) to (2-59) are strictly valid only in the limit of infinitely-large cavity dimensions. Corrections may be necessary if the wavelength is of the same order as one of the dimensions [39]; they are likely to be significant only in the very far infrared or, complementarily, at extremely low temperatures.

A primary total radiation thermometer measures temperature by application of Equation (2-59). However, absolute measurement of radiant power to better than 0.1% is exceedingly difficult; this would correspond, for example, to only 0.25 K at 100 K and 2.5 K at 1000 K, which are insufficient for primary thermometry. It is much simpler to measure ratios of radiant power in order to determine T. Suppose that a thermal sensor at temperature T_S and with total emissivity ε_S views a blackbody simulator (with effective emissivity ε_1) at temperature T_1 in such a way that it collects some fraction g of its hemispherical exitance. The purely geometrical quantity g is determined by the projected solid angle subtended at the sensor by the blackbody. Then the power Q_1' absorbed by the sensor per unit area will be, using Equation (2-59),

$$Q_1' = g\,F_\varepsilon\,n^2\,\sigma\,(T_1^4 - T_S^4) + Q_0' \qquad (2\text{-}60)$$

where F_ε is some function of ε_S and ε_1 (that in some cases is closely equal to the product $\varepsilon_1\,\varepsilon_2$) and Q_0' is the net radiant power exchanged with the surround. Similarly, if the sensor next views a second blackbody at temperature T_2, the power Q_2' absorbed is

$$Q_2' = g\,F_\varepsilon\,n^2\,\sigma\,(T_2^4 - T_S^4) + Q_0' . \tag{2-61}$$

In Equations (2-60) and (2-61) it is assumed that g and F_ε are the same for the two viewing conditions, and this is not difficult to achieve in practice. Similarly, the sensor temperature T_S and the exchange with the surround are assumed to be the same when viewing the blackbodies at different temperatures; this is not strictly true but no significant error is so incurred. Dividing one equation by the other,

$$\frac{Q_1'}{Q_2'} = \frac{g\,F_\varepsilon\,n^2\,\sigma\,(T_1^4 - T_S^4) + Q_0'}{g\,F_\varepsilon\,n^2\,\sigma\,(T_2^4 - T_S^4) + Q_0'} . \tag{2-62}$$

If one of T_1 or T_2 is known, the other can be calculated from Equation (2-62) from the measured ratio of sensor outputs if the various other quantities are somehow accounted for. Quinn and Martin [40] solved the problem in an elegant fashion using a calorimeter at $T_S \sim 2$ K as sensor and a radiation trap at 4.2 K as the bulk of the surround, with the complete system evacuated. For T_1 and $T_2 > 200$ K the terms T_S^4 and Q_0' become negligible and the equation takes the simple form

$$Q_1'/Q_2' = T_1^4/T_2^4 . \tag{2-63}$$

If either T_1 or T_2 is 273.16 K then the other is measured absolutely.

For high accuracy, corrections that must be considered include diffraction at the apertures, stray power that might reach the sensor by reflection from aperture edges, surround, etc., nonequality of ε_1, ε_2, ε_S (each of which is calculable [38]), and temperature gradients in the blackbody simulators.

2.6.1.9 Spectral Radiation Thermometer

According to Planck's law (Equation (2-57)) the spectral radiance of a blackbody at a particular wavelength varies only with the blackbody temperature. The monochromatic spectral radiation thermometer (or optical pyrometer) makes use of this principle by measuring ratios of spectral radiances in much the same way as it is used to define temperatures on the ITS-90 (Equation (2-23)) by

$$\frac{L_\lambda(T_1)}{L_\lambda(T_2)} = \frac{e^{c_2/\lambda T_2} - 1}{e^{c_2/\lambda T_1} - 1} .$$

One measures the ratio on the left hand side of Equation (2-23) and can then calculate T_1 if T_2 and λ are known. Thus an optical pyrometer, like many other primary thermometers, requires a known reference temperature in terms of which to measure temperatures. It is not possible to use the water triple point as a reference because suitably-stable radiation detectors (photomultipliers or silicon photodiodes) are insufficiently sensitive at the far infrared wavelengths where a blackbody at 273.16 K emits the bulk of its energy.

In a typical optical pyrometer an image of the blackbody aperture is focussed on to the sensitive surface of a photon detector by a (refractive or reflective) optical system that contains a narrow band interference filter or a monochromator to isolate the wavelength. For historical reasons connected with the obsolescent disappearing filament visual optical pyrometer, λ is frequently chosen to be near 650 nm and a photomultiplier is used as detector. More advantageously, since the maximum of $L_\lambda (T)$ shifts to the infrared as T decreases, use of a silicon photodiode allows measurements to much lower temperatures because the diode is sensitive at longer wavelengths.

Although called monochromatic, an optical pyrometer has a finite spectral bandwidth (typically between 1 and 20 nm). When it views a blackbody cavity at temperature T_1 its response R_1 is of the form

$$R_1 \propto \int_0^\infty \varepsilon_1 \, L_\lambda (T_1) \, s(\lambda) \, \tau_f (\lambda) \, \tau_0 (\lambda) \, d\lambda \tag{2-64}$$

where ε_1 is the effective emissivity of the cavity, $s(\lambda)$ is the spectral sensitivity of the detector, $\tau_f (\lambda)$ is the spectral transmittance of the interference filter or monochromator, and $\tau_0 (\lambda)$ is the spectral transmittance of all other optical components. Within the pass band, $\tau_0 (\lambda)$ is non-selective and can be taken outside the integral.

Thus we may write for the measured ratio R_1/R_2 of the pyrometer responses when it alternately views two blackbodies at temperature T_1 and T_2

$$\frac{R_1}{R_2} = \frac{\int_0^\infty \varepsilon_1 \, L_\lambda (T_1) \, s(\lambda) \, \tau_f (\lambda) \, d\lambda}{\int_0^\infty \varepsilon_2 \, L_\lambda (T_2) \, s(\lambda) \, \tau_f (\lambda) \, d\lambda} . \tag{2-65}$$

Since $L_\lambda (T)$ increases exponentially with T, the ratio R_1/R_2 can be far from unity; for example, for blackbodies at the freezing points of Al and Au and $\lambda = 900$ nm, $R_1/R_2 \sim 500$. Furthermore, it follows directly from Planck's law that an uncertainty ΔL in spectral radiance produces an uncertainty ΔT in measured temperature given by

$$\frac{\Delta L}{L_\lambda (T)} \sim \frac{c_2}{\lambda T} \frac{\Delta T}{T} . \tag{2-66}$$

Thus, in the example considered, if we wish $\Delta T < 10$ mK, then we must measure $L_\lambda (T)$ (i.e., R) to better than 1 part in 10^4. This places an extraordinary linearity requirement on the detector. Either we must independently ascertain that the detector is linear to this order over a dynamic range of 500 or, which is equivalent, measure the amount of non-linearity and apply a correction. Alternatively, one can reduce the radiance at the detector from the higher temperature source to be near that from the lower temperature source by interposing in the optical beam from the hotter source a filter of known transmittance τ_n that is non-selective in the pyrometer pass band. This requires that τ_n be known to better than 1 part in 10^4, which is a stringent requirement, especially when τ_n itself is small (~ 0.002 in the above example). Rotating sectored disks have been used as neutral filters. Another possibility is simply to use a pryrometer already calibrated on the ITS-90. So long as c_2 and T_{90} (Ag) are correctly assigned, the ITS-90 is coincident with thermodynamic temperatures above 1235 K. For the ITS-90 calibration a radiance multiplier that allows superposition of several equal radiances can be used instead of a neutral filter to establish the scale [41]. In this case, the pyrometer is calibrated against reference standard lamps. The accuracy of measurement of T will involve the uncertainties in the scale realization and in the lamp reliability.

For a measurement of this type the pyrometer detector acts essentially as a null detector. Bedford and Ma [42] have shown that the requirements on knowledge of $s(\lambda)$ are not stringent. The pass band of the selective filter ($\tau_f(\lambda)$) must be carefully determined, however, typically to within 0.1 nm for a 10 nm band width. Just as important, at all wavelengths outside the pass band where the detector has significant sensitivity, the filter must be blocked to the extent that the out-of-band transmittance at any wavelength (and especially on the long wavelength side) is less than 10^{-5} of the maximum of the pass band transmittance.

Historically, to simplify the solution of Equation (2-65) for T_1, the selective filter has been characterized by an effective wavelength λ_e so defined as to make the ratio of Equation (2-65) equal to the ratio of Equation (2-23). Methods for determining λ_e are described by, for example, Kostkowski and Lee [43]. It is now simpler and more accurate to solve Equation (2-65) directly by numerical integration and iteration.

For high precision measurements, corrections also have to be included for the effective emissivities of the blackbodies; effects of any scattered or reflected radiation as, for example, between optical components; proper orientation of the interference filter whose transmittance is extremely sensitive to angle of the incident radiation; and size-of-source effects. For the latter, if the radiating areas of the two sources are of different size, there may be an error caused by losses of different amounts of energy from the two geometric images by diffraction and scattering. The amount of this is best determined by measurement for a particular pyrometer [44].

2.6.1.10 Vapor Pressure Thermometer

A vapor pressure thermometer is one of a number of pseudo-thermodynamic thermometers. The underlying equation is derivable from thermodynamic considerations but it contains parameters whose values are not well enough known to give adequate temperature measurement accuracy. Instead, the required values are obtained essentially from calibration at several known reference temperatures; thereafter the thermometer produces a thermodynamically smooth set of temperatures. For any particular working substance, the range is limited from near the melting point to near the boiling point, but for many substances there is an enormous potential sensitivity. Beginning with the Clausius-Clapeyron equation in the form

$$\frac{\mathrm{d}P}{\mathrm{d}T} = \frac{L}{T(V_G - V_L)} \tag{2-67}$$

(where P is the saturated vapor pressure, L is the molar heat of vaporization and V_G (V_L) is the molar volume in the gaseous (liquid) state), using a virial expansion for PV, and integrating Equation (2-67), one can deduce the following relation between P and T:

$$\ln P = -\frac{L_0}{RT} + \frac{5}{2}\ln T + i_0 - \frac{1}{RT}\left(\int_0^T S_L \, dT - \int_0^P V_L \, dP\right) + \varepsilon(T) \qquad (2\text{-}68)$$

where L_0 is the heat of vaporization at 0 K, i_0 is the chemical constant ($= \ln[(2\pi m)^{3/2} k_B{}^{5/2}/h^3]$, where m is the atomic mass), S_L is the molar entropy, and $\varepsilon(T)$ is a quantity involving the virial coefficients. Equation (2-68) was used in conjunction with calibrations to derive the vapor pressure/temperature relationships [45] for ^{3}He and ^{4}He that are incorporated into the ITS-90, although the finally recommended equations are expressed in a different form.

For other substances, it is more usual simply to measure the vapor pressure at a series of temperatures and to fit an equation of the form

$$\ln(P/P_0) = \sum_{i=-m}^{m} a_i T^i . \qquad (2\text{-}69)$$

Such equations are available for many substances [46].

Some of the difficulties attached to accurate vapor pressure thermometry are common to gas thermometry – in particular, thermomolecular pressure effects can be troublesome and result in discrepancies between the true saturation vapor pressures and those indicated by the pressure sensor. Purity of the working substance must also be assured, and hydrostatic head pressures corrected for.

2.6.1.11 Magnetic Thermometer

The magnetic susceptibility χ of a paramagnetic salt is another pseudo-thermodynamic thermometer. The form of the equation relating χ to T can be derived from first principles but the numerical values of the three or four constants in the equation cannot be sufficiently accurately calculated for thermometric purposes and are obtained by calibration at reference temperatures. The equation then provides thermodynamically smooth interpolation and extrapolation. The technique is most often used in the approximate range 10 mK to 20 K, but has been successfully extended to above 80 K. Different salts are used in different ranges: cerium magnesium nitrate (CMN) has been the most extensively used, its range for primary measurements having an upper limit of about 2.6 K with powdered samples, and rather higher for single crystals, and a lower limit below 10 mK. Other useful salts are chromic methylammonium alum (CMA) from 0.3 to 30 K, manganous ammonium sulfate (MAS) from 0.9 to 83 K, and gadolinium sulfate from 1 to 83 K. Stability and repeatability to about 1 mK have been achieved at the higher, and to about 0.2 mK at the lower, temperatures. The susceptibility, and therefore the sensitivity, increases as T decreases.

The basis of the magnetic thermometer lies in the ordering of the magnetic dipoles of the paramagnetic salt that occurs at low enough temperatures in the presence of an applied magnetic field. For a salt to be suitable for thermometry, the temperature-dependent susceptibility should be associated solely with electron spin magnetism and not with the orbital angular momentum, so the latter will have to have been quenched. Under these conditions the first excited state normally lies much higher than the ground state, so is effectively unpopulated and can be ignored. CMN is an exception; it is the relative closeness of the first excited state and its increasing population density with temperature that limits the useful upper temperature to near 2.6 K.

According to the Curie law, the susceptibility of a system of non-interacting dipoles that has a degenerate ground state whose levels are split by an applied magnetic field varies inversely with temperature:

$$\chi = N_a J(J + 1) g^2 \mu_B^{2/3} k_B T = \frac{C}{T} \tag{2-70}$$

where J is the total angular momentum quantum number of the ground state, g is the Landé g-factor, μ_B is the Bohr magneton, and C is the Curie constant. This simple relation must be modified for a real salt to take account of dipole-dipole interactions and ground state splitting in the crystal field. The result is

$$\chi = \chi_V + \frac{C}{T - \left(\frac{4\pi}{3} - \varepsilon\right) C - \theta + \frac{\delta}{T}} \tag{2-71}$$

where χ_V is the temperature-independent susceptibility, $(4\pi/3 - \varepsilon)$ accounts for dipole-dipole coupling in the local field, ε is the demagnetization coefficient, θ results from ion exchange interactions, and δ/T is associated with Stark splitting of the ground state. The denominator of Equation (2-71) is sensitive to the geometry of the sample of salt – whether it is spherical or cylindrical, for example, and whether it is a single crystal or in powdered form. For a spherical sample, $\varepsilon = 4\pi/3$ and the second term disappears. These terms are also dependent upon the angle between the applied field and the crystal axis. For a detailed discussion of the theory, see Hudson [47]. Equation (2-71) is usually rewritten in the form

$$\chi = \chi_V + \frac{C}{T + \Delta + \frac{\delta}{T}}\,. \tag{2-72}$$

When a sample of the paramagnetic salt is placed within a set of mutual inductance coils, the coefficient of mutual inductance is linearly related to χ so that the bridge reading M may be written

$$M = M_0 + \frac{B}{T + \Delta + \frac{\delta}{T}}\,. \tag{2-73}$$

In Equation (2-73) M_0 and B, which depend upon χ_V and C respectively, also depend upon the coil design; Δ is a property of the salt and depends upon the shape of the sample;

δ is a property of the salt only. Calibration at four temperatures is necessary to determine the values of these four constants. CMN is nearly an ideal paramagnetic salt, having both θ and δ very near to zero. The constants of Equation (2-73) determined for one sample cannot be used generically; every salt sample must be individually calibrated.

Magnetic thermometers can also be based upon nuclear spin systems, the mutual interactions of which are about 1000 times weaker than those between electron spin systems. Consequently, the ordering temperatures for nuclear dipoles is much lower than for electronic dipoles and nuclear spin systems follow a Curie law of the form of Equation (2-70) down to about 10 μK. It is necessary that the spin system be in thermal equilibrium with the lattice. To be useful for thermometry, this spin-lattice relaxation time must be short and so metals (commonly platinum) are used. See Hudson [47] for details of nuclear paramagnetism.

2.6.1.12 Spectroscopic Techniques

Measurement of the temperatures of flames, heated gases, plasmas, and stellar objects is inconvenient or (more often) impossible with conventional sensors. For this a variety of spectroscopic techniques has been employed. They rely upon formulae that relate the intensities or profiles of atomic, ionic, or molecular spectral lines and bands to temperature, or upon Planck-Kirchhoff relationships between emission and absorption. The systems are gaseous and may be optically thin, in which case the average temperature in the line of sight is measured. All of the techniques require (and assume) the existence of local thermodynamic equilibrium (LTE) within the sampled volume. (For good discussions of LTE see [48–50]). The measured temperatures are in the range 500 K to 100000 K; an accuracy within a few percent is very good. Within the scope of this chapter we can do little more than list a few of these techniques. For good detailed descriptions see [49, 51–53].

The method of line reversal involves a simultaneous spectral radiance match between the system (usually a flame) and an auxiliary, calibrated, variable-temperature source. This source is viewed through the medium. A spectral line originating in the medium appears darker or brighter than the source depending as the medium temperature is lower or higher. At the point of reversal the temperatures are equal. Frequently the alkali metal resonance lines (such as sodium D) are used; if not naturally present, metal atoms can be seeded. If the medium path length is nonisothermal, the observed reversal temperature will approximate the temperature of the hotter region if it is nearer the sensor; otherwise some sort of average is obtained. For high speed processes this classical method has been generalized so that the actual point of reversal need not be observed, which would be prohibitively slow. Instead, the spectral radiances of source, medium, and medium plus source are independently observed; from these three quantities T can be calculated.

The two-path method involves a comparison of spectral radiances along two different paths through the medium, as by doubling the path with a mirror. From the ratio of these radiances the medium temperature is calculable through successive applications of Planck's law. Values of transition probabilities are not required. Again, it is commonly the radiance of a seeded material that is observed. In another variation, emission from the medium (at a laser wavelength) is measured simultaneously with absorption of a background laser beam by the medium.

There are a variety of methods based upon the absolute or relative intensities of atomic, ionic, or molecular lines, relative measurements being much the more common. The Boltzmann distribution relates temperatures to the ratio of intensities of two lines by

$$\frac{I_{ki}}{I_{nm}} = \frac{A_{ki}}{A_{nm}} e^{-(E_k - E_n)/k_B T} \tag{2-74}$$

where I_{ki} is the intensity of the line between levels k and i, A_{ki} is the corresponding transition probability, and E_k is the energy of level k. This requires knowledge of the A's and E's. Frequently, the intensities of several rotational lines of a branch are compared. The temperature is then obtained from:

$$\ln \frac{I_{ki}}{A_{ki}\, \nu_{ki}^4} = C - \frac{E_{\mathrm{k}}}{k_B T} \tag{2-75}$$

where ν_{ki} is the line frequency. A plot of the left hand side of Equation (2-75) versus E_k is a straight line with slope $1/T$.

As the temperature of a gaseous system increases, the increasing random thermal motion of the radiating particles produces a Doppler broadening of their spectral lines. The resulting line half-width $\Delta\lambda_D$ is related to temperature by

$$\Delta\lambda_D = 7.16 \times 10^{-7}\, \lambda \left(\frac{T}{M}\right)^{1/2} . \tag{2-76}$$

The numeric in Equation (2-76) arises from a constant involving the gas constant and the speed of light. Since the broadening arises from translational particle motion, it is a fair measure of the system temperature, but one must ensure that the broadening is not due to other causes. The method is insensitive, however, requires highly precise measurement of line widths, and so is used only in special circumstances.

A variety of ways of deducing gaseous temperatures through analysis of scattered radiation has been devised [51]. These make use of Thomson scattering, Rayleigh scattering, Raman scattering, or laser-induced fluorescence. Because scattered intensities are generally low, thermometry based upon scattering almost invariably involves the use of a high intensity laser to provide a sufficient number of photons in the scattered beam. At the same time this allows rapid ($\sim$nsec) measurement of the temperatures in very small volumes (<0.1 mm^3). Of all of these methods, the one that has gained the most prominence and is probably the most accurate (and expensive) has been given the appellation "coherent anti-Stokes Raman spectroscopy" (CARS) [53]. Because it involves coherence, the radiant beam to be analyzed emerges laser-like and can be completely collected, in contrast to incoherent processes where more than 99% of the scattered radiation may be lost. Furthermore, one can tune the system to resonance. Together, these produce a tremendous gain in signal level. In a typical system three high power laser beams are made to intersect in the desired location in the gas. Two of these of fixed frequency ν_1 are derived from the same beam (commonly a Nd:YAG laser); the second (commonly a YAG-pumped dye laser) has tunable frequency ν_2 where $\nu_2 < \nu_1$. These beams interact through the third order non-linear susceptibility; ie, the induced polarization exhibits a cubic dependence on the optical electric field strength of the target particles. When the frequency ν_2 is tuned until $(\nu_1 - \nu_2)$ is near the frequency of a Raman active resonance, an enhanced, coherent, laser-like, anti-Stokes signal is produced at frequency $\nu_3 = 2\nu_1 - \nu_2$. With sufficient ν_2 tuning, or alternatively with a broad-band laser, a whole anti-Stokes spectrum can be generated. Temperature is deduced from line intensities and widths, usually by comparing measured and computer-modelled theoretical spectra. The analysis is complex.

2.6.2 Secondary (or Practical) Sensors

For secondary sensors there is no theoretically derivable equation that relates the sensitive parameter to temperature or, if there is, it is too nebulously-based or agrees insufficiently well with experiment to provide a useful working relationship. Instead, an empirical interpolating equation (that may rely on theory for its general form) is used, the coefficients of which are obtained by calibration at a suitable number of known temperatures. These sensors are invariably calibrated in terms of practical temperatures as defined by the ITS-90. For simplicity in the discussion to follow we will designate temperatures by T but it must be remembered that they now mean T_{90} and not thermodynamic temperatures. Lack of space permits discussion of only a few of the almost unlimited number of physical properties that have been used as the basis of thermal sensors. In particular, there is no discussion here of such optical sensors as liquid crystals or fluorescence decay, nor of the application of infrared photon detectors in thermography.

2.6.2.1 Electrical Resistance

The variation of electrical resistance (R) with temperature has been used for almost 100 years as the basis of a thermometer. The method is simple, reliable, and sensitive. The sensitive parameter R is easily measured in terms of known reference resistors by the voltage drops across them when a fixed current is passed through them. Digital voltmeters or direct reading resistance bridges with discriminations to parts in 10^7 or 10^8 are readily available. Resistance sensors of all kinds are among the most-used thermometers and, of these, the platinum resistance thermometer is the most accurate temperature-measuring instrument available. Resistance thermometers are described in detail in Chapter 3; the discussion here is limited to a few remarks on the physical principles of their operation. The types of materials used can be broadly classed as metals and non-metals.

a) Metals

Theory is of limited help in predicting a specific relation between R and T. The electrical resistance in a pure metal arises from scattering of the conduction electrons by phonons and electrons and by impurities and lattice imperfections. The former can be shown to be roughly proportional to T at temperatures above about 0.2 θ_D (where θ_D is the Debye temperature), although in a transition metal such as platinum this variation is modified to become nearer a T^2 dependence. Below 0.2 θ_D the temperature dependence is approximately as T^5, approaching a constant value at very low temperatures where the approximately temperature-independent impurity and defect scattering dominates. This constant low-temperature resistance is called the residual resistance, which arises from scattering processes that would not be present in an ideally pure metal. If we express the resistance in the form

$$R(T) = R_1(T) + R_2 \tag{2-77}$$

where R_2 is the residual resistance, and if R_2 is truly temperature-independent, then the function

$$Z(T) = \frac{R(T) - R(T_1)}{R(T_2) - R(T_1)}, \tag{2-78}$$

where T_1 and T_2 are calibration temperatures, should allow correct thermodynamic interpolation for any specimen of a pure metal such as platinum. Equation (2-77) is an expression of Matthiessen's rule; $Z(T)$ is called the Cragoe function. Unfortunately, $Z(T)$ has been found to give only rough thermodynamic interpolation and, worse, to be a different function for every thermometer. Nevertheless, Equation (2-78) is from time to time utilized as an empirical interpolation function for platinum. At very high temperature one might expect an additional term in Equation (2-77) of the form $\exp(-E_V/k_B T)$ (where E_V is the energy of formation of a lattice vacancy) to allow for the increasing equilibrium concentration of vacancies in the metal lattice as T increases.

By far the most used pure metal resistance sensor is platinum although copper and nickel are used industrially to some extent. Interpolation is usually by simple polynomials that relate resistance ratio $W(T)$ to T, except where platinum is used as an ITS-90 defining instrument. There a much more complicated interpolation is needed (Equations (2-15) to (2-22)) because of the dual requirements that different calibrated thermometers produce the same values for measured temperatures to within a very close tolerance and that T_{90} be a very close approximation to T. The sensitivity $\mathrm{d}W/\mathrm{d}T$ of a platinum sensor is roughly 0.4%/K above 50 K, reaching a maximum near 80 K and then decreasing slowly at T increases. Below 50 K, it falls rapidly to about 0.01%/K at 10 K. Copper has a sensitivity of about 0.4%/K above 200 K and nickel of about 0.7%/K.

For very low temperatures where the sensitivity of platinum is low, metal alloy thermometers (notably RhFe) have proven effective. It has been shown [54] that the impurity contribution to the resistivity of rhodium of less than 1% iron has a large and positive temperature coefficient below about 30 K. The result of this is that the overall resistivity of the RhFe alloy decreases approximately monotonically as T decreases down to very low temperatures. Rusby [55] exploited this to develop the most reliable resistance sensor below about 30 K using an optimum iron concentration in rhodium of about 0.5%. Its sensitivity $\left(\frac{1}{R}\frac{\mathrm{d}R}{\mathrm{d}T}\right)$ is lower than that of platinum down to about 5 K, but below it is immeasurably higher. More important, its voltage sensitivity ($\mathrm{d}V/\mathrm{d}T$) for typical measuring currents is relatively constant and comparable to that of platinum down to 20 K, and then rises as the temperature is further decreased. Another dilute alloy thermometer with properties similar to RhFe is platinum with 0.5 atomic percent cobalt.

b) Non-metals

Electrical conduction in a doped semiconductor is strikingly different from in a metal. In a metal the number of conduction electrons per unit volume is large and relatively constant with temperature. In a semiconductor the number is small and very sensitive to temperature because electrons must be thermally excited into the conduction band or from one impurity atom to another in order to contribute to the conductivity. At the temperature where semiconductors are of interest as thermal sensors (< 30 K), intrinsic conduction plays little part. Conduction is almost wholly due to the ionization of impurity atoms, the number of electrons released to the conduction band being determined by the temperature and the number and kind of impurity donor and acceptor atoms. As T decreases, ionization of the donor atoms decreases in an exponential fashion and so the resistivity rises exponentially. Finally, ionization ceases and below about 10 K conduction is by direct electron transfer between impurity atoms where again the dependence of resistivity on temperature is exponential. Thus the resistance of a semiconductor below 30 K can be controlled by the nature of the doping. Unfortunately, except to indicate an exponential dependence of R on T^{-1}, theory is of little help for quantifying the resistance-temperature relation of semiconductors. Nevertheless, germanium with arsenic as a dopant has long been used as a low temperature thermometer. It is now largely supplanted by RhFe.

Although carbon is not a semiconductor, it exhibits an electrical resistivity variation with temperature that is similar to that of semiconductors. It behaves as if there were a small energy gap between the valence and conduction bands. Consequently, in the temperature range from about 0.05 K to 30 K, carbon (and graphite) has a high negative temperature coefficient of resistance. No theory has been developed to even qualitively relate R to T, partly because the characteristics of carbon are found to be heavily dependent upon the method and temperature of manufacture. Some of the carbon resistors used in electrical circuitry have been successfully used as sensitive, moderately reliable, inexpensive, cryogenic thermometers. Interpolation is strictly empirical.

Thermistors are polycrystalline mixtures of sintered metallic oxides (NiO, Mn_2O_3, Co_2O_3) or solid solutions ($MgCr_2O_4$ in Fe_3O_4) that behave essentially as semiconductors. As a result they have high negative temperature coefficients of resistance ($\sim 4\%/K$). They have proved successful in a variety of shapes as small, inexpensive, sensitive, fast-response temperature sensors, largely within the range -100 °C to 300 °C. Any one thermistor give better performance if its use is restricted to a narrow range, so thermistors are manufactured for specific use over small portions of the wider range. Resistance values are typically in the kilo-ohm range.

2.6.2.2 *Thermoelectric*

The physical principles of thermoelectric sensors stem from three thermoelectric effects that were discovered in the early part of the 19th century. The application of thermodynamics to these effects leads to several inter-relationships between them that underpin the operation of the sensors. Thermodynamics says nothing, however, about the magnitude of the effects and how they vary from one material to another. For this one must turn to the quantum theory of electrons in metals. Unfortunately, although the results are in many cases qualitatively in agreement with experiment, they are of no use for quantitative application to thermoelectric sensors.

The first thermoelectric effect, the Seebeck effect, concerns the conversion of thermal energy to electrical energy. An electric current will flow in a closed circuit composed of two different conductors when the two junctions are at different temperatures. Alternatively, if the circuit is opened, a Seebeck emf appears at the terminals. The polarity and magnitude of this Seebeck emf depend upon the temperatures of the junctions and upon the materials. For a small temperature difference, $\mathrm{d}T$, the resulting small emf, $\mathrm{d}E_{\mathrm{AB}}$, is proportional to $\mathrm{d}T$:

$$\mathrm{d}E_{\mathrm{AB}} = S_{\mathrm{AB}}\,\mathrm{d}T. \tag{2-79}$$

S_{AB} is called the Seebeck coefficient or, more commonly, the thermoelectric power or thermopower of material A relative to material B. The integral form of Equation (2-79) between temperatures T_1 and T_2 ($T_2 \geqslant T_1$) is

$$E_{\mathrm{AB}} = \int_{T_1}^{T_2} S_{\mathrm{AB}}\,\mathrm{d}T, \tag{2-80}$$

an equation that is basic to thermoelectricity.

The Peltier effect is the inverse of the Seebeck effect; it concerns the conversion of electrical energy to thermal energy. When an electric current flows across a junction of two dissimilar conductors, heat is either liberated or absorbed at the junction. The rate of generation or absorption of heat is proportional to the current and dependent upon the material of the conductors:

$$Q_\pi = \pi_{\mathrm{AB}}\,I \tag{2-81}$$

where Q_π is the quantity of heat liberated or absorbed per second when a current I amperes flows. The coefficient of proportionality, π_{AB}, is the Peltier coefficient or Peltier emf of material A relative to material B. By convention, π_{AB} is positive if heat is evolved when current flows from A to B.

It is important to realize that these two effects are completely reversible thermodynamically, in contrast to Joule heating and thermal heat conduction, which are not.

The Thomson effect concerns the heat content of a single conductor along which there is a temperature gradient when an electric current flows in it. Heat is liberated or absorbed in each portion of the conductor according as the electron flow is towards lower or higher temperatures. The rate of generation or absorption of heat is proportional to the product of the current I and the temperature difference $\mathrm{d}T$:

$$\mathrm{d}q_\sigma = \sigma_{\mathrm{A}}\,I\,\mathrm{d}T \tag{2-82}$$

where $\mathrm{d}q_\sigma$ is the heat liberated or absorbed per unit volume per second and σ_{A} is the Thomson coefficient of material A. The product $\sigma_{\mathrm{A}}\,\mathrm{d}T$ is sometimes called the Thomson emf. It follows that if the ends of the conductors are kept at temperatures T_1 and T_2 respectively, the total heat evolved per second is

$$Q_\sigma = \int_{T_1}^{T_2} \sigma_{\mathrm{A}}\,I\,\mathrm{d}T. \tag{2-83}$$

From Equation (2-83), if the single homogeneous conductor forms a closed circuit, equal and opposite emfs will be set up in the two paths between T_1 and T_2 and $Q_\sigma = 0$. This is known as the *Law of Homogeneous Conductors*: a thermoelectric current cannot be sustained in a single homogeneous conductor that forms a closed circuit solely by the application of heat.

Suppose now that we have a closed circuit consisting of two materials A and B with the two junctions at temperatures T_1 and T_2 ($T_2 > T_1$). By application of the first law of thermodynamics to the transport of a unit charge around the circuit, we find from Equations (2-81) and (2-83) that the total emf developed, E_{AB}, is

$$E_{AB} = \pi_{AB}(T_2) - \pi_{AB}(T_1) + \int_{T_1}^{T_2} \sigma_B \, dT - \int_{T_1}^{T_2} \sigma_A \, dT . \tag{2-84}$$

Equation (2-84) can also be written in the equivalent differential form (by setting $T_2 = T_1 + \Delta T$

$$\frac{dE_{AB}}{dT} = \frac{d\pi_{AB}}{dT} + (\sigma_B - \sigma_A) . \tag{2-85}$$

Equation (2-85) is the fundamental theorem of thermoelectricity. It expresses the Seebeck effect as the algebraic sum of the Peltier and Thomson effects.

Some important consequences for thermoelectric sensors are embodied in Equation (2-84):

1. *The Law of Magnus:* If conductors A, B are homogeneous, E_{AB} depends only upon the two junction temperatures T_1 and T_2. This follows because σ_A and σ_B are invariant along the respective conductors. This is the basis of all thermocouple operation, allowing the voltage developed to be related solely to the two junction temperatures independently of the details of the temperature gradients traversed by the thermocouple wires.

2. *The Law of Intermediate Conductors:* The algebraic sum of the emfs in a circuit composed of any number of dissimilar materials is zero if all of the circuit is at a uniform temperature. This follows because $E_{AB} = 0$ when $T_1 = T_2$, independent of the values of π_{AB}, σ_A and σ_B.

It is this law that allows the introduction of a device to measure the emf so long as all junctions to the devise are at the same temperature and the device itself is isothermal. It also allows fixed point calibrations of thermocouples by the wire method where a small piece of metal of known melting temperature is fastened between the thermocouple wires at the hot junction. As the junction temperature is slowly increased, the emf at the melting point can be observed. The melting metal introduces no extraneous emfs so long as the region of the thermocouple where it is fastened is isothermal. This law also allows additivity of emfs for different materials.

3. *The Law of Intermediate Temperatures:* If two dissimilar materials produce an emf E_1 when their junctions are at T_1 and T_2, and an emf E_2 where their junctions are at T_2 and T_3, then the emf generated when the junctions are at T_1 and T_3 will be $E_1 + E_2$. This follows from Equation (2-84), since

$$\int_{T_1}^{T_3} = \int_{T_1}^{T_2} + \int_{T_2}^{T_3} .$$

It is this law that allows the use of thermocouple tables based upon a reference junction at T_0 with a thermocouple whose reference junction is at some other temperature T_1, and hence the use of electrically-simulated reference junctions. It is also because of this law that extension leads can be used with thermocouples.

4. Although E_{AB} depends only on T_1 and T_2, the third and fourth terms on the right hand side of Equation (2-84) show that E_{AB} is actually generated in the region of the temperature gradient. When the conductors are homogeneous, all gradients paths from T_1 to T_2 produce the same result according to the Law of Magnus. When they are not homogeneous, however, different gradient paths can produce different emfs. This is fundamental to thermocouple thermometry; it leads to some of the basic limitations of thermocouples. No thermocouple wire, even if homogeneous initially, is likely to remain so with use. When it is inserted into a furnace, physical and chemical processes such as oxidation, defect formation, and impurity movement begin which dehomogenize the wires. The practical result is that the emf is not solely dependent on T_1 and T_2. An inhomogeneous thermocouple will produce different emfs in two furnaces, both at T_2, if the furnace gradients from T_1 to T_2 are different.

We can now apply the second law of thermodynamics to the change of entropy as unit charge is transported around the circuit. In doing this we assume that the application is valid for these reversible thermoelectric effects in the simultaneous presence of irreversible Joule heating and thermal conduction. The only real justification is that a rigorous treatment based upon the theory of irreversible thermodynamics leads to the same results [56]. On the assumption of reversibility, there must be no net change in entropy associated with the transport of the unit charge. Since by Equations (2-81) and (2-82) the change in entropy associated with the Peltier and Thomson emfs are π_{AB}/T and σ_A/T respectively, then the net change in the entropy of the surroundings is

$$\frac{-\pi_{AB}(T+\Delta T)}{T+\Delta T}+\frac{\pi_{AB}(T)}{T}-\frac{\sigma_B}{T+\dfrac{\Delta T}{2}}+\frac{\sigma_A}{T+\dfrac{\Delta T}{2}}=0 \tag{2-86}$$

where we have written T and $T+\Delta T$ for T_1 and T_2. After some algebraic manipulation Equation (2-86) leads to

$$\frac{\mathrm{d}}{\mathrm{d}T}\left(\frac{\pi_{AB}}{T}\right)=\frac{\sigma_A-\sigma_B}{T}. \tag{2-87}$$

Using Equations (2-79) and (2-85) with Equation (2-87) leads to

$$S_{AB}=\pi_{AB}/T. \tag{2-88}$$

Substituting Equation (2-88) in Equation (2-87), we can write

$$S_{AB}=\int_0^T\frac{\sigma_A}{T}\,\mathrm{d}T-\int_0^T\frac{\sigma_B}{T}\,\mathrm{d}T. \tag{2-89}$$

The first integral in Equation (2-89) is dependent only upon material A and the second only upon material B. If we define the absolute thermopower of material A (S_A) to be

$$S_A = \int_0^T \frac{\sigma_A}{T} \, dT , \tag{2-90}$$

then

$$S_{AB} = S_A - S_B \tag{2-91}$$

and

$$E_{AB} = \int_{T_1}^{T_2} (S_A - S_B) \, dT . \tag{2-92}$$

Once the absolute thermopower of any material is known, that of any other can be obtained from measurement of S_{AB}.

This is as far as thermodynamics can take us. A great deal of effort has been expended on attempts to develop an expression for the absolute thermopower of a metal. Considerations based upon both the heat capacity of a metal and, more rigorously, upon the effect of an electrical field and a temperature gradient upon the Fermi-Dirac function lead to

$$S = -\frac{\pi^2 k_B^2 T}{2 e E_F} \ \mu V/K \tag{2-93}$$

for a normal metal and

$$S = -\frac{\pi^2 k_B^2 T}{6 e (E_0 - E_F)} \ \mu V/K \tag{2-94}$$

for a transition metal, where e is the electronic charge, E_F is the Fermi energy, and E_0 is the highest energy level in the d-band. Although qualitatively correct for some metals, Equations (2-93) and (2-94) do not predict even the correct sign for others. For a detailed discussion of thermoelectric principles with specific orientation to thermoelectric sensors, see Pollock [57].

2.6.2.3 *Thermal Expansion*

Thermal expansion was the first physical property to be used as the basis of a thermal sensor. It may still today be considered the workhorse of industrial thermal sensing, being used in a wide variety of applications, especially those involving temperature control. We outline here the principles of the three commonest types of thermal expansion sensors.

a) Liquid-in-Glass

A liquid-in-glass sensor relies upon the differential thermal expansion between the glass of the thermometer bulb and the liquid in the bulb to drive a thin column of liquid back and forth in a capillary attached to the bulb in response to temperature changes. The capillary is provided with an attached graduated scale from which the position of the top of the liquid column, and so temperature values, can be read after suitable calibration. Much the commonest liquid is mercury. Mercury-in-glass thermometers can be used from about −35 °C to almost 600 °C (not necessarily on a single instrument), but the upper temperature is generally below 300 °C. In thermometers for use above 100 °C the volume above the mercury contains pressurized dry nitrogen to suppress distillation of the mercury. The coefficient of cubical expansion of mercury is eight to ten times that of the glasses used. To attain maximum sensitivity without deleterious effects on the movement of the mercury in the capillary, the volume of the bulb is usually about 6000 times the volume of a 1 °C length of the capillary. For measurement below the temperature of freezing mercury, organic liquids (alcohol, toluene, pentane, butane) are used with a possible lower limit of about −200 °C.

A typical mercury-in-glass thermometer consists of the bulb that contains most of the mercury; an attached capillary, with a very uniform bore, enclosed in a glass stem; sometimes one or more auxiliary reservoirs (or contraction chambers) along the capillary which serve to shorten the overall length of higher temperature thermometers; an expansion chamber at the end of the capillary; the graduated scale; and an auxiliary scale that includes the ice point temperature if the main scale does not.

Properly used, a mercury-in-glass thermometer should be immersed in one of three ways in the medium (usually liquid) whose temperature is required:

a) *totally immersed:* the top of the mercury column in the capillary is set coincident with the level of the top of the medium; for this the thermometer position must be adjusted whenever the medium temperature, and hence the position of the mercury meniscus, changes;

b) *partially immersed:* an engraved immersion line on the stem is set coincident with the level of the top of the medium;

c) *completely immersed:* the whole thermometer is immersed in the medium; for this, viewing of the scale through the medium and its container is necessary.

Partial immersion is the most common. With it, a correction to the observed reading must be made to account for the part of the mercury column that is above the immersion line, and so not in the medium, being at a different temperature than the medium. This *emergent stem correction,* Δt, is

$$\Delta t = Kn\,(t_1 - t_2) \tag{2-95}$$

where t_1 is the indicated temperature of the bulb, t_2 is the average temperature of the emergent mercury column, n is the number of degrees of the emergent column, and K is the differential volume coefficient of thermal expansion between mercury and glass (closely equal to 16×10^{-5}/°C). A similar correction applies in total immersion if the mercury column protrudes above the medium.

The level of the mercury is also sensitive to pressure, both due to external pressure and to the head of mercury in the capillary. External pressure higher than that at the time of calibration will slightly increase the indicated temperature. The correction (β) is small, sometimes as much as 0.1 K per atmosphere, and can be estimated from

$$\beta = 3 \times 10^{-7} \frac{R_e^2}{R_e^2 - R_i^2} + 1.1 \times 10^{-7}\ \text{K/Pa} \tag{2-96}$$

where R_e and R_i are the external and internal radii of the bulb. The first term on the right hand side in Equation (2-96) results from the effect of external pressure or from the head of mercury and the second from the compressibility of the mercury.

Proper use of a mercury-in-glass thermometer demands that ice point readings be taken regularly. When a thermometer is exposed to high temperatures the bulb undergoes inelastic expansion, resulting in a depression of the zero if the thermometer is returned quickly to the ice point. This expansion is reversible so that the ice point depression, which may reach 0.1 °C, is temporary (1 to 3 days). Additionally, there is an aging effect due to a long time contraction of the bulb volume that produces a slow rise in zero reading but at a decreasing rate over years. It affects all indicated temperatures equally so a correction measured at 0 °C can be applied to all readings. This ice point rise may be as large as 0.05 °C in the first year. For a detailed discussion of all aspects of liquid-in-glass thermometry, see Wise [58].

A mercury-in-glass thermometer can form a simple but effective temperature controller. Two platinum electrodes are introduced into the capillary, one fixed near the bottom of the scale and the other either fixed at some higher desired temperature control point or adjustable over a range of temperature points. When the mercury contacts the higher electrode, a circuit is closed that is used to operate some control circuit.

b) Filled-System

Under this heading we consider three widely-used industrial-type thermal sensors that have the common feature of translating a temperature change into a pressure change that is indicated by a Bourdon gauge. One of them (vapor pressure) is not really a thermal expansion device at all. A Bourdon gauge is essentially a tube of oval cross section bent into an arc of a circle. One end is sealed and attached to an indicting pointer. The other end is open to the pressure of the sensor. When the pressure in the tube exceeds the external pressure, the tube cross section changes from oval towards circular with consequent motion of the sealed end and hence of the pointer, the motion being approximately linear with pressure. The amount of movement of the free end varies inversely as the tube wall thickness and directly with the angle subtended by the arc through which the tube is bent. It also depends upon the cross-sectional shape. When the tube is bent into a helix or a spiral, the effective angular length is increased and, correspondingly, the amount of motion of the free end and of the pointer. None of these sensors is very accurate (a few percent), but they are rugged and have moderately good repeatability. All require calibration.

b.1) *liquid filled:* This is the counterpart of the liquid-in-glass thermometer. The liquid is confined in a metal bulb connected through a capillary to a Bourdon gauge, the bulb and capillary being completely filled. When the bulb temperature is increased, the positive differential expansion of the liquid relative to the bulb attempts to increase the volume of the liquid which, being confined, exerts a pressure on the Bourdon tube. The liquid is usually mercury or an organic such as xylene or ethanol, and a wide temperature span of the gauge is possible. Such a sensor can be used for temperature control, and for remote reading since there is no special limitation on the length of the capillary. In the latter case it may be necessary to compensate for temperature changes along the capillary which will also influence the Bourdon reading. One way is run an identical capillary (no bulb) alongside the main one to a second Bourbon gauge and to arrange for the pointer to register the difference in movement of the two Bourbon gauges. Thus, the pointer motion will indicate only bulb temperature changes.

b.2) *gas-filled:* This is the counterpart of the constant volume gas thermometer of Section 2.6.1.1. Its behavior is adequately represented by Equation (2-6). As in b.1), a bulb and capillary of constant volume are filled with an inert gas (usually nitrogen). Changes in bulb temperature produce corresponding changes in pressure which actuate the Bourdon gauge. Because the thermal capacity of the gas is less than of a liquid, this sensor has a faster response than a liquid-filled one. However, to reduce the effect of ambient temperature changes along the capillary, the bulb volume must be many times larger than that of the capillary. A dual capillary compensation such as described in b.1) will not work with a gas-filled system because any pressure change due to temperature change along the capillary will cause gas to flow into the bulb to equilibrate bulb and capillary pressures, whereas the pressure change in the compensating capillary would be transmitted to the Bourdon gauge.

b.3) *vapor pressure:* This is the counterpart of the vapor pressure thermometer of Section 2.6.1.10. In this system the bulb is partially filled with an organic liquid that has a moderately-high saturated vapor pressure. Changes in bulb temperature cause changes in vapor pressure which are sensed by the Bourdon gauge. For proper operation the liquid-vapor interface must lie within the bulb. Pressure changes are transmitted to the gauge by vapor when the bulb temperature is below that of the rest of the system and by liquid when the bulb temperature is higher. Different, suitably-filled units are used for these two cases. These systems have the advantage of being independent of bulb volume and of the temperature distribution along the capillary; they have the disadvantage of a non-linear response (equal temperature indications larger at higher temperatures) and of a limited temperature range for any particular liquid. They are not generally satisfactory for measurements near ambient temperature because of erratic behavior of the indicator due to liquid distilling into or out of the capillary. The difficulty can be overcome if a second, thermally-inert liquid is added in such a way that it fills the capillary and part of the bulb so as to act simply as a pressure-transmitting fluid.

c) Bimetallic

A bimetallic sensor makes use of the difference in thermal expansion coefficients of different metals. If strips of two such metals are bonded together into a cantilever or a helix with one end fixed and the strips are heated, one strip will expand more than the other, causing motion of the free end of the cantilever or helix. By suitable coupling, this motion can be made to drive an indicator. One of the metals is usually Invar which has a very low thermal expansion coefficient; the other may be brass, or iron or nickel alloys. A bimetallic sensor is robust, not very accurate, but more precise by up to a factor of 10. It is widely used for temperature control. It can span a wide temperature range, with a usual upper limit a little below 400 °C. At very low temperatures all metal expansion coefficients converge to a common value at nearly the same rate, so a bimetallic sensor becomes insensitive.

2.6.2.4 Diodes

Silicon, gallium-arsenide, or germanium p-n junction diodes can serve as sensitive, moderately-precise thermal sensors over a wide temperature range from about 1 K to 200 °C.

Their region of greatest reliability (to within about 10 mK) is below about 50 K; at higher temperatures the inaccuracy is nearer 0.1 K. The diodes cannot tolerate temperatures above about 200 °C. The forward bias characteristic of a silicon or germanium diode is such that practically no current flows below a particular voltage (V_F) called the forward conduction voltage. V_F is a measure of the minimum energy required by current carriers to cross the p-n junction. Above V_F the current flow increases approximately exponentially with V/T. Above V_F the forward voltage at constant current varies with temperature in one of two ways. From about 20 K to 150 °C the voltage decreases almost linearly with increasing temperature (the intrinsic region) with a coefficient (or slope) for silicon near −2 mV/K. Below 20 K the voltage increases exponentially with decreasing temperature (the extrinsic region) and is influenced by the dopant. The break between the two regions depends upon the current and is in the range 10 K to 30 K. The picture with gallium-arsenide is somewhat different. Above about 100 K the conduction is similar to that in silicon or germanium. Below 100 K the main conduction mechanism results from tunnelling of carriers through the junction barrier. The sensitivity is not constant below 100 K but passes through a minimum value near 20 K.

A silicon diode can also be incorporated into an integrated circuit with an amplifier in such a way as to produce either an output voltage or current that is proportional to temperature.

As practical sensors, all of these diodes must be calibrated at several temperatures and an appropriate curve fitted to the data.

2.6.2.5 *Quartz Crystal*

Quartz crystals are highly piezoelectric. Different resonant frequencies of vibration are produced depending upon the orientation of the cut faces relative to the crystallographic axes. Suitably cut, an extremely stable, temperature-independent resonant frequency is obtainable (and used for time keeping). With a different cut, the resonant frequency changes linearly with temperature at about 1000 Hz/K over a wide range from −80 °C to 250 °C. This allows a temperature resolution of 10 μK. As so far developed, temperature indications are obtained from beating the thermally-sensitive frequency against the thermally-inert frequency used as a reference oscillator. Because temperature is converted directly to frequency, the device is expected to be free from problems associated with lead resistance and pickup, so allowing the use of very long leads. Unfortunately, hysteresis occurs on thermal cycling that limits the achieved accuracy to about 0.01 K for wide temperature excursions. Reliability to 1 mK is possible when the sensor temperature changes less then 10 K. Currently available commercial instruments are very expensive. These sensors are discussed in detail in Chapter 7.

2.6.2.6 *Quadrupole Resonance*

A nuclear quadrupole resonance thermometer has many of the attributes demanded in both standards and general purpose thermometry. It has high resolution (~0.1 mK); wide range (10 K to 450 K); good reproducibility (~1 mK from 90 K to 400 K); good long-term stability; the sensitive parameter is independent of the shape and size of the thermometer material; temperature is determined from a frequency measurement, which is both easy and accurate; the temperature versus frequency relationship for a particular material should be universal for all samples of that material (of the same high purity). Experimental [59, 60] and commercial [61] thermometers have been shown to exhibit many of these ideal characteristics. It is difficult to understand why they have not come into wider use.

The basis of the thermometer is the variation with temperature of the nuclear quadrupole resonance frequencies of certain nuclei, the most widely studied being ^{35}Cl in $KClO_3$. The ^{35}Cl nucleus possesses an electric quadrupole moment which interacts with electric field gradients produced by valence electrons and surrounding ions in the $KClO_3$ lattice. Even in the absence of an applied magnetic field, these interactions are sufficient to cause a splitting of an atomic energy level into two components between which radio frequency transitions can be excited and detected by techniques similar to those of Section 2.6.1.11. The separation in energy of the two components corresponds to a frequency ν_0 given by

$$\nu_0 = \frac{e\,Q\,q_{zz}}{2h} \tag{2-97}$$

where e is the electronic charge, h is Planck's constant, Q is the electric quadrupole moment of the nucleus, and q_{zz} is the component of the electric field gradient sensor along the principal axis. For ^{35}Cl, the resonant frequency ν_0 is near 29 MHz. In a rigid lattice ν_0 is the frequency that would be observed independent of temperature, that is to say, in a real lattice near 0 K. In $KClO_3$ torsional vibrations of the ClO_3 ion produce fluctuations in the orientation of the electric field gradient tensor, with the result that q_{zz} and so the resonant frequency ν become temperature dependent. The resulting expression for ν is

$$\nu = \nu_0 \left[1 - \sum_i \frac{3\,A_i\,h}{4\,\pi\,\omega_i} \left\{\frac{1}{2} + \frac{1}{\exp\left(h\,\omega_i/2\,\pi\,k_B\,T\right) - 1}\right\}\right] \tag{2-98}$$

where ω_i is the angular frequency of the ith mode of lattice vibration and A_i is the reciprocal of the moment of inertia of the ith lattice mode. For $KClO_3$, Equation (2-98) gives a sensitivity of only about 0.1 kHz/K at 15 K but it rises rapidly to near 2.5 kHz/K at 80 K; the corresponding measurement capabilities are ±10 mK near 15 K but about ±1 mK above 50 K. Below 80 K Equation (2-98) has been shown to hold to within the accuracy of measurement when two modes of vibration are included; above 80 K there are increasing departures as the lattice expands and anharmonic oscillations dominate. The result is an even higher sensitivity that increases approximately linearly to near 7 kHz/K at 400 K. Above 80 K empirical formulae are used to relate ν to T. Several short range polynomials have been used, but Vorob'ev and Brodskii [62] suggest the following equation of the form of Equation (2-98) for the range 60 K to 300 K:

$$\nu(T) = \nu_0 - M\left[\sum_i \frac{h\omega_i/2\pi\,k_B}{\exp\left(h\omega_i/2\pi h T\right) - 1}\right] - y(T) \tag{2-99}$$

where $M = 3/2\,\dfrac{\nu_0\,k_B}{A_i\,\omega_i}$ and $y(T)$ is a quartic polynomial in T.

Vanier [59] has shown that internal heating of the $KClO_3$ sample is completely negligible and [60] that there is only a slight pressure dependence of T. At low temperatures, variations in atmospheric pressure have negligible effect; at room temperature and above the temperature decreases at about 6 mK per atmosphere.

Although $KClO_3$ becomes insensitive below 15 K, ferromagnetic (and antiferromagnetic) salts offer promise in this region. In this case the observed resonance is between components of nuclear levels that are split in the magnetic field produced by the atoms of the crystal lattice.

2.6.2.7 Radiation

The fundamental principles of radiation thermometers (frequently called pyrometers) are outlined in Sections 2.6.1.8 and 2.6.1.9. The use of radiation thermometers as practical sensors does not differ in principle from their use as primary instruments. Radiation thermometers form an extremely important group of thermal sensors and a large number and a wide variety are commercially available. The chief differences amongst these are their wavelengths of operation and the widths of the spectral bands to which they are sensitive. Radiation thermometers are discussed in detail in Chapter 5; we shall simply outline briefly a few of the basic principles associated with their use. For a comprehensive account of all aspects of radiation thermometry, see [63].

One significant difference between a practical and a primary pyrometer is the need for calibration. A primary pyrometer is most often simply a null device that compares an unknown to a known spectral radiance. A practical pyrometer must carry a calibration; it can be in terms of electric current versus filament radiance temperature in an internally-mounted tungsten ribbon lamp, as in the long-standing disappearing filament pyrometer, or directly in terms of the pyrometer sensor output voltage versus temperature, as in almost all infrared radiation pyrometers. Seldom is a pyrometer truly a total radiation pyrometer because, even if the sensor response is non-selective, the various windows and lenses are sure to be selective, especially in the far infrared. Monochromatic thermometers are seldom calibrated directly according to the ITS-90 definition (Equation (2-23)), but against some kind of transfer standard. The latter is almost always a tungsten strip-filament lamp if the pyrometer wavelength is near 650 nm, sometimes a lamp for longer wavelengths, more often a calibrated standard pyrometer for infrared wavelengths, or one or more blackbodies at known temperatures. Pyrometers with a wide spectral band require the latter.

Once calibrated, a monochromatic pyrometer always measures a target surface temperature in terms of the spectral radiance of the surface (see Equation (2-64)). For any real non-black surface, the spectral radiance is given by $\varepsilon_\lambda(T)\,L_\lambda(T)$. Herein lies the major stumbling block of radiation thermometry. The pyrometer cannot distinguish between changes in $\varepsilon_\lambda(T)$ and changes in T, so unless $\varepsilon_\lambda(T)$ of the surface is known, T cannot be measured. What a pyrometer measures is a radiance temperature (T_r) defined (roughly) by

$$\varepsilon_\lambda(T)\,L_\lambda(T) = L_\lambda(T_r)\,. \tag{2-100}$$

The radiance temperature is always smaller then the true temperature. Details of how to obtain T from measurement of T_r or, which is the same thing, of how to correct or compensate for $\varepsilon_\lambda(T)$ are given in Chapter 5.

It is sometimes suggested that relative values of $\varepsilon_\lambda(T)$ at different λ may be better known than $\varepsilon_\lambda(T)$ itself. If so, the emissivity correction may be reduced, or even eliminated by measuring the ratio of spectral radiances at two or more wavelengths. The fallacy is that the corresponding relationships between $\varepsilon_\lambda(T)$ and λ seldom hold accurately enough; we are really forced to know several values of $\varepsilon_\lambda(T)$ instead of only one. Nevertheless, multi-wavelength pyrometers have been found useful in certain special circumstances.

The response $R(T, T_0)$ of a radiation pyrometer to a target temperature may be expressed in the form of Equation (2-64) as

$$R(T, T_0) \propto \int_0^\infty \varepsilon_\lambda(T)\, L_\lambda(T)\, s(\lambda, T_0)\, \tau_f(\lambda, T_0)\, \tau_0(\lambda)\, \mathrm{d}\lambda \qquad (2\text{-}101)$$

where the quantities are as defined for Equation (2-64) and T_0 is the pyrometer temperature. An explicit dependence of $R(T, T_0)$ on T_0 is shown to emphasize that for precise measurements it may be necessary to control the pyrometer temperature.

Each of the quantities in the integrand in Equation (2-101) can affect the deduction of T from the measured response $R(T, T_0)$. Apart from the spectral emissivity $\varepsilon_\lambda(T)$, the most important of these is the sensor spectral sensitivity $s(\lambda, T_0)$. It must remain stable at its value at the time of calibration and must have a suitably-low temperature coefficient. It is this stability that determines the reproducibility of the output signal both as a function of incoming radiant flux and as a function of time. Sensor instability is not well understood but is usually the factor determining pyrometer operability. A full discussion of the characteristics of sensors used in pyrometers, and of the limitations caused by the other parameters in Equation (2-101) is given in Chapter 5.

2.6.2.8 Noise

The principles of thermal noise sensors are the same whether they are used for primary (Section 2.6.1.5) or secondary applications. Noise thermometers have proved extremely useful for measuring temperatures in the hostile environments of nuclear power plants, petrochemical plants, and oil-, coal-, and gas-fired power plants where more conventional thermometers suffer from a variety of deleterious effects. Temperatures from 300 °C to 1200 °C have been measured to within 0.5% and sometimes 0.1% in the long term. For such industrial measurements the practical difficulties associated with resistor materials, length of measuring cables, interference signals, shielding, etc. must be considered. A full discussion is given in Chapter 6.

2.6.2.9 Ultrasonic

Ultrasonic techniques have particular application to measurement of high temperatures (as in nuclear reactor fuels) where conventional sensors are inadequate. The principle is simple: to measure the speed of a sound wave propagating in a solid. In a thin wire of diameter much less than the acoustic wavelength the speed v_e of an extensional wave is

$$v_e = (\varepsilon / \rho)^{1/2} \qquad (2\text{-}102)$$

where ε is the modulus of elasticity and ρ is the density. Correspondingly, the speed v_ψ of a torsional wave is

$$v_\psi = K(G/\rho)^{1/2} \tag{2-103}$$

where K (< 1) is a shape factor and G is the shear modulus. The temperature dependence of v is given through ε or G. Both types of waves have been used as the basis of thermal sensors, the former more commonly. These sensors require calibration, after which temperature is deduced from a measured frequency.

Two measurement techniques have been successfully used [64]. The first detects ultrasonic resonances. One example of this is the quartz crystal resonator described in Section 2.6.2.5; its maximum temperature is limited to about 250 °C. For much higher temperatures, metals such as molybdenum or iridium, or ceramics such as sapphire, have been used. A typical probe is a cylinder of length $\lambda/2$ which is excited through a long transmission line by a package of sine waves with center frequency at about 120 kHz. The centre frequency is varied until resonance occurs as determined by the echo. Another variation that has worked well with sapphire to 2000 °C is to shape the sensor in the form of a tuning fork. Both of these are excited with extensional waves but the fork resonates in a bending mode. Because a frequency is being measured, high resolution is attainable.

The second technique (dubbed the pulse echo technique) measures the transit time of a single pulse through a thin wire. Rhenium or, better, thoriated tungsten has been successful. An acoustic pulse is injected into the transmission line and reflected from the end of the sensor or from discontinuities in the sensor produced, for example, by grinding the sensor wire into different diameter sections. This allows measurement of a temperature profile. The echo pulses are sensed by the transmitting transducer. With thoriated tungsten, round trip pulse times range from about 4.5 μs/cm at room temperature to about 5.5 μs/cm at 2700 °C. Usual time measurement techniques allow a time resolution of about 1 ns or about 0.2 K at 2000 °C for a 5 cm long sensor. In practice, a variety of experimental problems limit the achieved resolution to a few kelvins. Ultrasonic thermometers are rugged and can tolerate intense neutron irradiation, but are complex instruments. They have been used successfully for short times to above 2800 °C.

2.6.2.10 *Thermally Sensitive Cones, Paints, Pigments*

For rough-and-ready, one-shot temperature measurements where low accuracy is acceptable, substances that melt, soften, change color, or blacken at particular temperatures serve as temperature indicators. Pyrometric cones are much used in the ceramics industry. These are mixtures of silicate minerals that soften (and melt) at particular temperatures. They are formed in the shape of eccentric pyramidal cones; when the cone reaches its registration temperature the tip bends over to touch the surface. A series of such cones is provided that spans the temperature range from about 600 °C to 1900 °C at 20 °C to 50 °C intervals. The cone softening temperature is also dependent on the rate of heating. Under best conditions with controlled heating rates an accuracy within ±10 °C may be attainable.

Paints or pigments that change color within a temperature range of 1 or 2 °C can be applied to a surface to indicate the maximum temperature it has reached. The pigment is dispersed in an organic carrier which melts at the specified temperature. When the carrier melts, the pigment changes hue because it then scatters less light. Again, there is a whole series of such paints that indicate temperatures in steps of about 5 °C between 0 °C and 350 °C. They are used, for example, in the electronics industry to indicate if components have exceeded their operating range temperatures. The paints are also available in the form of adhesive tapes. Similarly, there are pellets that melt and tapes that blacken at specific temperatures.

2.7 Heat Flux Sensors

A heat flux sensor is essentially a thermal sensor that measures a heat flux in terms of a temperature difference. The heat flux can be conductive/convective across a wall or surface (see Section 2.5) or radiant, where the principles of radiometry apply. The sensors are adaptations of some of those already discussed. For radiation, the flux is commonly absorbed by a blackened surface, the temperature change of which is measured with the sensor. High sensitivity is required because the temperature change may be small. A thermopile is frequently used; it is simply a large number (8 to 80) of thermocouple junctions connected in series to give an amplified signal. One set of junctions is temperature controlled; the other is exposed to radiation from the absorber (which may or may not be attached directly to the hot junctions). Pyranometers and pyrheliometers are essentially thermopiles specially designed for measuring solar heat fluxes or earth radiative flux balances. Alternatively, a bolometer, which is a sensitive resistance thermometer, can serve. The change in resistance of one resistor exposed to the flux is compared to the change of a second which is unexposed. Or a pneumatic technique is possible. The radiant flux heats a small volume of gas contained in a cell of which one wall contains a flexible membrane (or mirror). The increase in gas pressure, due to heating causes motion of the membrane which is detected capacitively or optically. Pyroelectric sensors can be used. They are formed from ceramic crystals that exhibit spontaneous changes in electrical polarization when their temperature is changed. Electrical charges of opposite polarity are induced on opposite faces. The magnitude and polarity of the induced voltage is dependent upon the magnitude and direction of the temperature change. A pyroelectric sensor requires chopped incident radiation because it responds only to a changing temperature. The induced voltage dissipates when the crystal is at constant temperature.

An example of a device that measures the sum of convective and radiative heat flux is a temperature-controlled probe in which the sensor is a cylindrical metal plug. The front face of the plug has high emissivity through grooving and blackening, while the rear face is water cooled. An incident heat flux establishes a temperature gradient in the plug that is measured with a differential thermocouple whose junctions are suitably positioned in the plug. The construction is such as to minimize radial heat flow in the plug, so that the measured temperature difference created by longitudinal heat flow can be related to the incident flux.

2.8 References

[1] Quinn, T. J., *Temperature,* London: Academic Press, 1983, pp. 3-5.
[2] Preston-Thomas, H., *Metrologia* **27** (1990) 3-10.
[3] *Comptes Rendus des Séances de la Quinzième Conférence Générale des Poids et Mesures,* Sèvres: Bureau International des Poids et Mesures, 1976, pp. A1-A21, [English translation: *Metrologia* **12** (1976) 7-17].
[4] Bedford, R. E., Ma, C. K., Macready, W., Steski, S., *Metrologia* **23** (1987) 197-205.
[5] Brickwedde, F. G., Van Dijk, H., Durieux, M., Clement, J. R., Logan, J. K., *J. Research Natl. Bur. Standards* **64A** (1960) 1-18.
[6] Sydoriak, S. G., Roberts, T. R., Sherman, R. H., *J. Res. Natl. Bur. Standards* **68A** (1964) 547-588.
[7] Bureau International des Poids et Mesures, *Metrologia* **15** (1979) 65-68.
[8] Hill, K. D., Bedford, R. E., BIPM (1989), *Document CCT/89-5,* 1989, 17e Session, Comité Consultatif de Thermométrie.
[9] Hill, K. D., Bedford, R. E., BIPM (1989), *Document CCT/89-6,* 17e Session, Comité Consultatif de Thermométrie.
[10] *Supplementary Information for the ITS-90,* Sèvres: Bureau International des Poids et Mesures, 1990.
[11] *Techniques for Approximating the International Temperature Scale of 1990,* Sèvres: Bureau International des Poids et Mesures, 1990.
[12] Baker, H. D., Ryder, E. A., Baker, N. H., *Temperature Measurement in Engineering* Vol. I, New York: John Wiley 1953, ch. 7.
[13] Baker, H. D., Ryder, E. A., Baker, N. H., *Temperature Measurement in Engineering* Vol. I, New York: John Wiley 1953, ch. 8.
[14] Jakob, M., *Heat Transfer* Vol. II, New York: John Wiley, 1957, pp. 151-159.
[15] Baker, H. D., Ryder, E. A., Baker, N. H., *Temperature Measurement in Engineering* Vol. II, New York: John Wiley, 1961, ch. 5.
[16] Baker, H. D., Ryder, E. A., Baker, N. H., *Temperature Measurement in Engineering* Vol. II, New York: John Wiley, 1961, pp. 50-54.
[17] McLaren, E. H., Murdock, E. G., *Canadian Journal of Physics* **44** (1966) 2631-2652.
[18] McLaren, E. H., Murdock, E. G., *Canadian Journal of Physics* **44** (1966) 2653-2660.
[19] Kerlin, T. W., Shepard, R. L., Hashemian, H. M., Patersen, K. M., in: *Temperature, Its Measurement and Control in Science and Industry,* Vol. 5, Schooley, J. F. (ed.); New York: American Institute of Physics, 1982, pp. 1357-1366.
[20] Baba, T., Ono, A., Hattori, S., *Rev. Sci. Intrum.* **56** (1985) 1399-1401.
[21] Aziz, R. A., McCourt, F. R. W., Wong, C. C. K., *Molecular Physics* **61** (1988) 1487-1511.
[22] Schooley, J. F. *Thermometry,* Boca Raton, FA: CRC Press, 1986, pp. 115-138.
[23] Berry, K. H., *Metrologia* **15** (1979) 89-115.
[24] Plumb, H., Cataland, G., *Metrologia* **2** (1966) 127-139.
[25] Colclough, A. R., *Proc. Roy. Soc. London Ser. A.* **365** (1979) 349-370.
[26] Moldover, M. R., Trusler, J. P. M., *Metrologia* **25** (1988) 165-187.
[27] Gugan, D., Michel, G. W., *Metrologia* **16** (1980) 149-167.
[28] Colclough, A. R., *Temperature, Its Measurement and Control in Science and Industry,* Vol. 5, Schooley, J. F. (ed); New York: American Institute of Physics, 1982, pp. 89-94.
[29] Rice, O. S., *Bell System Tech. J.* **23** (1944) 282-290.
[30] Garrison, J. B., Lawson, A. W., *Rev. Sci. Instr.* **20** (1949) 785-794.
[31] Crovini, L., Actis, A., *Metrologia* **14** (1978) 69-78.
[32] Klein, H.-H., Klempt, G., Storm, L., *Metrologia* **15** (1979) 143-154.
[33] Kamper, R. A., Zimmerman, J. E., *J. Appl. Phys.* **42** (1971) 132-136.
[34] Soulen, R. J., Van Vechten, D., *Temperature, Its Measurement and Control in Science and Industry,* Vol. 5, Schooley, J. F. (ed); New York: American Institute of Physics, 1982, pp. 115-123.
[35] Borkowski, C. J., Blalock, T. V., *Rev. Sci. Instr.* **45** (1974) 151-162.
[36] Greywall, D. S., *Phys. Rev.* **B33** (1986) 7520-7538.
[37] Marshak, H., *J. Res. Natl. Bur. Standards* **88** (1973) 175-218.

[38] Bedford, R. E., in: *Theory and Practice of Radiation Thermometry,* De Witt, D. P., Nutter, G. D. (eds.); New York: John Wiley, 1988, ch. 12.
[39] Baltes, H. P., *Applied Physics* **1** (1973) 39–43.
[40] Quinn, T. J., Martin, J. E., *Phil. Trans. Roy. Soc. London* **316** (1985) 85–189.
[41] Erminy, D. E., *J. Opt. Soc. Am.* **53** (1963) 1448–1449.
[42] Bedford, R. E., Ma, C. K., *High Temperatures-High Pressures* **15** (1983) 119–130.
[43] Kostkowski, H. J., Lee, R., *NBS Monograph 41,* 1962, U.S. Goverment Printing Office, Washington, D. C.
[44] Ohtsuka, M., Bedford, R. E., *Measurement* **7** (1989) 2–6.
[45] Rusby, R. L., Swenson, C. A., *Metrologia* **16** (1980) 73–87.
[46] Bedford, R. E., Bonnier, G., Maas, H., Pavese, F., *Metrologia* **20** (1984) 145–155.
[47] Hudson, R. P., *Principles of Magnetic Cooling;* Amsterdam: North Holland, 1972.
[48] Drawin, H. W., *Z. Physik* **228** (1969) 99–119.
[49] Drawin, H. W., *High Temperatures-High Pressures* **2** (1970) 359–409.
[50] Lapworth, K. C., *J. Phys. E* **7** (1974) 413–420.
[51] Laurendeau, N. M., *Prog. Energy Combust. Sc.* **14** (1988) 147–170.
[52] Generalov, N. A., Lapp, M., Penney, C. M., Muntz, E. P., in: *Methods of Experimental Physics* **18 B,** Emrich, R. J. (ed.); New York: Academic Press, 1981, pp. 463–497.
[53] Verdiek, J. F., Shirley, J. A., Hall, R. J., Echbreth, A. C., *Temperature, Its Measurement and Control in Science and Industry,* Vol. 5, Schooley, J. F. (ed.); New York: American Institute of Physics, 1982, pp. 595–608.
[54] Coles, B. R., *Phys. Lett.* **8** (1964) 243.
[55] Rusby, R. L., *Temperature, Its Measurement and Control in Science and Industry,* Vol. 5, Schooley, J. F. (ed.); New York: American Institute of Physics, 1982, pp. 829–833.
[56] Domenicali, C. A., *Rev. Mod. Phys.* **26** (1954) 237–275.
[57] Pollock, D. D., *Thermoelectricity-Theory, Thermometry, Tool,* Philadelphia: ASTM, 1985.
[58] Wise, J. A., *Liquid-in-Glass Thermometry,* NBS Monograph 150, 1976, U.S. Government Printing Office, Washington, D. C.
[59] Vanier, J., *Metrologia* **1** (1965) 135–140.
[60] Vanier, J., *Temperature, Its Measurement and Control in Science and Industry,* Vol. 4, Plumb, H. H. (ed.); Pittsburgh: Instrument Society of America, 1972, pp. 1197–1212.
[61] Ohta, A., Iwaska, H., *Temperature, Its Measurement and Control in Science and Industry, Vol. 5,* Schooley, J. F. (ed), New York: American Institute of Physics, 1982, pp. 1173–1180.
[62] Vorob'ev, I. V., Brodskii, A. D., *Izmeritel'naya Tekhnika* Number 5, May (1970) 52–55.
[63] *Theory and Practice of Radiation Thermometry,* De Witt, D. P., Nutter, G. D. (eds.); New York: John Wiley, 1988.
[64] Tasman, H. A., Richter, J., *High Temperatures-High Pressures* **11** (1979) 87–101.

3 Resistance Thermometers

Luigi Crovini, CNR: Istituto di Metrologia "G. Colonnetti",
Torino, Italy

Contents

3.1 Introduction

After Georg Simon Ohm had introduced his famous laws in 1827, it became clear that the electric resistance of some metals increases with temperature in a quasi-linear fashion. However, this property was not exploited for temperature measurement until 1871, when Wilhelm von Siemens described the first resistance thermometer before the Royal Society in London. He proposed a new electrical pyrometer for metallurgical furnaces made from a 1-m length of 0.1-mm platinum wire bifilarly wound on a porcelain (or fireclay) former inside an iron tube. At that time pure metals were becoming available, industry was requiring better process control, and Charles Wheatstone had described his famous bridge. This bridge was adapted by von Siemens to the needs of resistance thermometry, minimizing the effect of the cable resistance.

The platinum resistance thermometer (PRT) was introduced for the first time as a precision instrument in 1887 by H. L. Callendar, who demomstrated that it would achieve the requisite stability and reproducibility if properly constructed and treated. Fourty years later, the Comité International des Poids et Mesures prescribed it for the International Temperature Scale [1]. Specific DC bridges for very accurate PRTs were first described by Smith [2] and by Mueller [3] and were used since with only few modifications in numerous laboratories until the advent of AC bridges in the late 1960s.

Nickel resistors were introduced not long after as a less expensive alternative to platinum. Even lower cost and smaller size were achieved with the thermistor. At first this device appeared in Germany in the early 1930s with the name *Heissleiter* (hot conductor) and in the USA where it was called thermistor. It was initially used in control applications and was made first from UO_2 and subsequently from $MgTiO_3$ (spinel).

Low-temperature applications of resistance thermometry are more recent and include the introduction of carbon radio resistors below 20 K [4], germanium thermometers [5], rhodium/0.5% iron thermometers [6, 7], and carbon-in-glass thermometers [8]. Resistance thermometry can today be performed with numerous different types of transducers. The PRT by far covers the widest range of temperature, ie, from liquid helium (4.2 K) to the freezing point of gold (ca. 1337 K).

Long-stem PRTs are generally used above −200 °C. Industrial platinum resistance thermometers (IPRTs) are often realized as long-stem thermometers and are coded from −200 °C to 850 °C [9]. One exception is the application to surface-temperature measurements.

Other metal resistance thermometers, chiefly nickel and copper, are mostly used for industrial process and domestic appliance control. Copper resistance thermometers are also used in oceanography [10]. Iridium thin-film sensors have recently been introduced for industrial applications, mainly surface-temperature applications [11].

Cryogenic applications entail numerous resistance thermometers, including PRTs. Rhodium-iron and platinum-cobalt alloys, carbon-resistance sensors, and germanium thermometers are just a few examples. Cryogenic applications are sometimes carried out in the presence of intense magnetic fields, which adversely affect the responses of all types, although with different intensity. Above 5 T no resistance thermometer can be recommended for very accurate measurements.

The most relevant types of resistance thermometers and their temperature ranges are presented in Figure 3-1. The full range presented is the most often used in practice and may require more than one design of thermometer to cover it all. This is the case, for instance, with thermistors.

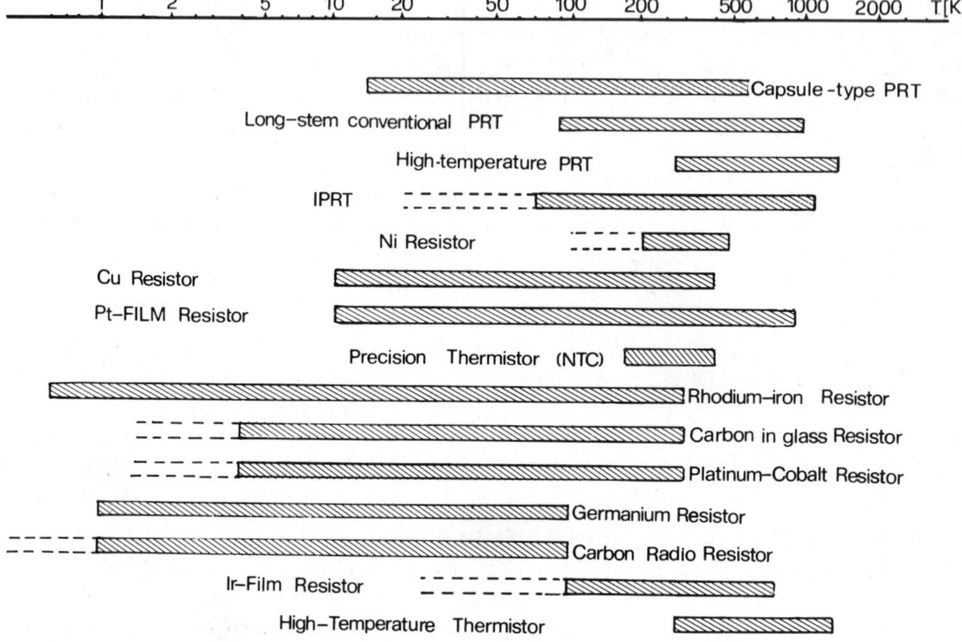

Figure 3-1. Ranges of resistance thermometers.

3.2 Principles of Resistance Thermometry

The current theories of electrical conduction, although well advanced and rather complex, provide only an approximate description of the temperature dependence of materials for resistance thermometry.

The principles of resistance thermometry also include the techniques of resistance measurement, which involve bridge and instrument principles, detector sensitivity, effect of lead resistances, and self-heating. Sensitivity, lead resistance and configurations, and self-heating are very important parameters for sensing element design. Other important considerations concern the insulating materials to support the sensing element.

3.2.1 Resistance Thermometers with Positive Temperature Coefficients

Positive temperature coefficients are peculiar to pure metals and some alloys. Some thermistors can also be manufactured with positive temperature coefficients, but their use is limited to control applications. In the case of metals one can note that:

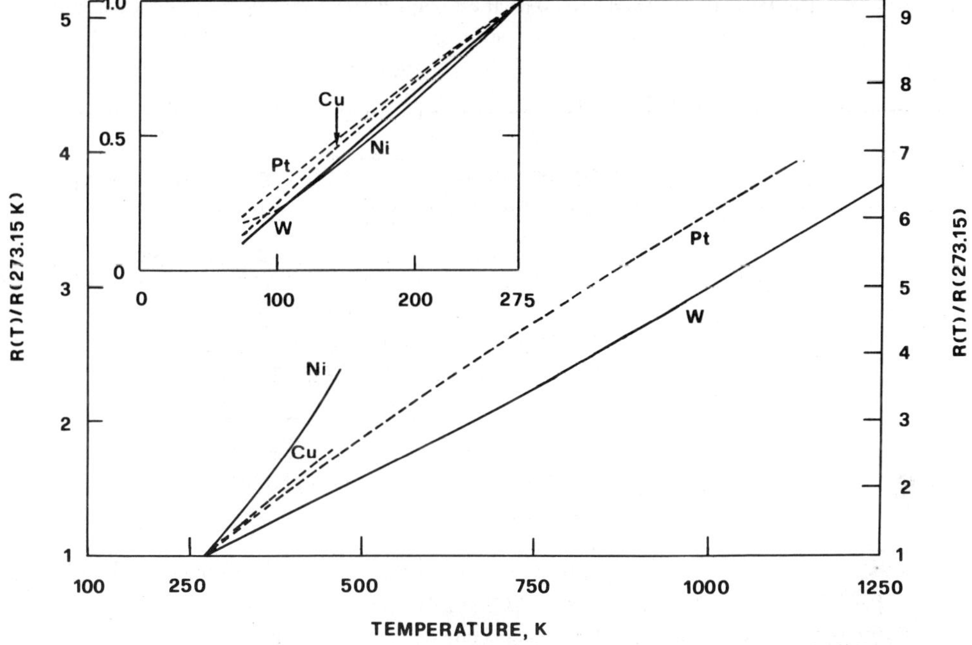

Figure 3-2. Resistance versus temperature of Cu, Ni, Pt, and W.

a) the electrical resistivity of the purest available metal falls between a typical value of about 10 μΩcm at room temperature and less than 10^{-3} μΩcm at liquid helium temperature;

b) that of concentrated alloys is higher by a factor of only ca. two over the same temperature range;

c) the addition of traces of an impurity to a pure metal always leads to an increase in resistivity;

d) the temperature coefficient of the resistivity of a concentrated alloy is much less than those of its constituents.

To understand in more detail the behavior of pure metals, it is possible to start with Figure 3-2, where resistance versus temperature relationships for copper, nickel platinum, and tungsten are reported. At room temperature, the coefficient $(1/R)$ $(\mathrm{d}R/\mathrm{d}T)$ ranges between 3.9×10^{-3} and 6.5×10^{-3} K^{-1}. Both copper and platinum have a coefficient close to 4×10^{-3} K^{-1}. All these metals have resistances that are approximately linearly dependent on temperature between ca. 200 and 500 K. For $T < 50$ K, the temperature coefficient decreases with T, becoming almost negligible below 4.2 K where the resistance reaches its minimum, or "residual", resistance. Considering non-superconducting pure metals, the residual resistance is determined by the amount of residual chemical impurities and lattice imperfections. Frequently the resistance ratio $R_{273.15\mathrm{K}}/R_{4.2\mathrm{K}}$ is taken as an index of purity. It increases after applying a specific heat treatment to remove lattice vacancies and other higher order defects. Another index of purity is the "alpha coefficient":

$$\alpha = (1/100\,°\mathrm{C})\,(R_{100} - R_0)/R_0 \tag{3-1}$$

where R_{100} and R_0 are the resistances at 100 and 0 °C, respectively. The greater is either of these indices, the higher is the purity. Very pure platinum yields $\alpha > 3.925 \times 10^{-3}$ °C^{-1}, but α is never greater than 3.9285×10^{-3} °C^{-1}. Of the above mentioned metals, platinum can be obtained with the highest purity in form of thin wires.

3.2.1.1 *References to the Theory of Electric Conduction*

Assuming the simple electron-gas model of Drude to describe the electric conductivity of metals, the mass and the charge of the electron being denoted by m and e, respectively, it can be shown (Dugdale [12], pp. 13 ff.) that the electric conductivity σ is given by

$$\sigma = n\,e^2\,\tau/m \tag{3-2}$$

where n is the number of electrons per unit volume and τ is the relaxation time. This is related to the rate at which the drift velocity of the electrons reaches a constant value following the application of an electric field. Under ordinary conditions the relaxation time of metals is of the order of 10^{-13} s and is the only parameter in Equation (3-2) that changes with temperature.

The basis of the present understanding of the conductivity of metals is the idea due to Bloch [13] that free electrons travel through the metal as plane waves modified by a function having the periodicity of the lattice. Consequently, the resistivity (inverse of σ) arises only from imperfections of the lattice. Foreign atoms, point defects, and grain boundaries all lead to additional scattering and, hence, to an increase in the resistivity. Additionally, the electrons are scattered by lattice vibrations through the emission or the absorption of a quantum of lattice vibrational energy, ie, a phonon. Each of these scattering mechanisms has its own relaxation time and its own rate of change of the drift velocity. The relaxation times are to a first approximation independent of each other. As rates must be added, the overall relaxation time is determined as follows from the individual times:

$$(1/\tau) = (1/\tau_1) + (1/\tau_2) + \ldots \tag{3-3}$$

The total resistivity can be considered as simply the sum of the resistivities due to separate mechanisms, as stated by Matthiesen's rule, which holds for small concentrations of foreign atoms and defects. Accordingly, only the electron-phonon scattering mechanism is temperature dependent, whereas foreign atoms and lattice defects cause nearly constant increments in the resistance. All resistance ratios between higher and lower thermal states of the same specimen decrease with increasing impurity level.

A full account on the current theory of electric conduction is not the purpose of this chapter, and it will suffice to list some of the conclusions:

(i) the resistivity of simple metals, such as Na, at ordinary temperatures ($T > \Theta_R$, Θ_R being a characteristic temperature of the metal related to its Debye temperature) exhibits a simple T dependence;

(ii) for the same metals and for $T \ll \Theta_R$ (very low temperatures) there results a T^5 dependence of the resistivity;

(iii) the more complex scattering mechanism of transition metals introduces a deviation from the simple T dependence that may be either positive (eg, nickel) or negative (eg, platinum);

(iv) at temperatures substantially higher than Θ_R a further term in T^2 is included on account of thermal expansion and of the increased amplitude of the lattice vibrations;

(v) at even higher temperatures there is a further increment to account for the effect of lattice vacancies. Under equilibrium conditions the resistivity increment $\delta\rho$ is given by

$$\delta\rho \propto \exp\,(-E_r/k_B\,\mathrm{T}) \tag{3-4}$$

where E_r is the energy of formation of a lattice vacancy, which in platinum is about 1.5 eV [14], and k_B is the Boltzmann constant;

(vi) with most transition metals electron-electron scattering becomes very important below 10 K, thus introducing a predominant T^2 term;

(vii) when very small amounts of magnetic metals are dissolved in certain non-magnetic metals, a resistance minimum occurs at very low temperatures; this is known as the Kondo effect. An anomalous behavior results with diluted alloys of iron in rhodium and of cobalt in platinum. Instead of a minimum there results a monotonic decrease in resistivity that is of great interest for low-temperature thermometry. The advantage in using one of these alloys is clearly shown in Figures 3-3 and 3-4.

A detailed theoretical approach to this matter can be found elsewhere (see, for instance, Quinn [15], pp. 167 ff.; Dugdale [12], pp. 164 ff., pp. 189 ff., and pp. 250 ff.).

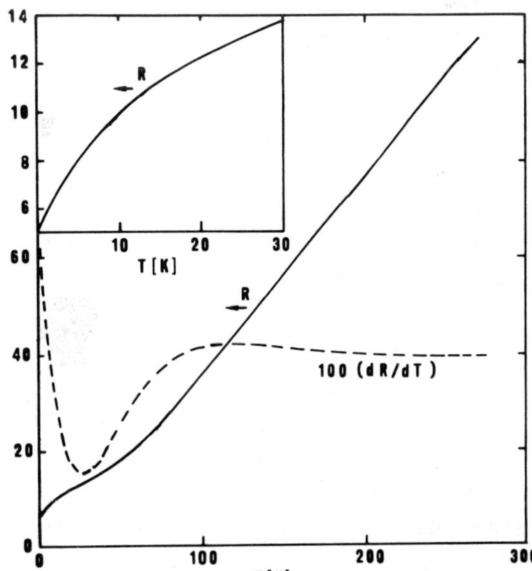

Figure 3-3.
Resistance (Ω) and sensitivity (Ω/K) of an RhFe thermometer.

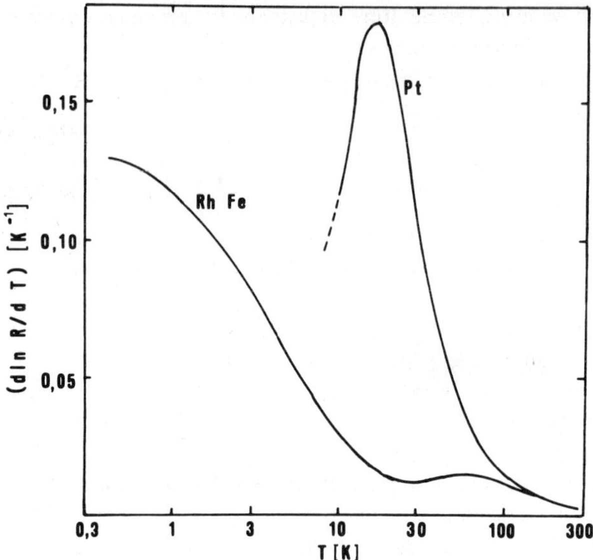

Figure 3-4. Sensitivity of RhFe and Pt resistance thermometers.

3.2.2 Resistance Thermometers with Negative Temperature Coefficients

Negative temperature coefficients are characteristic of semiconducting crystals, sintered semiconducting metal oxides and carbon resistors. Whereas the conductivity of metals is provided by the electron gas with a density that is almost independent of temperature, that of a semiconductor is due to the movement of electrons in the conduction band, with density n, and of holes (vacancies of electrons), with density p. At room temperature the energy gap E_g between the conduction and the valence band is much larger than $k_B T$ so that only a limited number of electrons are conducting the electricity. The conductivity increases with T, as predicted by Boltzmann statistics. Electrons, however, obey Fermi-Dirac statistics, so it is necessary to take into account the Fermi level μ and the fact that both electrons and holes respond to the electric field with effective masses m_e and m_h, respectively. The densities of conducting electrons and holes are thus

$$n = 2\,(2\pi\, m_e\, k_B\, T/h^2)^{3/2} \exp\left[-(E_g - \mu)/k_B\, T\right] \qquad (3\text{-}5)$$

$$p = 2\,(2\pi\, m_h\, k_B\, T/h^2)^{3/2} \exp\left[-\mu)/k_B\, T\right] \qquad (3\text{-}6)$$

where h is *the Planck constant.* For an intrinsic semiconductor (eg, pure silicon, germanium, and diamond) where $n = p$, one arrives at the following expression:

$$\sigma = 2e\left(\frac{2\pi\, k_B\, T}{h^2}\right)^{3/2} \times (m_e\, m_h)^{3/4} \times \left(\frac{e\, T_e}{m_e} + \frac{e\, T_h}{m_h}\right)^{-\frac{E_g}{k_B\, T}} \qquad (3\text{-}7)$$

where τ_e and τ_h are the relaxation times for the electron-phonon and the hole-phonon interactions, respectively. The mobilities $e\tau/m$ of electrons and holes are normally different. In contrast to metals, n and p are the dominating temperature-dependent factors in σ. Equation (3-7) is frequently written in one of the following forms:

$$\sigma = (1/A) \exp[-B/T] \tag{3-8}$$

or

$$\rho = A \exp[B/T]. \tag{3-9}$$

A exhibits a mild dependence on temperature because it includes $T^{-3/2}$ and also the mobilities have their own temperature dependence. With an energy gap between 0.1 and 1 eV, the strongest temperature dependence is introduced by the exponential term.

When a single-crystal intrinsic semiconductor is doped with foreign atoms having a valency one greater (donor) or one less (acceptor) than the host atom, extra electrons or extra holes are generated. These semiconductors are designated n-type and p-type, respectively, and have a higher conductivity and a lower temperature coefficient than has the intrinsic one. In their theoretical description one has to consider also the ionization energies of donors and acceptors and the energies required to have an electron jumping from one impurity atom to another. As these energies are drawn from $k_B T$, the temperature dependence is much more complex than that for an intrinsic semiconductor. Considering a germanium crystal doped with N_d donor atoms per unit volume and N_a acceptor atoms, it is possible to distinguish between four regions of conduction [15, 16]:

(I) *Above ca. 400 K,* where the predominant conduction is due to thermally excited electrons from the valence band, all impurity atoms having long since been ionized, and the material behaves just like an intrinsic semiconductor.

(II) *Below 400 K, but above 100 K,* where all active atoms, $N_d - N_a$, are ionized and their electrons are available in the conduction band and predominant over other thermally excited electrons. The conductivity depends on temperature only through the mobility of electrons, which decreases with temperature, and therefore the resistivity has a positive temperature coefficient.

(III) *Below ca. 100 K,* where the ionization of the effective donor atoms is no longer complete and the conductivity starts to depend exponentially on T, with a negative coefficient. Approaching 10 K very few of the donor atoms are ionized and hence the resistivity is as follows:

$$\rho = \{N_a/[C(N_d - N_a)]\} \exp[\beta/k_B T] \tag{3-10}$$

where virtually all of the temperature dependence is in the exponential term and C is constant.

(IV) *About 10 K,* the conduction due to ionized electrons is gradually replaced with that due to electron jumps. The expression for the resistivity is similar to Equation (3-10):

$$\rho = (1/D) \exp(\Gamma/k_B T). \tag{3-11}$$

Low-temperature germanium thermometers operate in regions III and IV. The transition may be "hard" or "soft", depending on the amount of mismatch in the activation energies β and Γ. With 1.8×10^{17} arsenic atoms per cm^3 as donors, of which 8% are compensated with gallium, the transition is soft and suitable for wide-range thermometry [16].

Most commercial semiconducting temperature transducers consist of compounds, mainly oxides or complex oxidic systems such as spinels or perovskites, rarely stoichiometric, and sintered in a controlled atmosphere from micrometer-size particles of the components. They appear under the generic name of "thermistors" (thermally sensitive resistors). Depending on the atmosphere during sintering, different combinations of n-type and p-type zones within a grain result. Grain boundaries also play an important role in electric conduction. Almost all commercial thermistors are made from combinations of metal oxides such as the oxides of manganese, nickel, cobalt, iron titanium, copper, lithium, magnesium, and chromium. Almost any combination of metal oxides will exhibit thermistor properties, but not all are stable. Manganese-nickel and manganese-nickel-cobalt systems are the most stable.

A qualitative insight into the conduction mechanism can be obtained from a comparison with single-crystal devices [17]:

Bond: Covalent (ME); covalent and ionic (OM).
Impurities: primarily chemical (ME); primarily physical (OM).
N-type: impurity results in excess valence electrons in covalent bonds (ME); deficit of oxygen results in an excess of ionized metal atoms in lattice-Frenkel defects (OM).
P-type: impurity results in a deficiency of valence electrons (holes) in covalent bonds (ME); excess of oxygen results in a deficiency of positively ionized divalent metal atoms in lattice-Schottky defects (OM);
Conduction: primarily intrinsic (ME); primarily extrinsic (OM);
Note: ME, monocrystalline elements such as germanium and silicon; OM, oxide-mixture semiconductors.

To a first approximation, all thermistors with negative temperature coefficients (NTC thermistors) follow the general law of semiconductors:

$$\rho = A \exp(B/T) \tag{3-12}$$

or

$$R_T = R_0 \exp \beta [(1/T) - (1/T_0)] \tag{3-13}$$

where β is a constant with typical values between 1000 and 5000 K and R_0 is the resistance at the reference temperature T_0, typically 25 °C. The incremental temperature coefficient is

$$(1/R)\,\mathrm{d}R/\mathrm{d}T = -\beta/T^2 \tag{3-14}$$

With β = 4000 K and at 300 K, the temperature coefficient is 4.4% K^{-1}, which is about ten times that of platinum. More extensive information on metal oxide semiconductor thermistors can be found elsewhere [17, 18].

3.2.3 Techniques of Resistance Measurement Applied to Thermometry

Precision resistance thermometry entails two distinct ranges of resistance values: from 0.1 to 1000 Ω for metal resistance thermometers and from 10 Ω to 1 MΩ for semiconductors and thermistors. Although partially overlapped, the two ranges may require different instruments. Metal resistance thermometers are, with few exceptions, low-resistance devices, thus requiring instruments of high resolution, efficient neutralization of lead resistance and parasitic thermal emfs, and operating with low current. Resistance measurements above 10 kΩ generally require electrical guarding to eliminate leakage currents. Both low- and high-resistance thermometers are frequently equipped with long leads crossing steep temperature gradients.

To minimize the temperature error caused by the dissipated power, thermistors are fed with 10–100 μA and platinum resistance thermometers with 0.5–10 mA. Self-heating effects not exceeding 0.01 °C are generally acceptable. The resulting systematic error can be corrected taking two measurements with two different currents, as explained later. Although high-resistance devices are not adversely affected by cable resistances and by parasitic emfs, thermistors and germanium resistance thermometers, when used over their full range, have low resistance and, thus, the electrical instrument must be adequate for both ranges. In the most general case the resistance measurement technique will require a resistance thermometer equipped with four leads, as shown in Figure 3-5.

Measuring circuits for resistance thermometry may be subdivided into potentiometric types, realizing a balance condition with no current flowing in the potential leads (P_1 and P_2 in

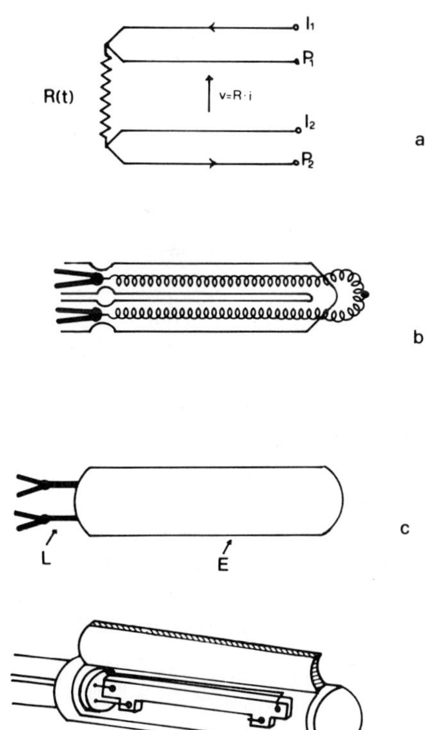

Figure 3-5.
Types of resistance thermometers: (a) four lead scheme; (b) platinum resistance element; (c) industrial platinum resistance thermometer element; L, leads, E, element (note: the elements are generally supplied with two leads; the four-lead arrangement should be realized by the user); (d) germanium resistance thermometer.

Figure 3-5 a), and bridge circuits, where, in principle, current flows at balance but may be made negligible for practical purposes. In the last 20 years potentiometric circuits have been finding more and more applications, for the introduction both of digital voltmeters and of analog-to-digital convertes of very high sensitivity, linearity, and input resistance, and for the adoption of this circuit in most AC bridges. Traditional DC bridges such as the Wheatstone, Siemens, Kelvin, Mueller and other types [19] have long been considered basic instruments in laboratory applications. They are today challenged by AC bridges, which offer the following advantages:

(a) extremely high and stable linearity achieved with inductive voltage dividers or ratio transformers, yielding linearities up to ± 1 in 10^8;

(b) very high sensitivity resulting from the use of high-resolution phase-sensitive detectors;

(c) complete neutralization of parasitic thermal emfs and minimization of the effect of lead resistances;

(d) automatic operation by means of self-balancing systems.

Many authors have described AC bridges [20–24]. The excitation frequency is sometimes chosen to be a multiple of the line frequency so as to increase the rejection of line interferences. The reactance of a resistance thermometer produces an appreciable effect above 500 Hz and an excitation frequency above this limit is not recommended. Another cause of systematic differences between DC and AC measurements is the coupling between the leads, particularly with long leads. Shielding is essential. The inductive coupling is considerably reduced by twisting together the current leads (I_1 and I_2 in Figure 3-5 a) and, separately, the potential leads P_1 and P_2. Some bridges require two coaxial cables for every resistor [21] and others four [20]. Some additional difficulties may arise with very long cables subjected to interferences from the mains. Saturation of the input stage of the phase-sensitive detector may result. AC-DC differences can also be ascribed to the sensing element or to its coupling to the surrounding medium: low-temperature PRTs used with a 375-Hz bridge showed differences of a few millikelvin [25]; industrial PRTs (IPRTs) may show differences of the same order of magnitude when used up to ca. 400 °C [25]. Between 400 and 960 °C and at frequencies above 300 Hz appreciable differences may be detected with sintered-alumina insulated IPRTs on account of frequency-dependent leakages [26]. Much lower leakages were detected in silica-insulated PRTs above 600 °C [26]. In all these cases the AC measurements are in error.

In contrast, germanium thermometers when measured with DC instruments are affected by a systematic error due to the low thermal conductance of their thin-wire gold leads combined with the high Peltier coefficient at the junction between the doped-germanium chip and the leads. This error may be as large as 0.2% [27, 28]. When they are measured with an AC bridge operating above a certain frequency, the error vanishes.

Advanced DC instruments are also available for the needs of resistance thermometry. The most accurate DC bridge is that developed by Kusters and McMartin [29] (see Figure 3-6 for the circuit diagram). The heart of the circuit is an adjustable ratio transformer with DC currents in both windings and equipped with a very sensitive flux detector to determine very precisely the condition of zero magnetic flux in the core. Thus, the current ratio is very accurately established by the winding-turn ratio. Consequently, a voltage drop generated by one current across an unknown resistance can be directly compared with that generated by the other current across a standard resistance. The linearity of this bridge is as good as that of

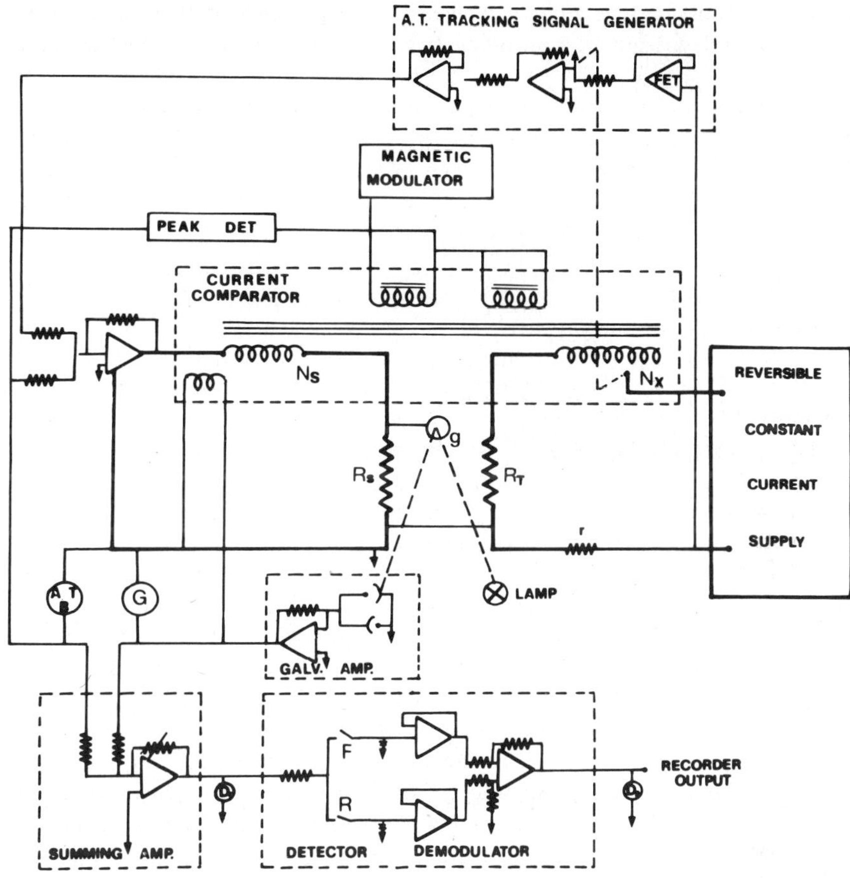

Figure 3-6. Kuster and McMartin birdge: it is based on a DC comparator providing a current ratio equal to the turn ratio N_s/N_x.

any AC bridge. Another DC bridge was described by Crovini and Kirby [30]. They used two isolated current loops with the current ratio established by high-accuracy resistors and high-sensitivity operational amplifiers, obtaining a very low voltage noise and offering battery operation.

Although DC bridges can hardly be operated automatically, automated data acquisition systems based on DC electronic ohmmeters are now available. They may include high-linearity analog-to-digital converters, an automatic zeroing system and an automatic switch to scan several thermometers; they may also be interfaced to a computer. Some models provide the autozero function for both the amplifier offset and the parasitic emf in the external circuit, including the sensor. Autocalibration with reference resistors and the current-pulse measurement technique, which virtually eliminates the self-heating effect, are also available. Microprocessor-based systems convert the resistance data into temperature data, which are directly displayed. Instruments of this type may have a resolution of 1 mK and an uncertainty of ±0.002%. Figure 3-7 shows an instrument of this type.

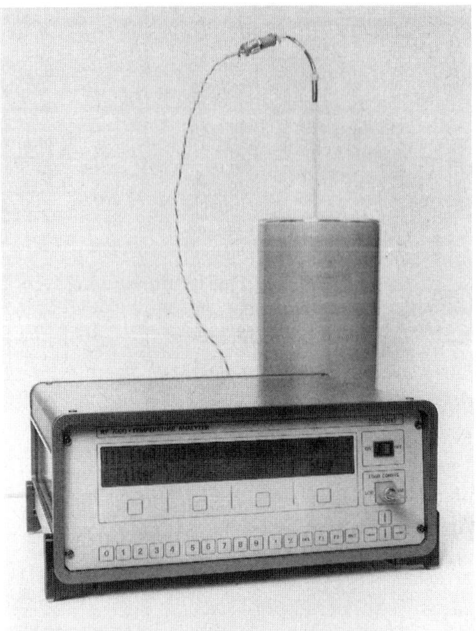

Figure 3-7.
Direct-reading instrument (courtesy of Franco Corradi).

3.2.4 Self-heating and Ultimate Sensitivity

One very common systematic error in resistance thermometry is caused by the Joule heating produced by the measuring current. The heat liberated within the sensing element flows either through the surrounding wall towards the zone whose temperature is to be measured or, to a much lesser extent, along the leads. Thus the resulting temperature difference is mostly concentrated across the interface between the thermometer and the unknown temperature zone. A smaller temperature difference is developed across the sheath and between the sensing element and the sheath. Figure 3-8 illustrates this concept for a standard platinum resistance thermometer (SPRT) in an ice-point bath.

The self-heating error will then depend on the thermal condition realized around the thermometer and, hence, must be determined either in situ or under exactly the same conditions of use. Neglecting radiation losses and assuming the sensing element and the adjacent part of the leads are fully isothermal, the stationary conditions are described by the following equations:

$$W_J = W_E + W_L \tag{3-15}$$

$$R(t)\,I^2 = (K_E + K_L)\,(t_M - t) \tag{3-16}$$

where W_J is the dissipated power, or $R\,I^2$, W_E and K_E are the thermal flow and the conductance towards the outside zone, respectively, W_L and K_L are the thermal flow along the leads

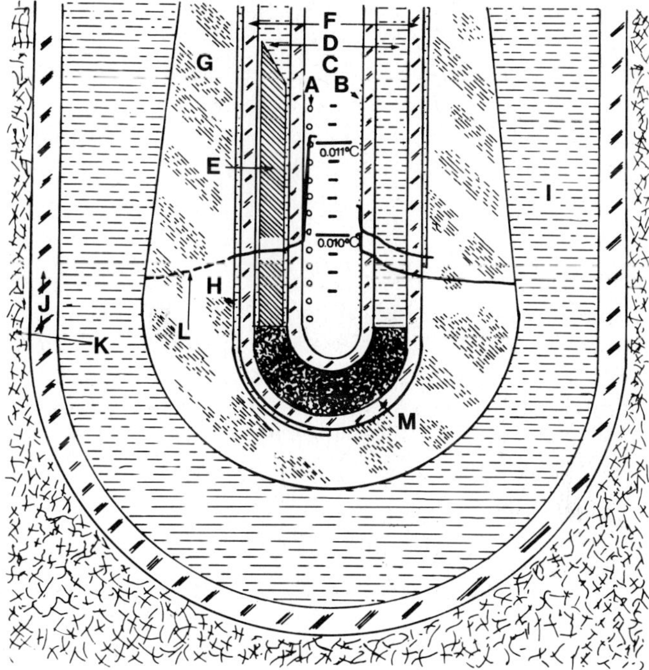

Figure 3-8. A standard platinum resistance thermometer in a cell for the triple point of water [33]: A, coiled-filament-type PRT element (see Figure 3-16); B, single-coil PRT element (see Figure 3-16 or 3-9); C, stem; D, water to provide the thermal exchange; E, aluminium or copper bushing (optional); F, glass well; G, ice mantle; H, remelted-water jacket; K, crushed ice; I, water phase; J, cell outer glass wall; L, temperature profile; M, flexible pad (optional).

and their thermal conductance, respectively, and t_M and t are the measured and the unknown temperature, respectively. The self-heating error Θ_{SH} is expressed as follows:

$$\Theta_{SH} = t_M - t \tag{3-17}$$

$$\Theta_{SH} = R I^2/(K_E + K_L). \tag{3-18}$$

Expressing with R_1 and R_2 the resistances measured for the same temperature but with two different currents I_1 and I_2 and assuming a linear relationship between the dissipated power and the consequent resistance increase, the self-heating is expressed as follows:

$$\Theta_{SH} = (R_2 - R_1)\, I^2/S_T\, R\, (I_2^2 - I_1^2) \tag{3-19}$$

where S_T is thermal sensitivity of the resistance thermometer and R is the average resistance. Measuring with I and then with $\sqrt{2}\, I$, the detected resistance increase is the same as produced by the self-heating of I (positive temperature coefficient detectors). With NTC detectors the same condition holds but with opposite sign. Equation (3-19) and its consequences hold for a dissipated power not exceeding a few milliwatts; preferably it should not exceed 1 mW.

A deviation bridge, or another equivalent instrument, enables the self-heating to be estimated if the transient output voltage when the measuring current is first switched on is recorded. The transient self-heating is also used to determine the response time [31].

The capability of a resistance thermometer to dissipate heat is reported in the technical literature with the parameter

$$W_D = W_J / \Theta_{SH} \tag{3-20}$$

commonly expressed in milliwatts per kelvin. Also the inverse parameter is prescribed (IEC publication 751 [9]). The data that are available are frequently given for bare elements and always under standard conditions, eg, flowing water at 1 m/s.

The IEC standard prescribes the determination in an agitated mixture of ice and water. Actual values of Θ_{SH} may differ from those given by the manufacturers for several reasons, including the particular assembly and medium. They may also vary with time. If the self-heating error is determined at a given time and accounted for, there always remains a residual uncertainty:

$$\delta\Theta_{SH} = \pm K_0 \, W_J / W_D \tag{3-21}$$

where K_0 depends on the particular case. In very sensitive measurements, as with microkelvin temperature controllers or in differential temperature measurements, $\delta\Theta_{SH}$ contributes to limit the ultimate sensitivity.

To clarify this point better, we can refer to a resistance thermometer connected at the input of an ideally noiseless amplifier. There, the only source of noise is the thermometer itself. Hence, the minimum detectable temperature change is

$$(\delta T_{min}) = \sqrt{(4\, k_B \, T R \, B_F)} \, / S_V \tag{3-22}$$

where $k_B = 1.380662 \times 10^{-23}$ J/K the Boltzmann constant, B_F is the frequency band width, and S_V is the thermal sensitivity of the voltage, or

$$S_V = S_T R I . \tag{3-23}$$

At constant S_v the sensitivity threshold increases with the square root of R. With this condition, the minimum uncertainty that is caused by the self-heating has the following magnitude:

$$\delta\Theta_{SH} = (K_0 / W_D) \, (S_V / S_T)^2 / R . \tag{3-24}$$

For a given type of thermometer both the minimum detectable temperature and the minimum self-heating uncertainty depend on the ratio $\sqrt{R / S_V}$. Adjusting it to a fixed value, thermometers of the same type, operating at the same temperature but with different resistances, provide the same temperature detectability.

In practice, however, it is not always possible to increase the size of the sensor as much as its resistance at a given reference temperature. At constant size, W_D would be inversely proportional to R; in this case, adjusting S_V to keep $\delta\Theta_{SH}$ constant with any R, the minimum detectable temperature would increase with the square root of the resistance. Moreover, every practical amplifier generates noise. When this is taken into account, Equation (3-22) is modified as follows:

$$(\delta T)_{min} = \sqrt{(A + CR + DR^2)} / S_V \tag{3-25}$$

where A, D, and C account for the amplifier noise voltage, the amplifier noise current and the term $4\,k_B\,TB_F$, respectively. It is possible to distinguish between three regions: at low R both the C and D terms are negligible and hence the minimum detectable temperature decreases with increasing square root of R, when the heat liberated is kept constant; at medium R (typically between 10 and 1000 Ω), when the equivalent noise resistance of the amplifier is of the same order of magnitude as the input resistance, or lower, the C term is no longer negligible and therefore (δT_{min}) will eventually remain constant; for even higher resistances (typically above 10^4 Ω) the D term comes into action, causing (δT_{min}) to rise with $\sqrt{R}$.

Concluding, W_J is set to a suitable value to provide the best sensitivity, typically between 50 and 500 μW. Consequently, the minimum detectable temperature is defined through Equations (3-23) and (3-25). The thermal conditions around the thermometer must be such as to generate an uncertainty $\delta\Theta_{SH}$ lower than $(\delta T)_{min}$, otherwise W_J has to be reduced. The optimum condition, implying that both terms are equal, is achieved with the following condition:

$$(K_0)_{OPT} = \sqrt{(4k_B TRB_F)}\;\; W_D\, R S_T^2 / S_V^3\,. \tag{3-26}$$

For instance, a PRT of 100 Ω at 273 K, exhibiting $S_T = 4 \times 10^{-3}\ K^{-1}$ and $W_D = 20$ mW/K, when measured with 1 mA and a band width of 1 Hz, yields

$$(K_0)_{OPT} = 6.14 \times 10^{-4}$$

$$(\delta T)_{min} = 3.1\ \mu K. \tag{3-27 a}$$

A thermistor of 10^4 Ω at 273 K, exhibiting $S_T = 0.04\ K^{-1}$ and $W_D = 0.6$ mW/K, when measured with 10 μA and a band width of 1 Hz, yields

$$(K_0)_{OPT} = 1.84 \times 10^{-3}$$

$$(\delta T)_{min} = 3.1\ \mu K. \tag{3-27 b}$$

It is worth noting that taking into account the amplifier noise, the detectability of the thermistor may be worse on account of the input noise current. Other self-heating data are presented in Table 3-1.

Table 3-1. Self-heating effect in various resistance thermometers. The given values are indicative of the behavior of different resistance thermometers. The manufacturers provide more accurate values for particular types.

Type	Temperature range (K)	Resistance (Ω)	W_D (mW/K)	Medium
Platinum: Ceramic encased	250–350	100 (0 °C)	20	still air
Platinum (SPRT): long stem	250–350	25.5 (0 °C)	50	still air
Thermistor: glass enclosed	220–550	10^4 (25 °C)	0.6	still air
Thermistor: glass enclosed	220–550	10^4 (25 °C)	3.0	still water
Germanium: Copper encased	10–30		0.5	metal block
Germanium: Copper encased	<2		0.02	metal block
Rh-Fe: Platinum encased	5–10	10 (10 K)	2	metal block

3.3 Resistance Thermometers

3.3.1 Platinum Resistance Thermometers

Platinum is the most precious metal used for thermometry. For a long time platinum resistance thermometers were expensive devices suited only to the realization of primary standards or other special laboratory applications. Owing to new manufacturing techniques that greatly reduce the amount of platinum that is required for a sensing element, platinum is today used also to make industrial thermometers at a cost comparable to that of many thermocouples and other devices. In the case of primary standards the cost of fabrication and of the preparatory work greatly exceeds that of the precious metal.

Platinum can be drawn in to very thin wires of high purity; wires with diameters of 0.05 mm or less and a mass-fraction purity better than 99.99% are today commercially available. From Table 3-2, where platinum and nickel properties are compared, it can be concluded that, for a given resistance value, platinum always offers the possibility of smaller volume and smaller thermal mass sensors. This property has been enhanced considerably with the adoption of thick-film and thin-film sensors that utilize a minimum amount of platinum.

The best stability is achieved when platinum is used in air or in a mixture of helium and a small amount of oxygen, such as is commonly used for low-temperature thermometers. Atmospheres containing hydrogen or carbon are poisonous for platinum and lead to unstable resistances. Silica particles in a reducing atmosphere again cause instability and brittleness.

Table 3-2. Properties of copper, nickel and platinum wires.

	Cu	Ni	Pt
Resistivity (20 °C, μΩ cm)	1.673	6.84	10.6
Density (20 °C, g/cm^{-3})	8.92	8.90	21.45
Coefficient of linear thermal expansion (10^{-6} K^{-1})	16.6	13	9
Length$^{(a)}$ (cm)	1173	287	185
Mass (mg)	205	50.1	77.9
Heat capacity$^{(b)}$ (mJ/K)	79	22	11

(a) Of a wire of 0.05 mm having a resistance of 100 Ω at 20 °C.
(b) Values at 25 °C.

The thermal expansion coefficient of platinum (9×10^{-6} °C^{-1}) is not very well matched by those of typical insulating materials such as silica or alumina. Thus, a strain-free design of the sensing coil of high-accuracy thermometers is unavoidable for eliminating hysteresis and enhancing stability. Industrial platinum sensors for use up to 400 °C may have platinum in tight contact with the insulator (eg, a thick platinum film on an alumina substrate) and still limit hysteresis to within ±0.25 °C [32].

3.3.1.1 The Traditional Standard Platinum Resistance Thermometer (SPRT)

The types of SPRTs that can commonly be found in metrological laboratories were developed just after the introduction of the first International Temperature Scale in 1927 and subsequently improved. The most traditional among them is the so-called "long stem" type shown in Figure 3-9. Different designs are used for the sensing element, such as those shown

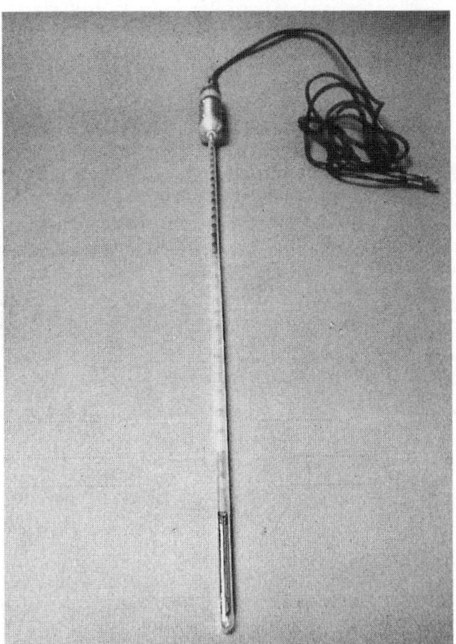

Figure 3-9.
Long-stem standard platinum resistance thermometer (courtesy of IMGC).

in Figures 3-5b and 3-10. The sensing platinum coil is wound on a former that at the same time provides excellent electrical insulation and freedom for the coil to expand and contract. The former material can be mica, borosilicate glass, or silica. The same material, in the form of discs and of either capillaries or plates, is used both to space and to isolate the outcoming leads. The thermometer is enclosed in a sheath of either borosilicate glass or fused silica, the latter being suitable above 500 °C. Alumina tubes have been used in some cases up to 1000 °C, but were eventually discontinued as they tend to crack upon thermal shock and to show excessive thermal conductance. With transparent sheaths it is often necessary to sand-blast or cover with colloidal graphite the outside surface over a band from about 3 cm above the tip and up to about 30 cm to stop the radiation piping in the sheath wall. Otherwise, radiation in both the visible and near-infrared regions would easily be lost through the glass, thus causing significant heat losses with immersion errors larger than 10 mK above 300 °C [33]. The typical size of the sheath is 0.7 mm in diameter and 45–80 cm in length.

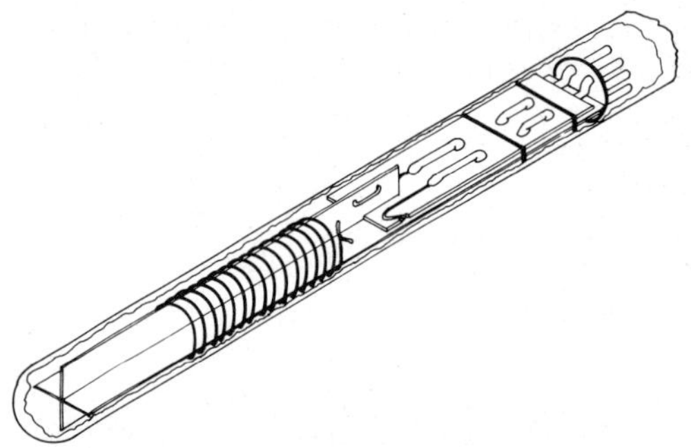

Figure 3-10.
Sensing element of a Meyers-type SPRT.

Mica insulators lose water of crystallization when exposed to high temperatures. Muscovite mica, $H_2KAl_3(SiO_4)_3$, is common but contains 5% in mass of water, which tends to be lost above 540 °C with alteration of its shape. Phlogopite mica, $H_2KMg_2Al(SiO_4)_3$, is recommended for operation above 500 °C as it does not begin to dehydrate until ca. 750 °C. A so-called "wet kick" may appear when the current of the thermometer is reversed, as a consequence of the moisture released by mica. This effect, which is caused by an exchange of energy with the water molecules when the electric field in the insulator is reversed, consists of a short-lived imbalance of the measuring bridge that causes the null detector to be kicked off its null position. A wet kick often implies a reduction of the electric insulation. Pumping off the gas in the sheath and replacing it with dry gas eliminates the effect. Care must be exercized in order to have the dew-point of the filling gas considerably below 0 °C. If this were not the case, some water would condensate when the thermometer is at about 0 °C, again causing a reduction in the insulation.

Sapphire and fused-silica insulators, albeit more expensive than those of mica, enable almost all of these problems to be solved and provide a much longer life at high temperature. The manufacturing cost of these insulators has recently been reduced by the introduction of laser and ultrasonic machining.

The resistance element is usually made from 0.07 mm diameter platinum wire and has an ice-point resistance of ca. 25.5 Ω. The platinum purity is specified by the requirement for $\alpha \geqslant 3.9250 \times 10^{-3}\,°C^{-1}$ or, as prescribed by the newly introduced International Temperature Scale of 1990 (ITS-90), by the requirement that $R(t)/R(0.01\,°C)$ as measured at the triple of mercury or at the melting point of gallium, be lower than 0.844235 or, higher than 1.11807, respectively. A purity of 99.999% by mass enables this requirement to be met after a preparatory annealing. The resistor is usually "noninductive", often bifilar, to reduce the pick-up of noise from stray fields and improve the performance in AC circuits. The length of the sensing resistance is usually between 2.5 and 4 cm. The fabrication process of the resistance element requires that the platinum wire be in a hard-drawn state. Details of the fabrication process are shown in Figures 3-11 and 3-12.

Figure 3-11.
Technique for formation of a platinum coil (courtesy of IMGC).

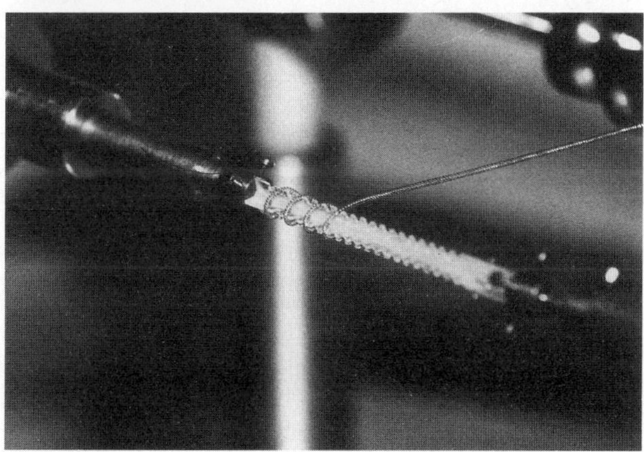

Figure 3-12.
Realization of a sensing element on a silica cross (courtesy of IMGC).

The thermometers are then connected to the leads, inserted into the sheath, filled with dry air at a pressure of ca. 30 kPa, and sealed after a preliminary annealing. The annealing temperature is usually between 450 and 500 °C and the duration may vary from manufacturer to manufacturer, but it should not be shorter than 12 h. To achieve the best performance of an SPRT, it must be submitted to both stabilization and maintenance annealing, as explained later.

SPRT sensing elements are equipped with four leads of either gold or platinum. Gold does not appear to contaminate platinum and is easy to work. Platinum has a lower thermal conductivity than gold and is immune to thermal emfs at the junctions with the sensing coil. To provide resistance uniformity with any temperature distribution, the four leads have to be taken from the same spool, with the same diameter and the same length, and run parallel. Analogous considerations apply to the cable prolonging the leads outside the stem of the thermometer.

Inside the stem, the lead wires are threaded through capillaries and discs. The discs ensure the right spacing of the leads, keeping them insulated, and provide a thermal tiedown with the sheath surface. For this reason, discs of sapphire or high-purity alumina are preferable. The immersion error, ie, the resulting systematic difference between the coil temperature and that to be measured, excluding the self-heating error, is virtually zero when no heat flow leaves the coil via the leads and the sheath. An effective thermal tiedown just above the sensing coil and the elimination of radiation losses enable this condition to be achieved. It also important that the coil be close to the inner surface of the sheath without, however, touching it.

The immersion error decreases with increasing immersion in an isothermal medium and eventually vanishes. A test of adequate immersion of a long-stem thermometer can be carried out in a metal freezing-point cell tracking the variation of the freezing temperature with the depth. According to the Clausius-Clapeyron equation, a linear gradient has to be observed with metals and other substances (eg, -7.3 μK/cm for the triple point of water, 22 μK/cm for tin freezing point, 27 μK/cm for zinc freezing point and 54 μK/cm for silver freezing point, the positive sign indicating an increasing temperature with increasing depth).

Figure 3-13 shows the behavior of two PRTs in freezing zinc [34].

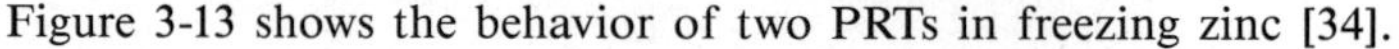

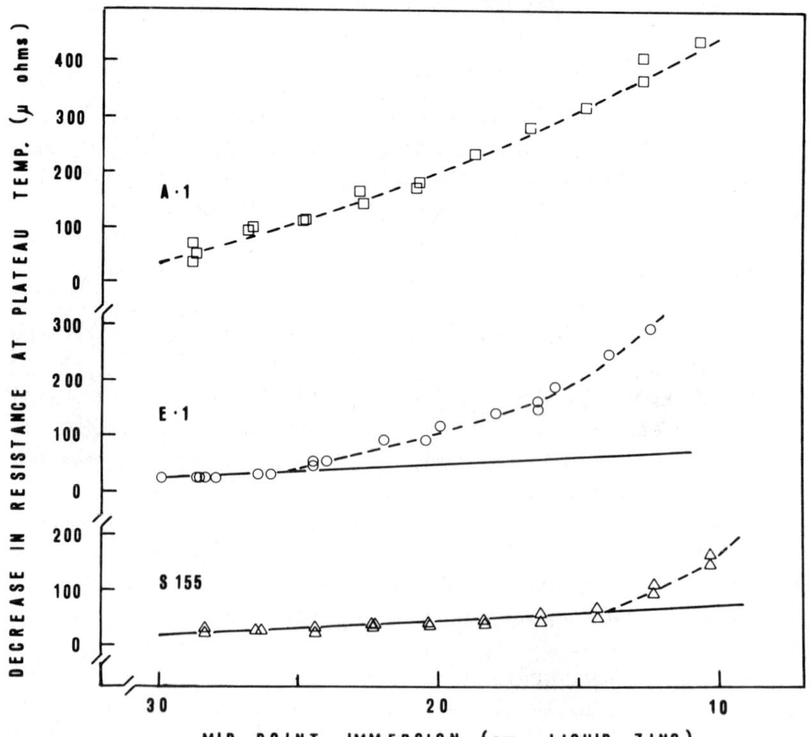

Figure 3-13. Immersion characteristics of three SPRTs in freezing zinc [34].

3.3.1.2 The Capsule SPRT

Low and very low temperatures cannot be accurately measured with long-stem thermometers. In most low-temperature apparatus one needs to enclose the thermometer within radiation shields and provide a thermal tiedown of the leads to the same temperature as the thermometer. Moreover, the thermometer must be perfectly gas-tight to operate in vacuo, when necessary. These and other requirements led to the adoption of the so called "capsule" design shown in Figure 3-14. The platinum wire characteristics are still the same as for long-stem types. An ice-point resistance of 25.5 Ω is mostly preferred, but 100-V thermometers are also available. The sensing resistance is inclosed in a platinum sheath made from hard-drawn tube of outside diameter 5 mm, wall thickness 0.1 mm and an overall length, including the glass seal, of about 6 cm.

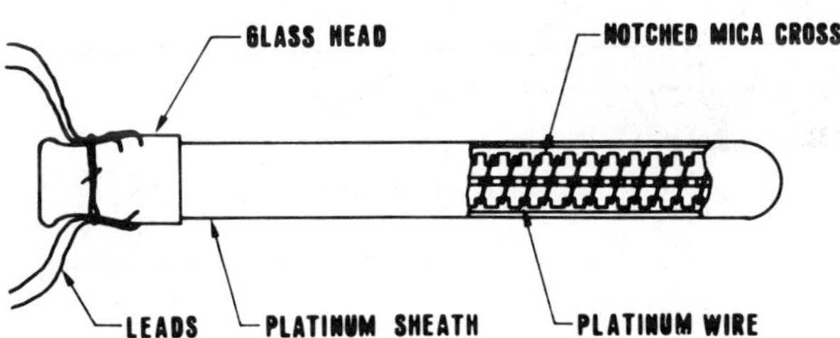

Figure 3-14.
Conventional capsule-type SPRT.

Four pure leads of platinum, welded to the coil ends, go across a lead-glass seal which closes the capsule. As reported previously, a helium atmosphere of about 30 kPa at room temperature is maintained inside the capsule, with a few kilopascals of partial pressure of oxygen to prevent the contamination of platinum when heated to 200 °C.

The thermometer is usually mounted, with a smear of vacuum grease to ensure thermal contact, in a copper sleeve or pocket which is in good thermal contact with the body whose temperature is to be measured. The quality of the contact and also the integrity of the capsule must be checked by measuring the self-heating effect. To reduce the heat conduction through the leads to a very low level, long fine wires are used outside the capsule, thermally anchored to the sample chamber and also to several other points within the cryostat, with a long length between the anchor points. Consequent errors may be reduced to less than 0.1 mK.

The self-heating error at room temperature for such a thermometer is about the same as that for long-stem types of equal resistance; it is of the order of 1 mK for a current of 1 mA. The helium filling of the capsule provides a good thermal conductance that decreases, however, with temperature. Nevertheless, as the resistance falls considerably faster, the heating effect at constant current falls with temperature in such a way as to be only about 0.01 mK at 14 K, always with 1 mA. Unusually large heating effects may indicate that the filling helium has been lost.

Low-temperature calorimetry often requires thermometers of low heat capacity. Miniature capsule SPRTs have been developed for this purpose [35]. The capsule is ca. 4 mm in diameter and 15 mm long (Figure 3-15). Miniature thermometers of this type have a larger heating effect than the regular types.

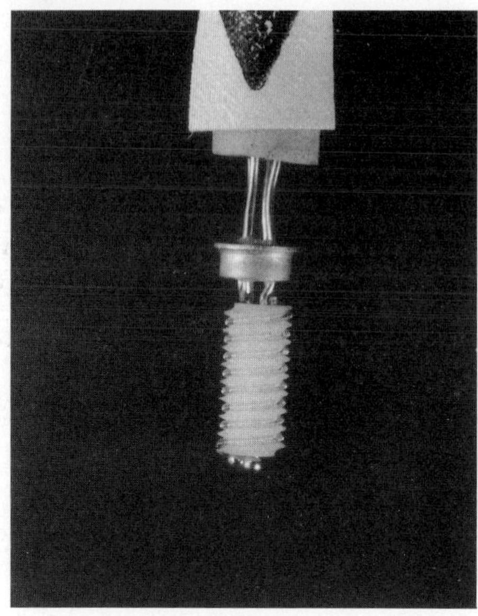

Figure 3-15.
Miniature capsule-type PRT (courtesy of Minco Products).

3.3.1.3 Heat Treatments

The resistance of platinum may change considerably as a consequence of cold work, quenching of excited vacancies or higher order point defects, and oxidation. In an SPRT the strain-free mounting leaves the platinum coil more exposed to the effects of accidental shocks and vibrations. These can force the wire to bend around its supports and produce strains that increase the resistance. For this reason all SPRTs must be treated with adequate care when used or transported in order to minimize shocks and vibrations. Special shock-absorbing cases have been developed to aid transportation [33]. It is recommended, however, that calibrated SPRTs are carried by hand.

Monitoring the resistance at the triple point of water (0.01 °C), or for lower accuracy the ice point, it can be determined when it is necessary to recondition the thermometer with a suitable heat treatment. Exposing it for short periods (eg, 0.5–1 h) to a temperature from 450 to 500 °C, 50% to 80% of the alteration can be removed. The remaining resistance increase does not generally alter the ratio between the resistance at any temperature and that at the triple point of water and, consequently, the calibration constants except R_0.

It is good practice to submit a new SPRT to a heat treatment at its maximum temperature of use or at 500 °C, whichever is the greater. One overnight exposure is often sufficient, but one must monitor R_0 to determine when the best stability has been achieved. Rapid cooling from above 500 °C, and particularly from above 650 °C, must be avoided as it causes the quenching of high-temperature excited lattice vacancies with a consequent increase in R_0. For this reason an SPRT must either be cooled slowly (at a rate of less than 100 °C/h) until 500 or 450 °C and then brought to room temperature, or be cooled quickly to room temperature and then exposed for 1 h to 500 or 450 °C.

Changes in resistance due to oxidation of the platinum will occur if the platinum wire surface is clean and the partial pressure of oxygen in the prevailing atmosphere is between 5 and

11 kPa [36]. A two-dimensional platinum oxide forms between −40 and 300 °C, with increasing R at rates up to the equivalent of 0.5 mK/hour initially, and decomposes rapidly at temperatures in the range 450–500 °C. A three-dimensional film of orthorombic β-phase PtO_2 forms above 300 °C, especially between 450 and 560 °C, and dissociates above 560 °C. At high oxygen partial pressures the resulting effect on R_0 is noticeable, eg, more than the equivalent of 40 mK for an exposure for 30 h to 450 °C at 83 kPa of oxygen [37].

Fortunately, the $R(t)/R_0$ ratios do not change very much and with some precautions they can be reproducible enough for high-accuracy thermometry. These are:

- always measure R_0 in the same oxidation state as R (Tt), preferably immediately after the determination of $R(t)$;
- in maintenance annealing do not keep the SPRT for long at 450–500 °C; exposures of 0.5–1 h are generally sufficient to remove quenched-in defects;
- repeat with sufficient periodicity the annealing at 450 °C when using a PRT between room temperature and 300 °C, in order to remove the accumulated two-dimensional oxide;
- use a reduced partial pressure of oxygen in the gas filling the sheath of high-temperature PRTs; 2 kPa is recommended [38].

Capsule-type PRTs are annealed at 450–500 °C at the factory and, when necessary, before recalibration. It is common experience that low-temperature capsule SPRTs, when properly operated at and below room temperature, do not appreciably change their R_0 values. However, when an annealing is required they cannot be annealed in a furnace as the glass seal would be destroyed. Electrical annealing is then used, passing a suitable current through the coil.

3.3.1.4 Interpolating Equations

The interpolating equations of the standard platinum resistance thermometer are specified by the text of the ITS-90 [39]. They are also reported in Chapter 2. Two reference functions, relating the ratio $R(T_{90})/R(0.01\,°C)$ to the temperature T_{90}/K for the ranges 13.8 K to 273.16 K and 273.15 K to 1235 K, respectively, represent the properties of two reference thermometers. A set of deviation equations enables one to account for the difference between a particular SPRT and the appropriate reference thermometer. These equations require calibrations at fixed points. For instance, the range 13.8 K to 273.16 K can be covered with the following deviation equation:

$$W - W_r = a(W-1) + b(W-1)^2 + \sum_{k=1}^{5} c_k \ln W^{k+2} \quad (3\text{-}28)$$

where W is the resistance ratio of the particular thermometer, W_r that of the reference thermometer, and a, b, and c_k are the calibration constants. An SPRT is to some extent an artifact, albeit very reproducible. Therefore, it is no surprise that some small difference may be found between different instruments fabricated in the same way and calibrated according to the same scheme. Should they be calibrated with ideally accurate fixed points they might nevertheless produce slightly different outputs for the same input temperature between fixed points. This is equivalent to saying that a scale is not unique in absolute terms, ie, that it suffers from non-uniqueness errors. It was demonstrated that the non-uniqueness uncertainty produced by Equation (3-28), as estimated at three-standard-deviation level, does not exceed ±0.5 mK [40].

Above 0 °C and up to the freezing point of aluminium (660.323 °C) a cubic deviation equation in (W − 1) is used. To reach the freezing point of silver (961.78 °C) a further quadratic term is added. This term is identical to zero for t_{90} ⟨660.323 °C. Above 0 °C, the non-uniqueness uncertainty has never been determined in a direct way. However, indirect tests show that it should not exceed ±1 mK [38].

3.3.1.5 High-Temperature SPRTs

Platinum rhodium versus platinum thermocouples are limited in accuracy to ±0.3 K between 600 and 1100 °C [41]. Industrial applications of thermometry need a temperature standard for this range that is at least five times more accurate. A maximum uncertainty of ±20 mK is required also on the grounds of advanced technical and scientific applications. The standard thermocouple prescribed by the International Practical Temperature Scale of 1968 (IPTS-68) is therefore totally inadequate.

The experiments with high-temperature platinum resistance thermometers (HTPRTs) began more than 25 years ago with the bird-cage type [42]. It was soon recognized that sufficient stability and reproducibility could only be achieved if both electrical leakages and the adverse effects at the platinum surface (eg, chemical contamination and scratches resulting from the friction with the support) were minimized with the adoption of a thicker wire than that of conventional SPRTs, with a consequent reduction of R_0 to 2.5 or 0.25 Ω.

Since then, a number of different designs of HTPRTs have been tried, some of which are illustrated in Figure 3-16 [42–46]. The results of several tests indicate that designs involving a single-layer bifilar helix or a coiled helix of platinum supported by a silica or sapphire cross, or by a silica blade, provide the best reproducibility. The wire diameter is between 0.2 and 0.5 mm.

The leads are separated from each other and from the sheath by carefully selected insulating components made of very high-purity silica or sapphire. Accurate cleaning must be performed before the final assembly of the thermometer. Electrical leakages through the sheath wall may cause the thermometer resistance to depend on the position of the stem with respect to the high-temperature region. An electric guard consisting of an isolated wire wound around the capillaries carrying the leads can be used to reduce this effect [44, 47]. This wire is laid down in such a way as to intercept all leakage paths and is maintained at an intermediate voltage with respect to the leads by means of an active circuit in the bridge.

The short-term stability of a HTPRT is determined when the thermometer is repeatedly heated at high temperature and then cooled to monitor R (0.01 °C) as shown in Figure 3-17 [48]. The repeatibility of this resistance, or more accurately that measured at the triple point of water, depends on the design of the sensing element and on the type of annealing that is applied after cooling. As shown by Equation (3-4), the resistivity increase due to lattice vacancies is noticeable above 600 °C. With rapid cooling, the quenched-in vacancies produce a resistance increase that is detectable at 0 °C, or at the triple point of water, and that is mostly eliminated by annealing for 0.5–2 h at a temperature between 450 and 500 °C [49]. A preliminary annealing for 1 h at 650 °C is required when the thermometer is cooled from above that temperature. Alternatively, the HTPRT must be cooled to 450 °C at a rate not faster than 50 K/h, annealed at this temperature for about 1 h, and then allowed to cool naturally to room temperature.

Chemical contamination seriously affects the long-term stability at high temperature. It was recently demostrated [50] that metal vapors of nickel, chromium, iron, and, possibly, copper permeate through the silica sheath at temperatures above 800 °C. It is therefore recommended that an HTPRT not be exposed to high temperatures for long periods without interposing a thick layer of graphite or ceramic between it and any metallic part.

Electrical leakages and chemical contamination increase so much with temperature that the use of HTPRTs as temperature standards cannot be recommended above 1000 °C. Also, platinum softening becomes a limiting factor above that temperature. The new International Temperature Scale (ITS-90) does, in fact, extend the scale of the platinum resistance thermometer up to the freezing point of silver (t_{90} = 961.18 °C).

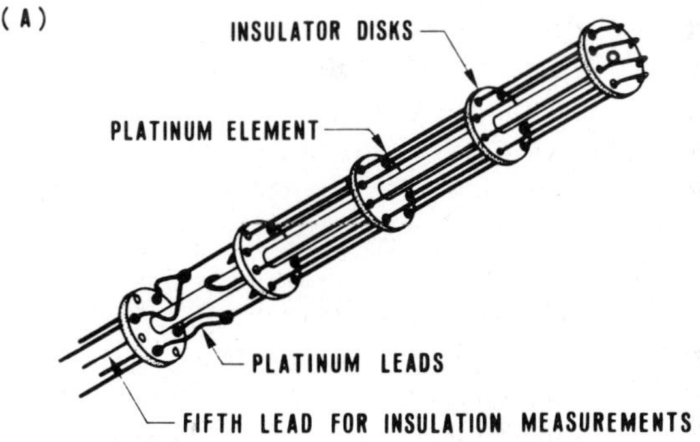

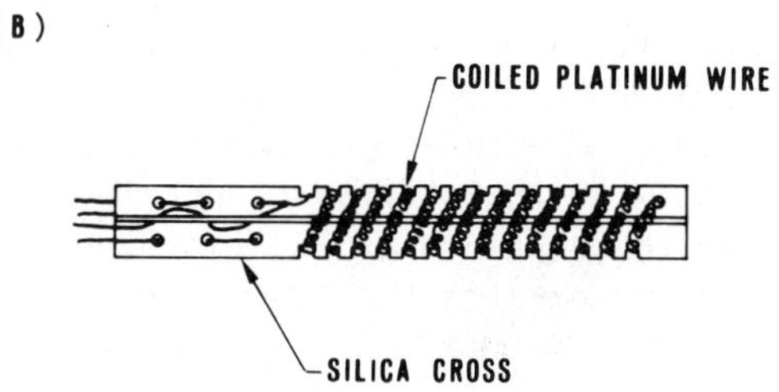

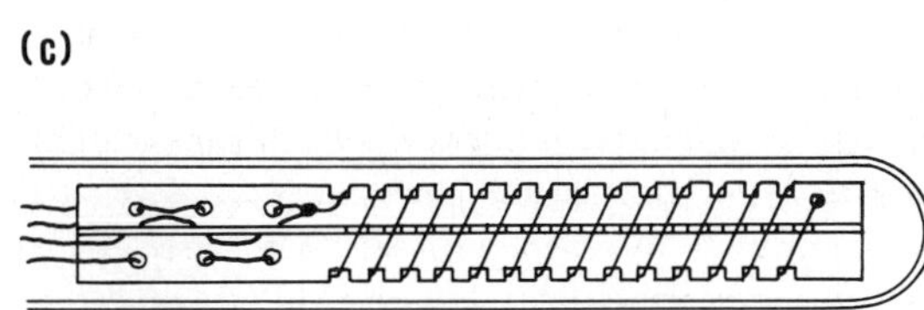

Figure 3-16.
Sensing elements for high-temperature platinum resistance thermometers: (A) bird-cage type with either silica or sapphire insulation; (B) coiled-filament type with notched insulating silica cross; (C) single-coil element with notched insulating silica cross.

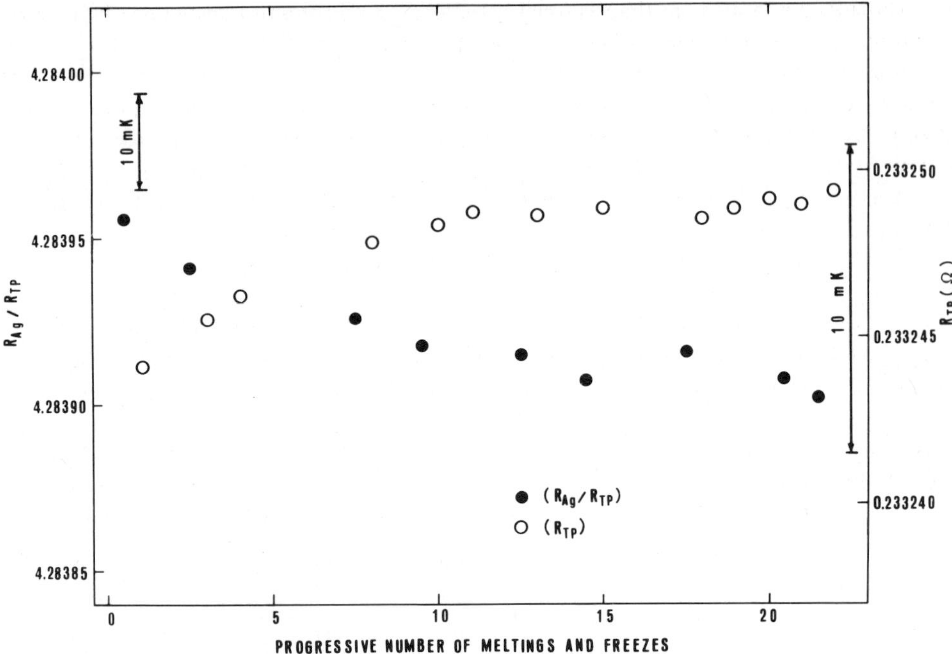

Figure 3-17. Reproducibility of an HTPRT (bird-cage type, 0.25 Ω in ice) on repeated exposure to the freezing point of silver [48]: R_{TP}, triple-point-of-water resistance; R_{Ag}/R_{TP}, measured ratio at the silver point.

3.3.2 Industrial Platinum Resistance Thermometers (IPRTs)

Practical applications of thermometry in industry, aeronautics and space, environmental science, geothermal research, and other research and testing activities impose different design criteria for a PRT compared with those recommended for an SPRT. Primarily, a PRT of this kind must withstand shocks and vibrations much more than an SPRT and comply with a standard calibration graph (or table) to within specified tolerance limits. Additionally, its sensing element should have a smaller size and smaller time constants than those of SPRTs. In some cases its shape is chosen in order to perform a particular type of measurement (eg, surface-temperature measurements). All these thermometers are called industrial platinum resistance thermometers, although they are not exlcusively used in industry. In the USA they are also referred to as RTDs (resistive temperature detectors).

Basically, there are two types of IPRT sensors: those made of platinum wire and those consisting of a thick or thin film of platinum. Wire-wound IPRTs are generally made using thinner wires than SPRTs, typically from 0.01 to 0.05 mm in diameter. In order to withstand a more severe mechanical environment it is generally necessary to support the winding partially, if not fully. Sensors with partially constrained coils have been promoted by a few manufacturers as providing a good trade-off between low hysteresis and good mechanical ruggedness. Very few models, however, are made according to this criterion, most types of IPRTs in fact having their coils embedded in a medium that reasonably matches the thermal expansion of platinum or that provides a certain degree of compliance (eg, alumina powder).

Still more rugged are the elements with their coils glued to an insulating support with cement or enamel, or embedded in glass. They exibit a rather large hysteresis, however, when used in the range from room temperature and below to at least 400 °C. The difference between the same temperature measured on heating or on cooling may be of the order of a few tenths of a kelvin. In contrast, elements with partially constrained coils exhibit under the same conditions differences of few tens of millikelvins, or even a few millikevins [32].

Ruggedness and low cost are combined in platinum-thick film thermometers. They too exhibit a comparatively large hysteresis and, in general, are not recommended for use above 450 °C. They are often made by the screen-printing process in which a platinum ink is printed on an alumina substrate, producing a thickness from 5 to 10 μm, and then fired. Subsequently, they are trimmed to obtain the required resistance, coated with a glaze and fired again. Thin films are produced by a sputtering technique, with a thickness of about 1 μm. With this thickness the electrical properties of the film, after recrystallization, are almost the same as for the bulk material.

Figure 3-18 shows a ceramic-insulated wire-wound platinum resistance element.

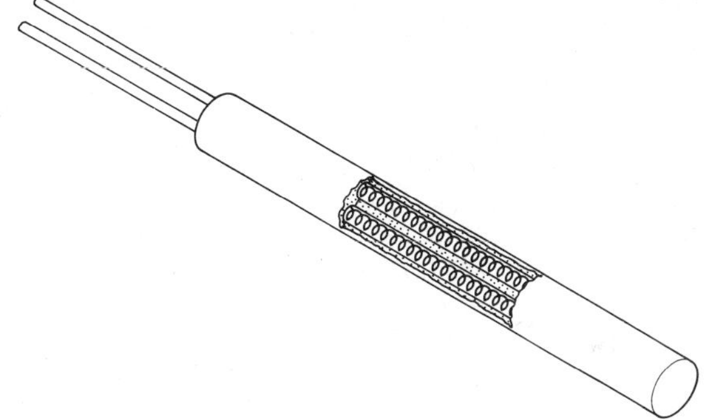

Figure 3-18.
Ceramic-insulated, wire-wound IPRT element.

3.3.2.1 The International Standard for Industrial Platinum Resistance Thermometers

The International Electrotechnical Commission (IEC) Publication 751 [9] is a world-wide accepted standard for industrial platinum resistance thermometers. This code prescribes a resistance ratio R (100 °C)/R (0 °C) nominally equal to 1.385, thus recommending the use of platinum with traces of other metals (eg, gold). Such a platinum wire is retained as more stable in an industrial environment, although no evidence, has ever been produced to prove this assumption. In the early days of resistance thermometry, it was almost impossible to draw thin wires of platinum with such a purity as to produce a ratio higher than 1.385. Also today elements having a coil embedded in glass or a film deposited on a substrate do not attain ratios much higher than 1.385 as a consequence of either contamination during fabrication or of a different thermal expansion coefficient of the platinum with respect to the insulating material.

The IEC Publication envisages two accuracy classes for IPRTs, namely A and B, for temperature ranges of −200 to 650 °C and −200 to 850 °C, respectively. They prescribe different manufacturing tolerances those for Class A being tighter than those for Class B. A typical IEC IPRT has a resistance of 100 Ω at 0 °C, but other resistance values are also permitted.

A reference table based on the following is provided:

$$R(t) = R_0 [1 + At + Bt^2 + Ct^3 (t - 100\,°\mathrm{C})], \tag{3-29}$$

with

$R_0 = 100\ \Omega$
$A = 3.90802 \times 10^{-3}\ °\mathrm{C}^{-1}$
$B = -5.802 \times 10^{-7}\ °\mathrm{C}^{-2}$
$C = -4.27350 \times 10^{-12}\ °\mathrm{C}^{-4}$ for $r < 0\,°\mathrm{C}$
$C = 0$ for $t > 0\,°\mathrm{C}$.

The manufacturing tolerances are shown in Figure 3-19. The tolerance at 0 °C is only a fraction of that at higher or lower temperature. Thus, a selection for a tighter tolerance at 0 °C does not cause a proportional reduction in the tolerances at other temperatures.

The constants of Equation (3-29), as given above, are referred to the IPTS-68. A new standard complying with the ITS-90 is in preparation. The ensuing differences would not exceed 0.13 °C up to 650 °C and 0.36 °C up to 850 °C.

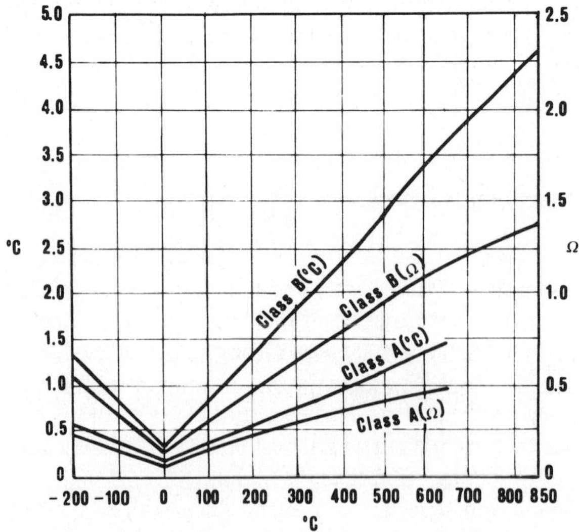

Figure 3-19. Manufacturing tolerances of industrial platinum resistance thermometers (IEC Publication 751 [9]).

3.3.2.2 Calibration of an IPRT

Some IPRTs exhibit such a good repeatibility and long-term stability that they may be calibrated individually. Figure 3-20 shows the results obtained by cycling 18 commercial IPRTs (wire-wound type) between room amd liquid oxygen temperatures. Half of this sample exhibits an R_0 stability better than ± 4 mΩ, with $R_0 = 100$ Ω. Similar results are expected with the ITS-90.

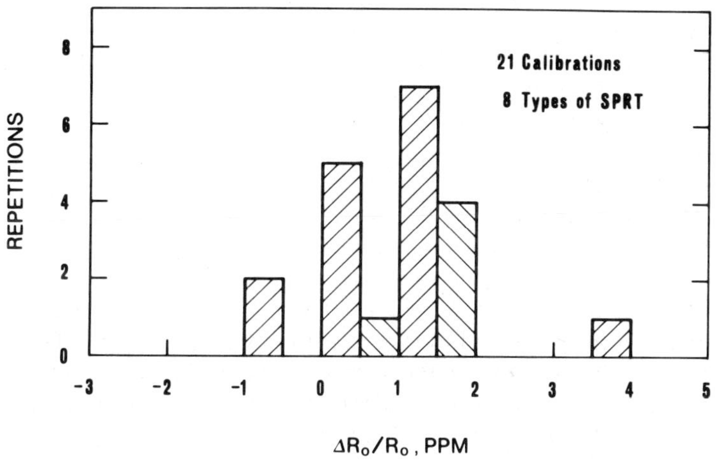

Figure 3-20. R_0 stability of 18 commercial IPRTs [51].

Using Equation (3-29), a calibration can be carried out with a minimum of three points between 0 and 850 °C, or of four points for the range −200 to 850 °C. A redundant number of calibration points may be fitted to Equation (3-36) by means of a least-squares approximation method to reduce the random calibration uncertainty. There remains a systematic error with respect to the IPTS-68 that may be as large as 0.05 K between 0 and 450 °C and 0.2 K at −200 °C [51]. Similar results are expected with the ITS-90.

When operating from 0 to 250 °C, a much better accuracy can be achieved with the following equation [52]:

$$R(t) = R_0(1 + a\,t + \mathrm{b}\,\mathrm{t}^2 + \mathrm{c}\mathrm{t}^3) \tag{3-30}$$

where *a, b* and *c* are calibration constants.

To reduce the systematic uncertainty to within ±10 mK between −50 and 450 °C, the following two equations can be recommended [51]:

$$R(t) = R_0(1 + At' + Bt'^2) \tag{3-31}$$

$$t = t' + g\,\frac{t}{100\,°\mathrm{C}}\left(\frac{t}{t_1} - 1\right)\left(\frac{t}{t_2} - 1\right)\left(\frac{t}{630.73\,°\mathrm{C}} - 1\right) \tag{3-32}$$

where the constants A, B, and g are determined with calibrations at t_1, t_2, and a third temperature, generally intermediate between the other two.

For the range 77–273 K an interpolating scheme yielding an uncertainty of ±50 mK was proposed by Besley and Kemp [53]. It requires only two calibration points, namely 90 and 273 K.

3.3.2.3 Assembly Techniques

Typical long-stem IPRTs are shown in Figure 3-21. Several operations must be performed to assemble then, or other equivalent types, namely:

- the selection of the sensing element;
- the construction of the stem;
- the assembly of the sensing element with three or four leads and their insulation;
- the sealing of the stem; and then
- the annealing of the thermometer.

A sensing element is selected primarily for its maximum temperature of use and its size. Other important parameters are the self-heating coefficient and the response time. Note that the response time provided by the manufacturer can be used only to compare the particular element with others. The significant response time is that measured with the assembled thermometer in real conditions of use. Also, the resistance to high pressure and the electrical insulating resistance are important selection parameters. When all other technical parameters

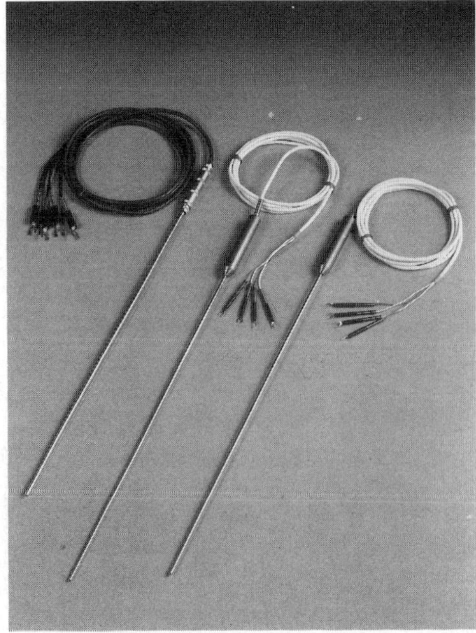

Figure 3-21.
Long-stem IPRTs: (a) precision thermometers (courtesy of Termofas CTC); (b) thermometer for process measurement and control (courtesy of Sensycon).

are equivalent, the choice of a particular constructional type (wire-wound, thick-film, etc.) is made on the grounds of accuracy versus cost considerations.

The stem has to be designed in order to comply with the requirements of the particular application. Specifically, the oxidation or corrosion resistance has to be taken into account, in addition to the mechanical strength, the vibration isolation, the response time, and the heat losses through the stem. A suitable choice of the stem material ensures chemical compatibility: stainless steel, nickel alloys, sintered alumina or other equivalent ceramic materials, and hard glass (fused silica) can be chosen to solve particular problems. For room- and low-temperature applications the sheath sensing tip is sometimes made from copper to improve the thermal transfer to the sensor. For high-temperature and high-pressure applications, low-creep material must be selected in order to preserve the geometrical integrity of the sheath and the sensing element.

A number of stem shapes have been introduced in order to combine mechanical strength, vibration isolation, and good response time [54]. Stems with thinner tips and with increasing cross section from the tip to the mounting flange often provide a good compromise. With a high level of vibration in the surrounding structure, vibration-absorbing mounting fixtures may be required.

The best, but most expensive, leads for medium and high temperatures are gold and platinum wires. Their cost can, however, be minimized by making them thin and not longer than required to reach room temperature. One important advantage of these two metals is the very low level of parasitic thermal emfs. The same property is offered at low temperature by annealed wires of pure copper. Higher resistances and thermal emfs but much lower cost are typical of alloy wires, particularly wires of nickel alloys. When assembling the sensing element with the leads care must be exercized to weld them to those emerging from the sensor case without shortening the latter if not for the minimum amount which is necessary to produce a good weld. In this way the original calibration will not be changed. If the two junctions are arranged in such a way that they will have the same temperature, no thermal emf will be created in the thermometer circuits.

Silica or sintered-alumina capillaries and compressed alumina or magnesia powder are the most frequently used materials as lead insulation. Since insulating powders are hygroscopic, gas-tight stems must be used with them. The same applies to any PRT operating below room temperature.

To reduce further the effect of vibrations and to improve the response time, the gap between the sensor outer surface and the stem inner surface must be reduced to a minimum. In some cases this gap may be filled with either alumina powder or silicone oil, depending on the temperature range.

The type of heat treatment which a manufacturer applies to a product is generally not known in detail. When an IPRT is to be calibrated individually, it is advisable to submit an assembled IPRT to a preparatory annealing for at least 12 h (overnight) at 450–500 °C. To use an IPRT above 500 °C it is also advisable to stabilize it at 50 °C above the maximum temperature of use.

Very good electrical insulation is required for accurate temperature measurements with IPRTs. For very accurate measurements the insulating resistance between two opposing leads must be of the order of 10^5 times the sensor resistance, eg, 30 MΩ at 600 °C. This resistance, however, cannot be measured directly without cutting the wire in the sensing element. To provide an index of the insulation it is recommended to measure the insulating resistance between any lead and the metallic sheath. This resistance can be assumed to be of the same order of magnitude as that between leads. Minimum values for 100-Ω thermometers at 300, 500, and 850 °C are 10, 2, 0.5 MΩ, respectively, (10 V DC) [9].

3.3.3 Other Metal Resistance Thermometers

3.3.3.1 Copper, Nickel, and Iridium Resistance Thermometers

(a) Copper

A copper resistance offers the most linear temperature dependence. The International Organization for Legal Metrology (OIML) has recommended the adoption of the following simple equation [55]:

$$R(t) = R_0(1 + \alpha t) \tag{3-33}$$

with $\alpha = 4.260 \times 10^{-3}\,°C^{-1}$ for temperatures from 0 to 100 °C.

Copper resistances, however, have two disadvantages that strongly limit their use:

- the resistivity of copper at 0 °C is less than one sixth of that of platinum with about the same temperature coefficient; a copper resistance of the same size as one made of platinum exhibits only one sixth of its sensitivity;
- copper begins to oxidize above 100 °C and deteriorates rapidly above 180 °C.

Copper wires, however, offer some advantages, eg, when they are directly wound onto copper parts (eg, windings of electrical machinery) they do not suffer from strains due to differential thermal expansion and they provide a very fast response, with time constants of about 0.1 s, and very low self-heating.

(b) Nickel and Nickel Alloys

The high temperature coefficient of nickel (ca. $6.6 \times 10^{-3}\,°C^{-1}$) has made this metal very popular for temperature measurement and control below its Curie point (ca. 376 °C). Its resistivity, however, is lower than that of platinum and the non-linearity of its resistance versus temperature characteristic is higher. In principle, with equal size it should provide the same sensitivity as platinum. In practice, the sensitivity is lower because nickel wires cannot be drawn as thin as those of platinum. Nickel resistance thermometer sensors of 100 Ω at 0 °C are always more massive, albeit less expensive, than those of platinum. Table 3-3 presents the resistance versus temperature relationship recommended by the OIML [56]. A considerably smaller size is obtained with thin-film nickel resistors, which have temperature coefficients of $(5.5\text{–}6.2) \times 10^{-3}\,°C^{-1}$.

A linear resistance thermometer for the range of −296 to 25 °C can be made connecting in series thin-foil grids of nickel and manganese. The device is commercially available and is affected by a linearity error not exceeding ±2 °C in the above range. The temperature coefficient is $0.82 \times 10^{-3}\,°C^{-1}$.

Alloys of about 70% nickel and 30% iron have a high temperature coefficient (ca. $5 \times 10^{-3}\,°C^{-1}$) and good oxidation resistance. The resistivity is about three times that of

Table 3-3. Resistance of a nickel thermometer of 1 Ω at 0 °C according to the International Organization of Legal Metrology (OIML).

°C	0	10	20	30	40	50	60	70	80	90
0	1.000	1.056	1.113	1.171	1.230	1.291	1.353	1.417	1.482	1.549
100	1.617	1.687	1.759	1.838	1.909	1.987	2.067	2.149	2.231	–

pure nickel so that smaller and more rugged sensors can be obtained for use up to 600 °C. This alloy, however, does not provide a very good stability and the response to temperature is very nonlinear.

Lastly, NiCr (Chromel P) thin films, 2-nm thick, are used from 1 to 4 K with substantial insensitivity to magnetic fields [56].

(c) Iridium Resistance Thermometers

Iridium is today more expensive than platinum and not as ductile. For both reasons iridium thin wires are uncommon. Therefore, an iridium resistance thermometer was not commercially available until the advent of thin-film technology. An iridium film matches the thermal expansion coefficient of its alumina substrate much better than any other available combination. A special process [11] enables 1-μm thick iridium films to be produced having positive temperature coefficients of ca. 4.3×10^{-3} °C^{-1}. The temperature coefficient increases slightly with temperature, in such a way that a linear temperature sensor can be prepared by simply placing a constant resistor in parallel with the iridium resistor. For instance, an iridium resistor having $R_0 = 103.69$ Ω in parallel with a fixed resistor of 2807 Ω results in a sensor having a sensitivity of 0.3956 Ω/°C and a linearity error not exceeding ±0.3 °C from 0 to 400 °C [11]. Using flat alumina substrates, iridium sensors are particularly suited to surface-temperature measurements.

3.3.3.2 Rhodium-Iron and Platinum-Cobalt Thermometers

A low-temperature resistance thermometer based on an alloy of rhodium and 0.5 at % of iron was first suggested by Coles [6] and subsequently realized by Rusby [7]. It is known that the addition of a small proportion of a magnetic impurity to certain metals leads to the formation of a local magnetic moment and, consequently, to a resistance minimum at low temperature. Instead, the addition of about 0.5 at % of iron to rhodium produce a resistance anomaly with a positive coefficient at low temperature. This new thermometer is a useful means to provide temperature scales in the range 0.5–30 K.

The process of making the wire starts with chemically depositing the iron in fine rhodium powder. The powder is then dried, sintered, swaged, and hot drawn. Finally, the wire thereby obtained is strain-annealed in hydrogen at 1100 °C. The sensing element is of the double-capillary type (see Figure 3-5b), sealed in a platinum capsule like a traditional PRT (Figure 3-14). Figure 3-3 presents the fractional sensitivity of Rh-Fe thermometers in comparison with that of a PRT and Figure 3-4 shows the voltage sensitivity. The temperature may be expressed in term of the resistance as follows:

$$T = \sum_{j=1}^{n} a_i (AR + B)^i \quad (3\text{-}34)$$

where the constants A and B are chosen so as to limit the term in parentheses within ± 1 [56, 57]. A fitting in the range 0.5–27 K with a standard deviation not exceeding 0.2 mK is obtained with $n = 11$. Up to 20 K, n can be reduced to 8. Between 27 and 273 K, Equation (3-40) with $n = 10$ still produces a fitting with a standard deviation of 1 mK, but R must be replaced with ln Z, where

$$Z = (R - R_{4.2\,\mathrm{K}})/(R_{273.15\,\mathrm{K}} - R_{4.2\,\mathrm{K}})\,. \tag{3-35}$$

Rh-Fe thermometers are affected by a noticeable transverse magnetoresistance below 10 K and with fields larger than 2 T. For instance, at 4.2 K and with 3 T the effect is equivalent to 1 K.

Diluted Pt-Co alloys have similar properties. The alloy of platinum and 0.5 at % of cobalt was studied at the National Research Laboratory for Metrology (Tsukuba, Japan) over the range 4–303 K [58].

3.3.4 Thermistors

3.3.4.1 Negative Temperature Coefficient (NTC) Thermistors for the Range −100 °C to 300 °C

NTC bead thermistors provide the best stability and interchangeability in this range. They are manufactured to approach the accuracy of PRTs at lower cost. Usually, they are made by first taking a pair of platinum alloy wires, which will eventually form the leads, stretched at the appropriate distance apart and then applying small blobs of the mixed oxides in a binder at intervals along the pair of wires. Subsequent sintering at about 1300 °C produces a series of thermistors having embedded leads. After cutting the leads, the individual thermistors can be coated with glass to protect the thermistor from the environment, and also to give strength to the assembly.

Disc thermistors also have good stability, albeit not as good as bead types. They are made by pressing the powder mixture and subsequently heating to about 1100 °C. Electrical contacts are realized by means of silver deposits on opposite faces of the disc, obtained by means of a spraying or screen-printing technique. This type of contact, however, is considerably less stable than that of bead types. Finished disc thermistors may be epoxy coated.

For both types, the temperature response depends on the ingredient mixing, the sintering time, the sintering temperature and other factors. Fine trimming is accomplished by grinding away a small portion of each thermistor. When mixtures of nickel and manganese oxides and a small amount of a binder are used, the more massive disc thermistors are expected to display a more uniform statistical unit-to-unit distribution of material [59]. Interchangeability of disc thermistors of ±0.1 °C for temperature spans of 50 or even 100 °C is guaranteed by some manufacturers. However, no interchangeability is guaranteed between different manufacturers. The reference resistance of an NTC thermistor is often given at 25 °C. Its value ranges from 1 to 30 kΩ and beyond, but 5 and 10 kΩ are the most common.

A typical glass-coated bead thermistor with a diameter of about 2.2 mm is characterized by a time constant in still air of 16 s and in flowing water at 1 m/s of 0.4 s and a dissipation constant of, 0.8 and 4 mW/°C, respectively [17]. This means that a 10-kΩ resistor of this type is heated 0.13 °C above the surrounding medium by a current of 100 μA, but only 25 mK in flowing water. Therefore, the current to be used in measurements in air and also in calibrations should not exceed 20 μA in order to exploit fully the accuracy of this thermistor.

The best approximation of the resistance versus temperature characteristic, as determined with individual calibrations, is provided by polynomials of the forms

$$\ln R = A_0 + A_1 (1/T) + \ldots + A_n (1/T)^n \tag{3-36}$$

$$(1/T) = a_0 + a_1 \ln R + \ldots + a_n [\ln R] A^n \,. \tag{3-37}$$

It was demonstrated [17] that a third-degree polynomial of either form provides a very good approximation.

Investigations on the stability of bead and disc thermistors have been carried out by several authors [59, 60, 61]. They demonstrated that selected bead thermistors may have drift rates not greater than 5 mK over 150 days at up to 100 °C. Disc thermistors under the same conditions may change by several tenths of a kelvin.

3.3.4.2 NTC Thermistors for Above 300 °C and Below −100 °C

Thermistors for above 300 °C are made of oxides of rare earth elements, which are more refractory than nickel and manganese oxides and possess a higher activation energy. Therefore, their resistance decreases more slowly with T, thus still providing a good sensitivity at high temperature. Mixtures of these oxides are used up to 1000 K. Zirconia-based thermistors are used at still higher temperature [18]. Their temperature response is adequately described by the general Equation (3-13).

Thermistors for cryogenic use [62] are mostly made from non-stoichiometric iron oxides which exhibit very low activation energy. They are characterized by high sensitivity, eg, 15% at 20 K and 300% at 4.2 K, and narrow operating ranges. They also have a low sensitivity to magnetic fields.

3.3.4.3 Positive Temperature Coefficient (PTC) Thermistors

Thermistors made from barium/strontium titanate in perovskite-type oxides present a large resistance anomaly at a temperature related to the Ba/Sr ratio. It consists in a sudden resistance increase of three order of magnitude, or even larger, in a range of few tens of kelvins and seems to be connected with the ferroelectricity of this type of crystal [63].

An obvious applications of PTC thermistors is in the realization of heat switches. They can be manufactured with the transition temperature anywhere from 15 to 115 °C. They have also been applied to the determination of small temperature differences, as required by heat-flux sensors [64].

3.3.5 Low-Temperature Semiconductor Thermometers

3.3.5.1 Germanium Thermometers

Practical germanium resistance thermometers for the range from 1 to 20 K are made from doped germanium single-crystal chips operating in regions III and IV (see Section 3.2.2). They always require individual calibrations with numerous points. No extrapolation of the calibration graph is possible.

Initially, germanium slices were clamped to the side of a calorimeter using mica for electrical insulation. However, high stability could not be attained owing to the strong piezoelectricity of germanium. Today, the capsule-type thermometer shown in Figure 3-5 d [16] ensures the optimum conditions of strain-free mounting and thermal coupling. Actually, heat is exchanged mostly through the leads and also through the helium in the capsule. Hence it is very important to effect good thermal anchoring of the leads outside the capsule. Recently, a technique was described for the construction of planar structures resulting from the diffusion of arsenic in a gallium-doped substrate [65]. It should yield a better uniformity of the production of thermometers and enable the electrical characteristic to be adjusted.

The germanium thermometer cannot be used as an interpolating instrument in primary scales, as initially proposed, owing to its instability [66–68]. The main source for instability is the different thermal expansion of the gold leads and of the germanium chip. It is possible, however, to select thermometers that have a repeatibility of 5 mK at 20 K on prolonged thermal cycling.

For high-accuracy applications the following interpolating equations are recommended [15]:

$$\ln R = \sum_{i=1}^{n} a_i (\ln T)^i \tag{3-38}$$

or

$$\ln T = \sum_{i=1}^{n} b_i (\ln R)^i . \tag{3-39}$$

The calibration points must be uniformly distributed over a logarithmic scale, that is, there must be as many points between 1 and 2 K as there are between 10 and 20 K. It is recommended that an increased number of points be placed at both ends of the range and, possibly, a few just outside both limits.

Lastly, it is worth mentioning that germanium thermometers exhibit a large magnetoresistance and are therefore not recommended for use in intense magnetic fields.

3.3.5.2 Carbon Thermometers

Carbon radio resistors have long been used as cheap medium-accuracy, low-temperature thermometers. They are small, very sensitive to temperature, and fairly intensitive to magnetic fields. They also exhibit a small heat capacity, fast response, a slight pressure dependence (0.3 K for 6.7 MPa at 20 K), and thermal instability on ageing. Clement and Quinnell [4] first introduced the use of cylindrical carbon radio resistors from Allen Bradley below 20 K. These resistors are still in production today, but changes in manufacturing techniques since then have led to a modified electrical characteristic [69]. Resistances between 80 and 330 Ω (at room temperature) and wattage ratings from 0.1 to 0.5 W are currently in use.

Another variety is obtained by forming a colloidal suspension of carbon particles with an appropriate organic vehicle, painting a film of the suspension on the sample, and then curing. Both the sensitivity and resistivity increase with decreasing particle size [70]. A suitable interpolating equation for the range 3–60 K, and with an interpolation accuracy of ±30 mK, is

$$\ln R = AT^{-m} + B \tag{3-40}$$

where A, B, and m are constants [71, 72].

A further development is the carbon-impregnated porous glass thermometer [73]. This type of sensor covers the range 1–300 K, providing higher sensitivity and better stability at low temperature than traditional carbon resistors. With a suitable choice of the dependent and the independent variables of the interpolating polynomial and with an individual calibration the uncertainty can be limited to within ±5 mK over the whole range [74].

3.4 Calibration Techniques

Standard platinum resistance thermometers are calibrated at fixed points. Rhodium-iron and germanium thermometers, and sometimes industrial platinum resistance thermometers and thermistors, may also be calibrated at fixed points. In all other cases resistance thermometers are calibrated by comparison with an SPRT, a standard thermocouple, a gas thermometer, or a vapour-pressure scale (eg, ^{3}He and ^{4}He).

3.4.1 Fixed-Point Calibration

An extremely constant and reproducible temperature is realized in the following cases:

(a) *Melting point:* transition from the solid to the liquid phase under normal pressure conditions (ie, 101325 Pa). The melting point is pressure dependent.

(b) *Freezing point:* reverse transition with respect to the melting point.

(c) *Triple point:* transition entailing the coexistence of solid, liquid, and vapor phases of a pure substance. According to Gibb's rule this point is invariant, ie, both temperature and pressure are constant and reproducible.

(d) *Vapour-liquid transition:* this involves the coexistence of vapor and liquid phases of a pure substance. It is often designated as boiling point when it is realized at normal pressure.

(e) *Sublimation point:* transition involving the coexistence of vapor and solid phases at normal pressure.

(f) *Superconducting point:* transition from the normal to the superconducting phase at zero magnetic field.

Numerous pure substances can in principle be used for realizing fixed points. However, many of them do not ensure the necessary stability and reproducibility and some others require a complex procedure and special laboratory facilities. It is then impractical to realize as many fixed points as are required to calibrate all types of thermometers. They are subdivided into primary and secondary fixed points [75]. The fixed points for very low temperatures are specified by the 1976 Provisional 0.5–30 K Temperature Scale (EPT-76) [73] and by the ITS-90 [39]. An extensive list of secondary fixed points is provided by both the International Union of Pure and Applied Chemistry [76] and the Consultative Committee for Thermometry of the CIPM [77]. Detailed information on the realization and the use of fixed points between −50 and 450 °C is also available [76].

The use of triple point of water cells is of great importance for the correct maintenance of SPRTs and, in general, of high-accuracy PRTs. Frequent determination of their resistance at this temperature enables the stability of the thermometer to be checked, and hence also the validity of its calibration. Moreover, small changes consequent on the bridge calibration or on annealing can easily be detected and accounted for.

3.4.2 Calibration by Comparison

Low-temperature calibrations (typically below 90 K, but also up to 300 K) are carried out in a vacuum-jacketed copper block which provides a temperature stability and uniformity to within ±10 mK (for details, see Quinn [15]). Above −90 °, but sometimes from −150 °C and up to 250–300 °C, a temperature-controlled stirred liquid bath is the most common means of calibration. Isopentane may be used from −150 °C and Freon from −90 °C. Silicone oil is used continuously up to 250 °C and intermittently up to 300 °C. Liquid baths may be uniform and stable to within ±50 mK and, in some cases, to within ±10 mK. For still higher temperatures, baths of molten salts (up to 550 °C) and of fluidized beds of alumina powder (up to 600 °C, or even to 1000 °C) may be used for comparisons to within ±0.2 °C or better. Metal-block electrical furnaces may be used from just above room temperature to slightly above 1000 °C.

Nickel-plated or stainless-steel-jacketed copper blocks are recommended for good temperature uniformity up to at least 500 °C, while Inconel-600 blocks are recommended from about 500 to 1100 °C. Extremely good temperature uniformity, stability, and repeatibility can be obtained with a gas-controlled heat-pipe furnace. For instance, a water heat pipe with a helium-ballast control can be used from room temperature to 150 °C with temperature stability and uniformity within a few tenths of a millikelvin [78]. A sodium heat pipe, when operated in the same way, can be used from 650 to 1000 °C with temperature stability and uniformity of a few millikelvins [79]. Moreover, its temperature can be changed by means of its pressure control in steps of a few tenths of a kelvin in a matter of few seconds, still keeping the same stability and uniformity, and eventually reset to its initial value within a few millikelvins.

In any intercomparison, the standard must be an interpolating instrument prescribed by the International Temperature Scale of 1990 or another reference thermometer derived from the former. The ITS-90 introduces the use of an interpolating gas thermometer from 4.2 to 25 K and of ^{3}He and ^{4}He vapor-pressure scales from 0.65 to 5.5 K. In both cases complex and delicate equipment is required [15].

3.4.3 Other Tests

Other tests, in addition to calibration, may be performed to qualify a resistance thermometer. Often they are carried out only on some specimens from the production of a particular type of transducer and the results thereby obtained can legitimately be extended to the whole production. This is the case, for instance, with the determination of the response time, the self-heating effect, the insulating resistance, the immersion error, and the effects of vibration and pressure. The practical realization of this and other tests on IPRTs is described in IEC Publication 751 [9]. The techniques and the apparatus recommended there can easily be adapted to other types of resistance thermometers.

One important but sometimes overlooked test is that realized with thermal cycling. In this way, both the stability of the resistance at the reference temperature (zero stability) and the hysteresis can be determined. Industrial resistance thermometers, including IPRTs [32] and thermistors, often exhibit hysteresis when operated in a wide temperature range. This can mostly be explained in terms of the difference in thermal expansion between the sensing material and its supporting insulator, or the electrodes attached to it. To a smaller extent, physico-chemical processes also play a role. Hysteresis may not be detected in ordinary calibrations but may be active in practical applications of the thermometer. To detect it, it is necessary to calibrate the thermometer, first continuously increasing the temperature and then with a continuously decreasing temperature, never cooling the thermometer to room temperature before the cycle is concluded.

3.4.3.1 Response Time

The response time of resistance thermometers is measured either by plunging them in a liquid flowing at a well known speed and temperature, or by stepwise heating a stream of gas flowing across them. Several types of apparatus have been devised for this purpose. However, they have the limitation of determining the response time under standard conditions and, most of the time, near room temperature. It is known that in the final applications the thermometer may experience very different conditions of temperature, fluid speed, and properties from those in the standard test. A technique for the *in situ* determination of the response time is, therefore, very desirable.

The determination of the response time of an installed sensor enables a control system performance to be assessed or a sensor malfunction to be detected. With industrial fluids, for instance, a deposit of insoluble material may cause fouling of the sensor, affecting both the response time and the immersion error. An indirect method of measurement entails the loop-current step response (LCSR): a current is applied stepwise to the sensor, thus producing a measurable overheating, which is recorded over a sufficient length of time. This technique is applied to both resistance thermometers and thermocouples. Cylindrical resistance thermometers meet the basic measurement requirements very closely, namely, that (1) the sensor heat transfer must be predominantly in one direction (eg, radially) and (2) there must be insignificant heat capacity between the sensing element and the center line. They imply that for an outside applied temperature step all the resulting heat flows towards the sensing element, whereas for a current step the liberated Joule heat flows in the opposite direction. In this case, it can be demonstrated [31] that the responses to both types of step can be described with the same time constants and, additionally, that the transfer function relating the outside temperature change to the resistance change in the sensor has no zeros in its numerator. Consequently, as it is possible to extract the significant time constants from the recorded response of an LCSR test carried out with the sensor under exactly the same conditions of use, the thermometer transfer function can be calculated and used to estimate the thermometer response time, whichever is defined.

The IEC recommends that the time required to cover half the temperature step be taken as the response time. It is also recommended that in no case should the response time be referred to as a time constant, because no IPRT can be represented as a simple first-order system.

3.5 Laboratory and Scientific Applications

3.5.1 Melting Point Determinations

The determination of the melting or freezing points of substances is an important analytical tool. From both the melting and the freezing curve the total purity of a substance can be assessed. Such a technique is applied, for instance, to the determination of the purity of organic substances melting between 0 and 450 °C [76]. They are frozen in a glass cell with a suitable stirrer and using a long-stem PRT to record the freezing curve [80]. To detect impurities at level 10^{-5} of the total mass, or even less, the temperature measurement has to be carried out with resolution to within ±1 mK.

With a small amount of substance (ca. 1 cm^3), the freezing and the melting points can be determined with a PRT and a special apparatus making use of a vacuum-jacketed copper block, where the PRT is placed, and a miniature cell containing the sample and linked to the block by means of a differential thermocouple [81].

3.5.2 Tracking Temperature Gradients

The precise determination of a temperature profile is a common task in many experiments and in the preparatory work on some industrial processes. It can be carried out, on a surface, by means of a radiation thermometer and, inside a medium, using either a stationary multijunction differential thermocouple or a movable transducer. For the last case, it is required that the sensor be conveniently small and of fast response and that it should not respond to changes in the temperature distribution over its stem above the sensing area. Both PRTs and thermistors posses these characteristics, the former in the range −200 to 1000 °C and the latter from −100 to 200 °C.

Selected IPRT elements, four-lead connected and assembled in a fused-silica stem, can be used up to 900 °C [82]. Special PRT element can be used up to 1100 °C [83]. Typically, a temperature resolution of ±1 mK up to 500 °C and of ±10 mK up to 1100 °C can be achieved. When determining the temperature distribution in a furnace or a liquid bath, the thermometer is first inserted with the mid-point of its senstive element at the desired initial position and sufficient time is allowed to equilibrate it. After having taken the initial reading, it is displaced by the desired amount and after a new equilibrium has been reached another reading is taken. This procedure is repeated as many times as necessary to track the temperature profile. At the end, it is good practice to replace the thermometer in its initial position to check that the temperature has not changed during the measurement sequence.

Tracking temperature gradients over a great depth is often required in oceanography and geophysical prospecting. In both cases high-accuracy PRTs or other resistance thermometers are used. The range of temperature in the oceans is only from −2 °C to 35 °C and, except for a few special areas and near the surface, an even narrower band from −2 °C to 15 °C. There are, however, some difficulties, notably the great depth, the high-pressure environment, the corrosive and mechanical action of sea water, and the fast response required to give a complete temperature profile to a depth of several kilometers in a reasonable length of time. A thermometer made from a copper wire has a time constant of ca. 40 ms, and it is accurate to within ±10 mK [10]. Alternatively, a slow but very accurate PRT can be used in combination with a microbead thermistor, the latter providing the high-frequency components of the temperature changes [84]. Fast platinum and copper sensors exhibit a pressure sensitivity of about −0.04 K/km depth. Slower jacketed sensors do not have a pressure sensitivity.

3.5.3 Dew-Point Measurements

The temperature at which water vapor begins to condense, the dew-point, is measured to infer the water-vapor partial pressure in a given environment and, hence, to determine its absolute or relative humidity. A mirror surface, cooled by means of a Peltier heat pump, is often used to accomplish this determination. The current in the Peltier heat pump (single-or double-staged) causes the heat to flow either in the mirror and, through it, towards a sink at room temperature, or out of the mirror, depending on the current direction. Adjusting the current, a film of condensate (either dew or frost) is formed on the mirror and detected by means of an optical system generating a feedback signal to the pump power supply. When equilibrium is reached, the heat flow is minimal and the mirror temperature is the dew-point of the sur-

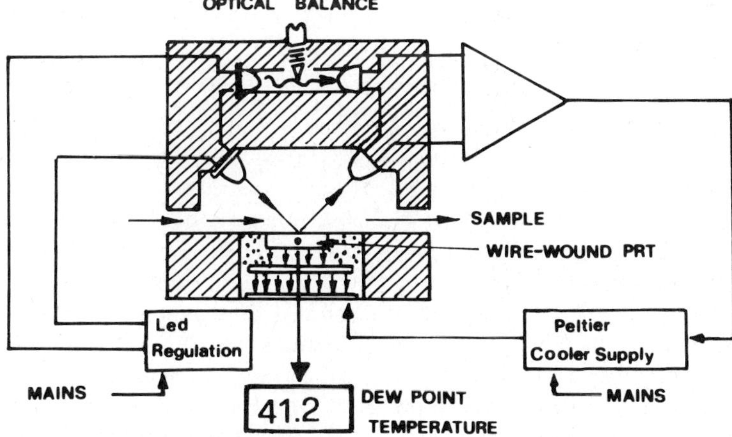

Figure 3-22. Schematic diagram of a dew-point hygrometer.

rounding gas. This is measured by means of a miniature PRT element placed in the mirror copper body. The best instruments of this type afford resolutions of ±10 mK and accuracies from ±0.05 to ±0.2 K for dew-point ranges from −80 to 90 °C. Figure 3-22 shows the operating principle of a dew-point hygrometer.

3.6 Industrial Applications

The use of resistance thermometers in industry is steadly increasing. Improved manufacturing techniques and the introduction of film technology have made resistance thermometers cheaper than traditional mechanical temperature transducers and of comparable cost to that of thermocouples, but of better accuracy. Specifically, process measurement and control problems requiring accuracies of ±0.5 K or better from −200 to 600 °C can today be solved with IPRTs with no significant additional cost with respect to thermocouples. Tests of heat exchangers, hydraulic machinery, and heat pumps, often requiring that temperature differences lower than 10 °C are measured to within ±1%, are better accomplished with resistance thermometers. Also other mechanical measurements, such as flow and fluid speed measurements, need accurate temperature determinations, frequently performed with good-quality IPRTs or thermistors.

The temperature control of liquid baths and environmental chambers is carried out with resistive temperature sensors. Fast response and good sensitivity and stability in a modest temperature span (50–100 °C) are required. Both thermistors and film PRTs are well suited to this application.

Lastly, a large numer of aeronautical and space applications involve temperature measurements with resistance thermometers. They include total temperature measurements, air conditioning, and temperature control of astronauts' suits and the determination of the temperature of ball bearings of landing gears by means of tip-sensitive sensors.

3.6.1 Total Temperature Probes

The total temperature, T_t, of a gas stream is defined as the kelvin temperature assumed by the gas when brought to rest adiabatically. It is linked to the temperature of the flowing gas, ie, the static temperature T_s, by the following equation:

$$T_t = T_s\{1 + 0.5\,[(c_p/c_v) - 1]\,M^2]\} \qquad (3\text{-}41)$$

where M is the Mach number of the gas stream and c_p and c_v are the constant-pressure and the constant-volume heat capacity, respectively. To measure accurately the total temperature the kinetic energy of the gas must be fully converted into thermal energy in the region surrounding the sensor, with negligible losses. The determination of the total temperature is required to calculate the other thermodynamic parameter such as T_s and enthalpy, provided that the Mach number is known. Any sensor in a fast gas stream measures a temperature T_m intermediate between T_t and T_s. When the gas flow transverse to the long axis of a cylindrical sensor, $T_m - T_s$ is about 65% of $T_t - T_s$, ie, the "recovery factor" is 0.65. With air ($\Gamma = 1.4$) at ambient pressure and 300 K and at a speed of 100 m/s, $T_t - T_s$ is ca. 5 K. An accurate temperature measurement under these conditions would require a recovery factor larger than 0.95.

Such measurements are required outside an aircraft in flight and in wind tunnels. They are also required with high-speed flows within ducts, or at the intake of a compressor of a jet engine. In all these cases platinum resistance thermometers are often preferred. The sensors are designed to ensure a fast response (Figure 3-23) and are housed in receptacles providing a recovery factor very close to unity [15, 54]. Figure 3-24 shows a sensor for measurements in ducts.

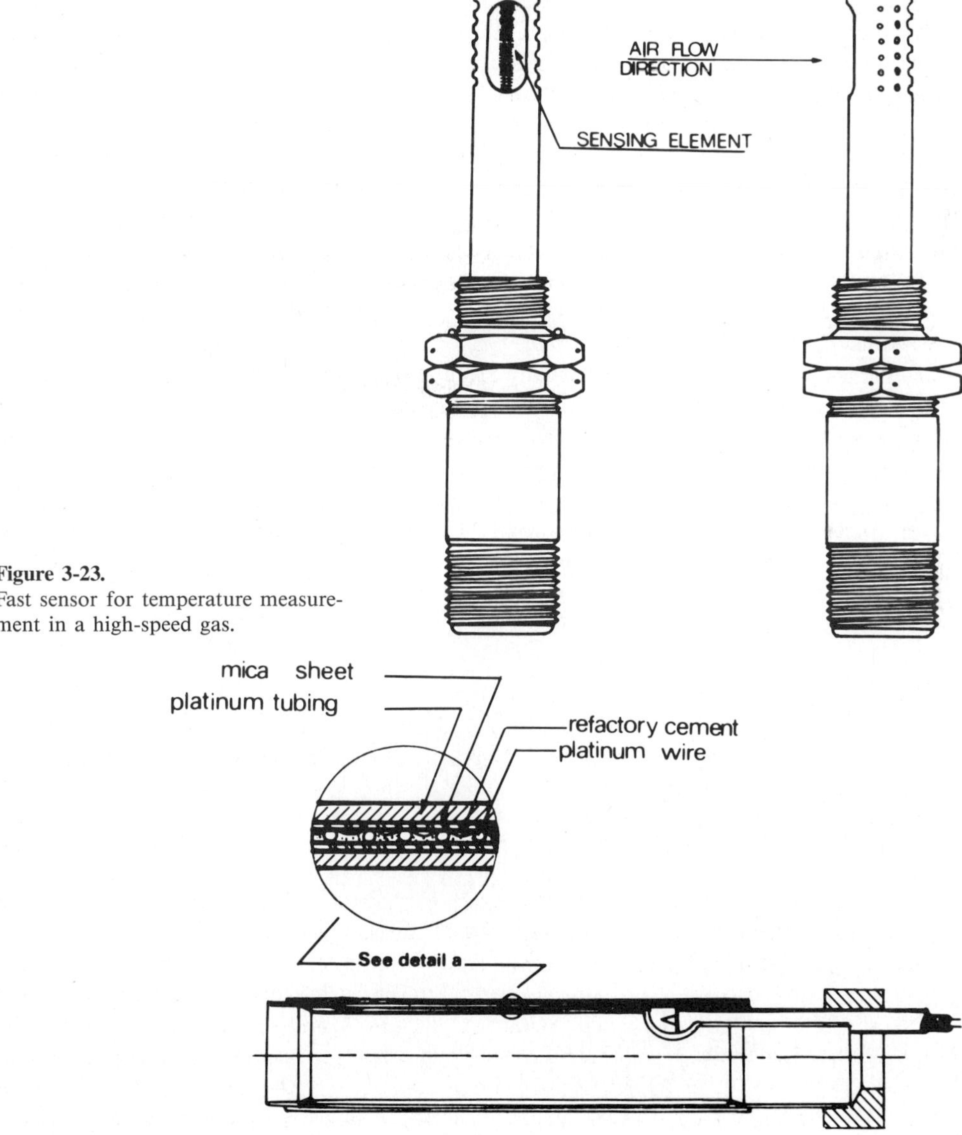

Figure 3-23.
Fast sensor for temperature measurement in a high-speed gas.

Figure 3-24. Total-temperature sensor to be applied in ducts.

3.6.2 Surface-Temperature Measurements

Ideally, a surface-temperature sensor must exchange heat exlcusively through the interface with the surface whose temperature must be measured. Consequently, the sensing element must be flat and a good heat-conducting electrical insulator must be placed between it and the surface, whereas heat insulation must be provided for the other sides. These conditions are fulfilled in practice with some approximation and, additionally, heat is exchangd through the leads.

Flat resistance thermometers are realized with both wire and film sensing elements of platinum, iridium, nickel, and nickel-iron alloy. The substrate can be either rigid (typically alumina) or flexible (Kapton, Mylar, or even silicone rubber). Then the thermal expansion coefficients of the sensing metal and the substrate are well matched, the rigid sensor exhibits the best stability and the lowest hysteresis. It is available in several different styles, including bolt-on, clamp-on, screw-on, and cement-on devices, for temperatures between −200 and 500 °C (exceptionally up to 800 °C). Some sensors can be spot-welded to the underlying surface. Flexible sensors, instead, provide good thermal contact also with moderately curved surfaces or can be gently compressed between two surfaces, but they can be used only between −150 and 200 °C. Silicone-rubber sensors are not recommended for use below −50 °C. Figure 3-25 shows some of these sensors.

For the best result, the sensing surface must adhere in an optimum way to that under measurement. A flexible pad of a thermally conductive polymer or a smear of silicone grease may help in a few instances and, particularly, with measurements in vacuo. Radiation losses from the other side of the sensor can be minimized by covering it with a reflective metal foil. Either polished aluminium or copper foils can be used, or even platinum caps for the highest

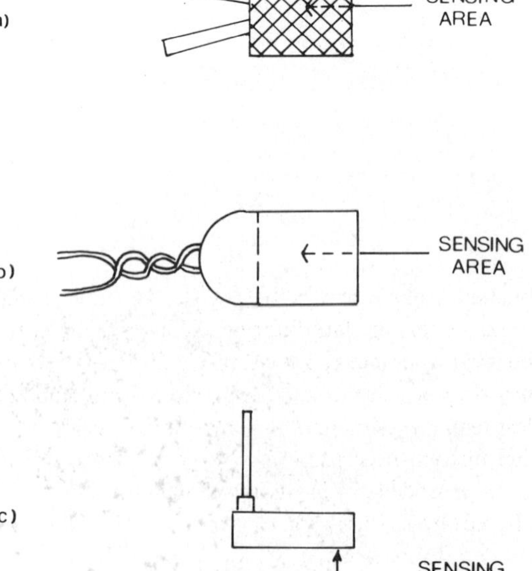

Figure 3-25.
Flexible sensors for surface-temperature measurements.

temperatures. It is also recommended to keep the lead wires in thermal, but not electrical, contact with the underlying surface for a few centimeters from the sensor.

Cylindrical sensors may also be used for surface-temperature measurements if some precautions are adopted. Figure 3-26 shows some examples.

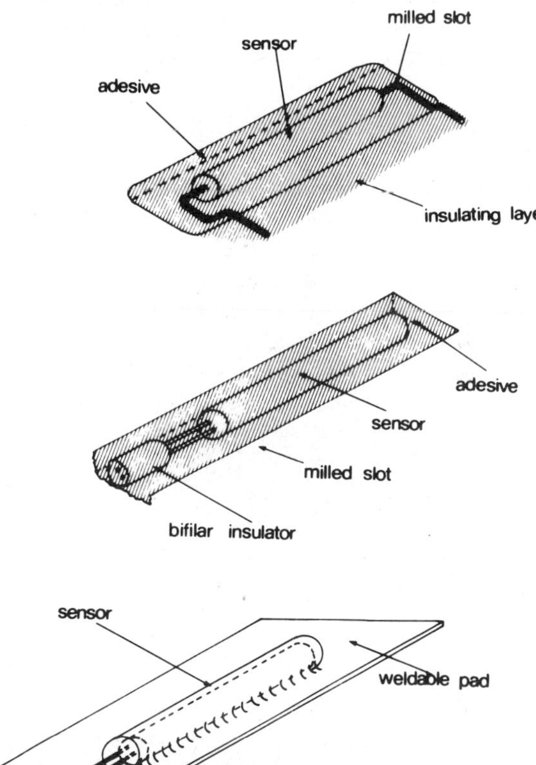

Figure 3-26.
Techniques for use of cylindrical sensors for surface-temperature measurements.

3.7 Conclusions

Resistance thermometry is a well established technique up to 1000 °C. In some cases, however, resistive temperature sensors have not yet left the laboratory stage as they are either too expensive or not produced with tight enough tolerances to meet the requirements of industrial applications. In few other cases they do not exhibit satisfactory long-term stability. One or more of these reasons limit the widespread use of many low-temperature devices and, to a much lesser extent, of high-accuracy thermistors. It is reasonable to expect that a few of these problems will be solved before long. Improved reproducibility should be one of the prime concerns of future developments. In contrast, the great advances achieved in the development of resistance-measuring instrumentation enable sensors with comparatively low sensitivity to be utilized.

The number of applications of industrial resistance thermometers should increase when the users increase their confidence in these sensors. The temperature range for their application will not be extended to either higher or lower temperatures, except marginally. Other metals, such as iridium, may be used more widely.

Thin-film technology will be used more extensively with both metals and thermistors. The combination of resistive sensors and electronic circuits on a single chip is ripe for development, although the semiconductor manufacturers seem much more in favour of semiconductor-junction sensors for use in this context. Other combinations of integrated sensors may also be developed, such as resistive temperature sensors and pressure and flow sensors.

With the introduction of the International Tempeature Scale of 1990, ITS-90, both the use and further improvement of high-temperature standard platinum resistance thermometers will increase significantly.

3.8 References

[1] Burgess, J. K., *J. Res. Natl. Bur. Stand.,* **1,** 635.
[2] Smith, F. E., *Phil. Mag.* **24** (1912) 541-565.
[3] Mueller, E. F., in: *Temperature, its Measurement and Control in Science and Industry Vol., 1,* Wolf (ed.); New York: Reinhold, 1941, p. 162.
[4] Clement, J. R., Quinnell, E. H., *Rev. Sci. Instrum.* **23** (1952) 213-217.
[5] Kunzler, J. E., Geballe, T. H. Hull, G. W., in: *Temperature, its Measurement and Control in Science and Industry,* Vol. 3, Herzfeld, C. M. (ed.); New York: Reinhold, 1962, Part 1, 391-398.
[6] Coles, B. R., *Phys. Lett.* **8** (1964) 243-244.
[7] Rusby, R. L., in: *Temperature Measurement 1975, Inst. Phys. Conf. Ser.,* No. 26; Billings, B. F., Quinn, T. J. (eds.); London Institute of Physics, 1975, pp. 829-833.
[8] Lawless, W. N., *Rev. Sci. Instrum.* **43** (1972) 1743-1747.
[9] International Electrotechnical Commission, *Industrial Platinum Resistance Thermometer Sensors, Publication 751,* 1983, Genève: Bureau Central de la Commission Electrotechnique Internationale.
[10] Dauphinee, T. M., in: *Temperature its Measurement and Control in Science and Industry* Vol. 5, Schooley, J. F. (ed.); New York: American Institute of Physics, 1982, Part 2, pp. 1317-1325.
[11] Diehl, W., in: *Sensors '83,* Vol. 1; Buckinham: Network Events, 1983, 28-41.
[12] Dugdale, J. S., *The Electrical Properties of Metals and Alloys;* London: Arnold, 1977.
[13] Bloch, F., *Z. Phys.* **52** (1928) 555.
[14] Berry, R. J., *Rev. Sci. Instrum.* **39** (1962) 106-112.
[15] Quinn, T. J., *Temperature;* London: Academic Press. 1983.
[16] Blakemore, J. S., *Rev. Sci. Instrum.* **33** (1962) 106-112.
[17] Sapoff, M., Siwak, W. R., Johnson, H. C., Slepian, J., and Weber, S., in: *Temperature, its Measurements and Control in Science and Industry* Vol. 5, Schooley, J. F. (ed.); New York: American Institute of Physics, 1982, Part 2, pp. 875-887.
[18] Sachse, H. B., *Semiconducting Temperature Sensors and their Application,* New York: Wiley, 1975.
[19] Luppold, D. S., *Precision DC Measurements and Standards;* Reading, MA: Addison-Wesley, 1969.
[20] Thompson, A. M., Small, G. W., *Proc. IEE* **118** (1971) 1662-1666.
[21] Cutkowski, R. D., in: *Temperature, its Measurements and Control in Science and Industry* Vol. 5, Schooley, J. F. (ed.); New York: American Institute of Physics, 1982; Part 2, pp. 711-714.
[22] Kirby, C. G. M., in: *Temperature, its Measurement and Control in Science and Industry* Vol. 5; Schooley, J. F. (ed.); New York: American Institute of Physics, 1982, Part 2, pp. 715-718.
[23] Crovini, L., Marcarino, P., Frassinetti, G., *IEEE Trans.* **IM-34** (1985) 337-341.
[24] Compton, J. P., in: *Temperature, its Measurement and Control in Science and Industry* Vol. 4; Plumb, H. H. (ed.); Pittsburgh: Instrument Society of America, 1972, Part 2, pp. 951-961.

[25] Crovini, L., Marcarino, P., Actis, A., Dematteis, R., *Report on the Comparative Measurements of the Resistance of IPRT's with Three Bridges (DC: 28.5 Hz; 318 Hz),* IMGC internal report S/240, 1985, IMGC, Turin.
[26] Crovini, L., Actis, A., in: *Temperature Measurement 1975, Inst. Phys. Conf. Ser.,* No. 26, Billings, B. F., Quinn, T. J. (eds.); London: Institute of Physics, 1975, pp. 398-408.
[27] Swenson, C. A., Wolfendale, P. C. F., *Rev. Sci. Instrum.* **44** (1973) 339-341.
[28] Kirby, C. G. M., Laubitz, M. J., *Metrologia* **9** (1973) 103-106.
[29] Kusters, N. L., McMartin, M. P., *IEEE Trans.* **IM-19** (1970) 291-297.
[30] Crovini, L., Kirby, C. G. M., *Rev. Sci. Instrum.* **41** (1970) 493-497.
[31] Kerlin, T. W., Shepard, R. L., Hashemian, H. L., Petersen, K. M., in: *Temperature, its Measurement and Control in Science and Industry* Vol. 5, Schooley, J. F. (ed.); New York: American Institute of Physcis, 1982, Part 2, pp. 1357-1366.
[32] Curtiss, D. J., in: *Temperature, its Measurement and Control in Science and Industry* Vol. 5: Schooley, J. F. (ed.); New York: American Institute of Physics, 1982, Part 2, pp. 803-812.
[33] Riddle, J. L., Furukawa, G. T., Plumb, H. H., *Platinum Resistance Thermometry, NBS Special Publication 126,* 1973, U.S. Department of Commerce, Washington, DC.
[34] McLaren, E. H., *Can J. Phys.* **37** 422-432.
[35] Lucas, D. A., in: *Temperature, its Measurement and Control in Science and Industry* Vol. 4. Plumb, H. H. (ed.); Pittsburgh: Instrument Society of America, 1972, Part 2, pp. 963-969.
[36] Berry, R. J., Zhang Jipei, in: *Temperature Measurement in Industry and Science; Beijing: China Academic Publishers, 1986, pp. 157*-163.
[37] Berry, R. J., *Surface Science* **76** (1978) 415-442.
[38] Bureau International des Poids et Mesures: *Supplementary Information for the IPTS-68 and the EPT-76,* Sèvres: BIPM.
[39] Preston-Thomas, H., *Metrologia* **27** (1990) 3-10.
[40] Kemp, R. C., in: *Temperature Measurement in Industry and Science;* Beijing: China Academic Publishers, 1986, pp. 85-90.
[41] McLaren, E. H., Murdock, E. G., *The Properties of Pt/PtRh Thermocouples for Thermometry in the Range 0*-1100 °C. NRCC 17407 and 17408, NRC Ottawa, 1979.
[42] Evans, J. P., Burns, G. W., in: *Temperature, its Measurement and Control in Science and Industry* Vol. 3, Herzfeld, C. M. (ed.); New York: Reinhold, 1962, Part 1, pp. 313-318.
[43] Sawada, S., Mochizuki, T., in: *Temperature, its Measurement and Control in Science and Industry* Vol. 4, Plumb, H. H. (ed.); Pittsburgh: Instrument Society of America, 1972; Part 2, pp. 919-926.
[44] Evans, J. P., in: *Temperature, its Measurement and Control in Science and Industry* Vol. 5, Schooley, J. F. (ed.); New York: American Institute of Physics, 1982, Part 2, pp. 771-781.
[45] Long Guang, Hongtu, Tao, in: *Temperature, its Measurement and Control in Science and Industry* Vol. 5, Schooley, J. F. (ed.); New York: American Institute of Physics, 1982; Part 2, pp. 783-787.
[46] Evans, J. P., Tillet, S. B., *National Bureau of Standards Technical Notes 1183,* Washington, DC: Natl. Bur. Stand. (U.S.), 1984.
[47] Evans, J. P., *J. Res. Natl. Bur. Stand* **89** (1984) 349-373.
[48] Marcarino, P., Crovini, L., in: *Temperature Measurement 1975, Inst. Phys. Conf. Ser.,* No. 26, Billings, B. F., Quinn, T. J. (eds.); London: Institute of Physics, 1975, pp. 107-116.
[49] Berry, R. J., in: *Temperature, its Measurement and Control in Science and Industry* Vol. 4, Plumb, H. H. (ed.); Pittsburgh: Instrument Society of America, 1972, Part 2, pp. 937-949.
[50] Marcarino, P., Demalteis, R., Gallorini, M., and Rizzio, E., *Metrologia* **26** (1989) 175-181.
[51] Actis, A., Crovini, L., in: *Temperature, its Measurement and Control in Science and Industry* Vol. 5, Schooley, (ed.); New York: American Institute of Physics, 1982, Part 2, pp. 819-828.
[52] Connolly, J., in: *Temperature, its Measurement and Control in Science and Industry* Vol. 5, Schooley, J. F. (ed.); New York: American Institute of Physics, 1982, Part 2, pp. 815-817.
[53] Besley, L. M., Kemp, R. C., *Cryogenics* (1983) January, 26-28.
[54] Johnston, J. S., in: *Temperature Measurement 1975, Inst. Phys. Conf. Ser.,* No. 26, Billings, B. F., Quinn, T. J. (eds.); London Institute of Physics, 1975, 80-90.
[55] Organization Internationale de Métrologie Legal (OIML), *Thermometres Electriques à Resistance de Platine-Cuivre-Nichel (à usage technique),* Project of International Recommendation, Mai 1973, Sécretariat-Rapporteur OIML.
[56] Griffin, E. L., Mochel, J. M., *Rev. Sci., Instrum.* **45** (1974) 1265-1267.

[57] Rusby, R. L., in: *Temperature, its Measurement and Control in Science and Industry* Vol. 5, Schooley, J. F. (ed.); New York: American Institute of Physics, 1982, Part 1, pp. 820–833.
[58] Shiratori, T., Mitsui, K., Yangishawa, K., Kobayashi, S., in: *Temperature, its Measurement and Control in Science and Industry* Vol. 5, Schooley, J. F. (ed.), New York: American Institute of Physics, 1982, Part 1, pp. 839–843.
[59] La Mers, T. H., Zurbuchen, J. M., Trolander, H., in: *Temperature, its Measurement and Control in Science and Industry* Vol. 5, Schooley, J. F. (ed.); New York: American Institute of Physics, 1982, Part 2, pp. 865–873.
[60] Wood, S. D., Mangum, B. W., Filliben, J. J., Tillet, S. B., *J. Res. Natl. Bur. Stand* **83 (A)** (1978) 247–265.
[61] Mangum, B. W., in: *Temperature Measurement in Industry and Science;* Beijing: China Academic Publishers, 1986, pp. 170–175.
[62] Schlosser, W. F., Munnings, R. H., in: *Temperature, its Measurement and Control in Science and Industry* Vol. 4, Plumb, H. H. (ed.); Pittsburgh: Instrument Society of America, 1972, Part 2, pp. 919–926.
[63] Heywang, W., *Solid State Electronics* **3** (1961) 51–58.
[64] Wendt, R. E., Jr. in: *Temperature, its Measurement and Control in Science and Industry* Vol. 5, Schooley, J. F. (ed.); New York: American Institute of Physics, 1982, Part 2, pp. 911–914.
[65] Swinehart, P. R., in: *Temperature, its Measurement and Control in Science and Industry* Vol. 5, Schooley, J. F. (ed.), New York: American Institute of Physics, 1982, Part 2, pp. 835–838.
[66] Besley, L. M., Plumb, H. H., *Rev. Sci. Instrum.* **49** (1978) 68–71.
[67] Besley, L. M., *Rev. Sci. Instrum.* **49** (1978) 1041–1045.
[68] Besley, L. M., *Rev. Sci. Instrum.* **51** (1980) 972–976.
[69] Weinstock, H., Parpia, J., in: *Temperature, its Measurement and Control in Science and Industry* Vol. 4, Plumb, H. H. (ed.): Pittburgh: Instrument Society of America, 1972, Part 2, pp. 785–790.
[70] Terry, C., *Rev. Sci. Instrum.* **39** (1968) 925.
[71] Anderson, A. C., in: *Temperature, its Measurement and Control in Science and Industry* Vol. 4, Plumb, H. H. (ed.); Pittsburgh: Instrument Society of America, 1972, Part 2, pp. 773–784.
[72] Ricketson, B. W., in: *Temperature Measurement 1975, Inst. Phys. Conf. Ser.,* No. 26, Billings, B. F., Quinn, T. J. (eds.); London: Institute of Physics, 1975, pp. 135–143.
[73] BIPM, *Metrologia* **15** (1979) 65–68.
[74] Ricketson, B. W., Grinter, R., in: *Temperature, its Measurement and Control in Science and Industry* Vol. 5, Schooley, J. F. (ed.); New York. American Institute of Physics, 1982, Part 2, pp. 845–852.
[75] CIPM, *Metrologia* **13** (1976) 7–17.
[76] Ambrose, D., Crovini, L., in: *Recommended reference materials for the realization of physicochemical properties.* Marsh, K. N. (ed.); Oxford: Blackwell Scientific Publications, 1987, pp. 163–218.
[77] Bedford, R. E., Bonnier, G., Maas, H., Pavese, F., *Metrologia* **20** (1984) 145–155.
[78] Busse, C. A., Labrande, J. P., Bassani, G., in: *Temperature Measurement 1975, Inst. Phys. Conf. Ser.,* No. 26, Billings, B. F., Quinn, T. J. (eds.); London: Institute of Physics, 1975, pp. 428–438.
[79] Marcarino, P., Bassani, C., in: *Proc. 2nd Symp. on Temperature Measurement in Industry and Science, Shul (GDR), IMEKO TC12;* pp. 235–244.
[80] Taylor, W. J., Rossini, F. D., *J. Res. Natl. Bur. Stand. 32* (1944) 197–212.
[81] Crovini, L., Marcarino, P., Milazzo,G., *Anal. Chem.* **53** (1981) 681–686.
[82] McAllan, J. V., in: *Temperature, its Measurement and Control in Science and Industry* Vol. 5, Schooley, J. F. (ed.); New York: American Institute of Physics, 1982, Part 2, pp. 789–794.
[83] Crovini, L., Actis, A., Galleano, R., *High Temp.-High Press.* **18** (1987) 697–705.
[84] Brown, N. L., in: *IEEE International Conference on Engineering in the Ocean Environment (Ocean 74);* New York: IEEE, 1974, Vol. 2, pp. 270–278.

4 Thermocouples

HERBERT VANVOR, Hanau, FRG

Contents

4.1 Fundamentals

4.1.1 Theory

Thermocouples are temperature sensors which find application in almost all fields of industrial production. Depending on the choice of materials, they cover the range from −270 to 2000 °C. Their simple construction allows sensors to be made which are able to provide accurate, reliable data under extreme mechanical, chemical and electrical conditions. The accuracy achievable depends on the temperature being measured, the temperature range, the choice of the two materials in the thermocouple and the accuracy in the determination of the electromotive force (emf) produced (the measurement circuit). Near ambient temperature it can be within the range 0.1–0.2 K, where as at very high temperatures it can be of the order of ±5–10 K.

A thermocouple consists of two conductors (A and B) of different compositions, which are made up into an electrical circuit as shown in Figure 4-1.

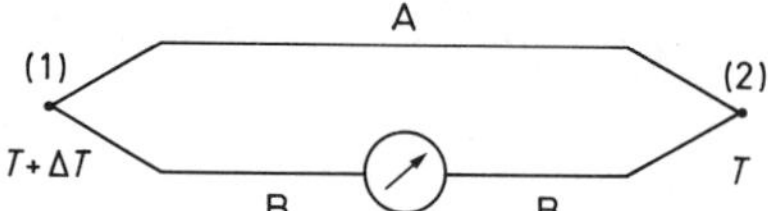

Figure 4-1. Basic circuit of a thermocouple.

In 1821 Johann Seebeck discovered that a small current flowed through a circuit, such as that in Figure 4-1, if junction 1 was brought to a different temperature (eg, higher) than junction 2. The electrical potential created by the temperature difference is known as the Seebeck emf (electromotive force) or the thermolectric potential. The magnitude of the thermal emf depends on the composition of the conductors chosen (A and B) and the temperature difference ΔT, and its polarity depends on the sign of ΔT. In order to perform temperature determinations, one of the two junctions (eg, 2) is maintained at a known and constant temperature (eg, 0°C).

A thermocouple can be regarded as a device for the conversion of thermal energy into electrical energy. By ignoring the irreversible losses such as Joule heat and thermal conductivity losses, the thermoelectric potential produced can be determined from the algebraic sums of the Peltier potentials at junctions A and B together with the Thomson potentials in materials A and B (legs).

In 1834, Peltier discovered that in a circuit according to Figure 4-2 temperature increase and decrease effects occurred at junctions 1 and 2 which depended on the intensity and direction of the current I (Peltier effect). The transport of electrons across the junctions between the two materials requires work to be done by or on the system. This makes the thermal energy

Figure 4-2. Peltier effect.

of the electrons either higher or lower, which leads to heating or cooling of the junction. If P is a constant of proportionality, the heat flow Φ given by

$$\Phi = PI\,. \tag{4-1}$$

P is called the Peltier coefficient in $WA^{-1} = V$.

Thomson recognized that the heat content of a single conductor with a current flowing through it changes when there is a temperature gradient along it, as shown in Figure 4-3. The flow of heat is proportional to the current I and to the temperature gradient ΔT:

$$\Phi = \sigma \cdot I \cdot \Delta T \tag{4-2}$$

where σ is the Thomson coefficient in $WA^{-1}K^{-1} = VK^{-1}$.

Figure 4-3. Thomson effect.

The emf $E = \frac{\mathrm{d}E}{\mathrm{d}T}\Delta T$ of a thermocouple, with legs A and B, temperature 1 of measuring junction $T + \Delta T$ and temperature 2 of the reference junction T, is equal to the algebraic sum of the Peltier and Thomson potentials [1, 2]:

$$E = \frac{\mathrm{d}E}{\mathrm{d}T} \cdot \Delta T = P(T + \Delta T) - P(T) + \sigma_A \cdot \Delta T - \sigma_B \cdot \Delta T \tag{4-3}$$

and

$$\frac{\mathrm{d}E}{\mathrm{d}T} = \frac{P(T + \Delta T) - P(T)}{\Delta T} + \sigma_A - \sigma_B\,. \tag{4-4}$$

When ΔT is small ($\Delta T \rightarrow 0$) Equation (4-4) becomes

$$\frac{\mathrm{d}E}{\mathrm{d}T} = \frac{\mathrm{d}P}{\mathrm{d}T} + (\sigma_A - \sigma_B) \tag{4-5}$$

which can be transformed into

$$P = T\frac{\mathrm{d}E}{\mathrm{d}T} \quad \text{and} \quad \sigma_A - \sigma_B = T\frac{\mathrm{d}^2E}{\mathrm{d}T^2} \tag{4-6}$$

and hence on integration

$$\frac{\mathrm{d}E}{\mathrm{d}T} = \int_0^T \frac{\sigma_A - \sigma_B}{T} \cdot \mathrm{d}T = \int_0^T \frac{\sigma_B}{T}\,\mathrm{d}T - \int_0^T \frac{\sigma_A}{T}\,\mathrm{d}T \tag{4-7 a}$$

or

$$\frac{\mathrm{d}E}{\mathrm{d}T} = \frac{\mathrm{d}E_B}{\mathrm{d}T} - \frac{\mathrm{d}E_A}{\mathrm{d}T} \tag{4-7 b}$$

$dE_A/dT = \int \frac{\sigma_A}{T} dT$ and $dE_B/dT = \int \frac{\delta\sigma_B}{T} dT$ are known as the absolute thermoelectric powers (ATP) of the individual legs of the thermocouple and each applies only to one leg, see Figure 4-4. Some typical ATP-data of metals are given in Table 4-1.

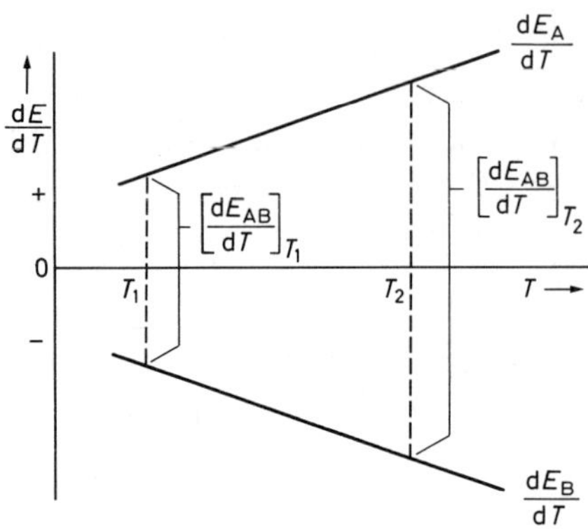

Figure 4-4.
Thermal power of a thermocouple consisting of legs A: positive ATP and B: negative ATP.

It is possible to describe the ATP in terms of the "theory of thermoelectricity" developed by Mott and Jones [3]; A detailed presentation can be found in ASTM 492-1. The results of these calculations are not generally amenable to direct analysis, since a series of simplifying assumptions are always necessary. In the case of binary alloys their thermoelectric properties are only slightly changed if a small amount of the added metal dissolves in the parent metal. Considerable changes occur when solution is almost complete. In the case of multicomponent alloys, the changes that can occur are no longer calculable in practice. If the alloying elements are only partially soluble in the metal matrix, hysteresis phenomena can occur when the temperature is changed.

The use of the ATP tables is very laborius as it is necessary to integrate the ATP-temperature curves to yield a thermal emf curve. Thermal emf curves with respect to a standard thermocouple arm (eg, platinum) are much more appropriate. It is possible to produce platinum wire (with a diameter of, for example, 0.5 mm) containing very few impurities which is suitable for use as a reference leg. Purer platinum exhibits a negative thermal emf with respect to a less pure sample. Comparison of thermal potentials at, for example, 1200 °C makes it possible to detect impurity concentrations of the order of a few ppm with relative ease. Platinum with a purity of at least 99.9995% is normally employed as a standard. International cooperative comparisons of platinum wires in the various national standards institutes (NIST, NPL, PTB, etc.) have led to the development of a platinum which yields a negative thermal emf with respect to all other samples of platinum wire. It is known as platinum 27 since the mean temperature coefficient α of its electrical resistance between 0 and 100 °C was 1.3927 ($\alpha = (R_{100} - R_0)/100\ R_0$; where R_{100} = resistance at 100 °C and R_0 = resistance at 0 °C). By general agreement this platinum is now generally employed as the reference for the citation of the thermal potentials of individual materials (see Table 4-2).

For the application of Table 4-2 the convention holds that the thermoelectric potential of a thermocouple is given by the thermal potential of leg 1 minus that of leg 2. Some illustrative examples are given below.

Table 4.1: Absolute Thermoelectric Power (ATP) of Metals.

T (K)	Cu	Fe	Ni	NiAl	NiCr	Ag	Au	Pt	PtRh 90/10	PtRh 87/13	Pd	W	Mo
100	1.19					0.73	0.82	4.29			2.0		
200	1.29					0.85	1.34	−1.27			−4.85		
300	1.83					1.51	1.94	−5.78	−0.14		−1.0	1.07	5.57
400	2.34	−9.6	−22,51	−19.6	21,0	2.08	2.46	−7.83	−1.0	−0.90	−13.0	4.44	8.52
500	2.83	7.0	−24.31	−19.4	20.8	2.82	2.86	−9.89	−2.01	−1.75	−16.03	7.53	11.12
600	3.33	3.90	−26.26	−20.6	20.0	3.72	3.18	−11.66	−3.49	−3.00	−19.06	10.29	13.27
700	3.83	1.3	−25.93	−21.7	19.2	4.72	3.43	−13.31	−4.42	−4.05	−22.09	12.66	14,94
800	4.34	0.9	−26.24	−22.8	18.4	5.77	3.63	−14,88	−5.60	−5.02	−25.12	14.65	16.13
900	4.85	−2.8	−27.35	−24.52	17.0	6.85	3.77	−16.39	−6.91	−6.25	−28.15	16.28	16.86
1000	5.36		−28.85	−26.07	15.5	7.95	3.85	−17.86	−8.23	−7.46	−31.18	17.57	17.16
1100	5.88		−30.36	−27.57	13.4	9.06	3.88	−19.29	−9.56	−8.75	−34.21	18.53	17.08
1200	6.40		−31.92	−29.0	12.9	10.15	3.86	−20.69	−11.00	−10.20	−37.24	19.18	16.65
1300	6.91						3.78	−22.06			−40.27	19.15	15.92
1400								−23.41			−43.3	19.60	14.94
1500													

Table 4-2: Thermal emfs (mV) of various metals against Pt (reference temperature 0 °C).

°C	Al	Au	Ag	Bi	C	Ca	Cd	Co	Cu	Fe	Ge	Hg	Ir	K	Li
−200	0.45	−0.21	−0.21	12.39			−0.04		−0.19	−3.10	−46.0		−0.25	1.61	−1.12
−100	−0.06	−0.39	−0.39	7.54			−0.31		−0.37	−1.94	−26.62		−0.35	0.78	−1.00
0	0	0	0	0	0	0	0	0	0	0	0	0	0	0	0
100	0.42	0.78	0.74	−7.34	0.70	−0.51	0.90	−1.33	0.76	1.98	33.9	−0.60	0.65		1.82
200	1.06	1.84	1.77	−13.57	1.54	−1.13	2.35	−3.08	1.83	3.69	72.40	−1.33	1.49		
300	1.88	3.14	3.05		2.55	−1.85	4.24	−5.10	3.15	5.03	91.8		2.47		
400	2.84	4.63	4.57		3.72			−7.24	4.68	6.08	82.3		3.55		
500	3.93	6.29	6.36		5.15			−9.35	6.41	7.00	63.5		4.78		
600	5.15	8.12	8.41		6.79			−11.28	8.34	8.02	43.9		6.10		
700		10.13	10.75		8.84			−12.88	10.49	9.34	27.9		7.56		
800		12.29	13.36		11.01			−14.00	12.84	11.09			9.12		
900		14.61	16.20		13.59			−14.49	15.41	13.10			10.80		
1000		17.09			16.51			−14.20	18.20	14.64			12.59		
1100					19.49			−12.98					14.48		
1200								−10.68					16.47		

°C	Mg	Mo	Na	Ni	Pb	Pd	Rh	Sb	Si	Sn	Ta	Th	W	Zn	NiAl
−200	+0.37		1.00	2.28	0.24	0.81	−0.20		63.13	0.26	0.21		+0.43	−0.07	2.39
−100	−0.09		0.29	1.22	−0.13	0.48	−0.34		37.17	−0.12	−0.10		−0.15	−0.33	1.29
0	0	0	0	0	0	0	0	0	0	0	0	0	0	0	0
100	+0.44	1.45		−1.48	0.44	−0.57	0.70	4.89	−41.46	0.42	0.33	−0.13	1.12	0.76	−1.19
200	+1.10	3.19		−3.10	1.09	−1.23	1.61	10.14	−80.58	1.07	0.93	−0.26	2.62	1.89	−2.17
300		5.23		−4.59	1.91	−1.99	2.68	15.44	−110.09		1.79	−0.40	4.48	3.42	−2.89
400		7.57		−5.45		−2.82	3.91	20.53			2.91	−0.50	6.70	5.29	−3.64
500		10.20		−6.16		−3.84	5.28	25.10			4.30	−0.53	9.30		−4.43
600		13.13		−7.04		−5.03	6.77	28.88			5.95	−0.45	12.26		−5.28
700		16.35		−8.10		−6.41	8.40				7.87	−0.21	15.60		−6.18
800		19.87		−9.35		−7.98	10.16				10.05	+0.22	19.30		−7.08
900		23.69		−10.69		−9.72	12.04				12.49	+0.87	23.36		−7.95
1000		27.80		−12.13		−11.63	14.05				15.20	1.73	27.80		−8.79
1100		32.21		−13.62		−13.70	16.18				18.70	2.80	32.60		−9.58
1200		36.91				−15.89	18.92				21.41	4.04	37.78		−10.34

°C	NiCr 90/10	Au-Fe 0.05 At%	Cu-Be	Cu-Ni (const.)	Manga-nin	Brass	P-Bronze	Spring steel	NiCr 18/8	NiCr 80/20	NiFeCr 60/24/16	CuSnZn 95/4/1	CuNi 75/25	AgCu 90/10	Soft solder 50Sn 50Pb
−200	−3.36			5.35											
−100	−2.20			2.98											
0	0	0	0	0	0	0	0	0	0	0	0	0	0	0	0
100	2.81	−0.17	0.67	−3.51	0.61	0.60	0.55	1.32	0.44	1.14	0.85	0.6	−2.76	0.80	0.46
200	5.96	−0.32	1.62	−7.45	1.55	1.49	1.34	2.63	1.04	2.62	2.01	1.48	−6.01	1.90	
300	9.32	−0.44	2.81	−11.71	2.77	2.58	2.34	3.81	1.76	4.34	3.41	2.60	−9.71	3.25	
400	12.75	−0.55	4.19	−16.19	4.25	3.85	3.50	4.84	2.60	6.25	5.0	3.91	−13.78	4.81	
500	16.21	−0.63		−20.79	5.95	5.30	4.81	5.80	3.56	8.31	6.76	5.44	−18.10	6.59	
600	19.62	−0.66		−25.47	7.84	6.96	6.30	6.86	4.67	10.53	8.68	7.14	−22.59	8.94	
700	22.96			−30.18					5.93	12.91	10.78				
800	26.23			−34.86					7.37	15.44	13.06				
900	29.41			−39.45					8.99	18.11	15.50				
1000	32.52			−43.92						20.91	18.10				
1100	35.56														
1200	38.51														

Example 1: *Gold-palladium at 1000°C*

Au-Pt 17.09 mV at 1000°C

Pd-Pt −11.63 mV at 1000°C

Au-Pd = 17.09 − (−11.63) = 28.72 mV at 1000°C

Example 2: *Silver-copper at 500°C*

Ag-Pt 6.36 mV at 500°C

Cu-Pt 6.41 mV at 500°C

Ag-Cu = 6.36 − 6.41 = −0.05 mV at 500°C

Cu is positive with respect to Ag. According to the usual convention (positive leg listed first) the element should be referred to as Cu-Ag.

Example 3: *Nickel-copper nickel (constantan) at 1000°C*

Ni-Pt −13.62 mV at 1000°C

CuNi-Pt −43.92 mV at 1000°C

Ni-CuNi = −13.62 − (−43.92) = 30.3 mV at 1000°C

The data in Table 4-2 have been taken from several literature sources.

4.1.2 Properties of a Thermocouple

Deviations of the readings of thermocouples of the same type may occur because the thermal emf is very dependent on the composition of the materials and on any additives or impurities which they may contain. The deviations of the thermal emfs of combinations included in IEC 584-1 (Fe-CuNi, NiCr-Ni, etc.) can be up to 1−3 K and even up to 0.75% of the actual temperature in the upper range of temperatures.

It is possible to summarize the properties of thermocouples for practical applications by means of three simple laws.

4.1.2.1 Circuits of Homogeneous Materials

A thermal emf cannot be set up in a circuit made of a homogeneous conductor even if the temperature is altered over the whole conductor or over just part of it.

When applied to a thermocouple this means that if it is made up of two different homogeneous conductors then only the measurement junction and the reference junction have an effect on the thermal emf. The temperature variation along the homogeneous conductors between the measurement and reference junctions has no effect [4].

Example:

A load to be heated in a continuous oven is passed through the oven on a conveyor belt. The temperature cycle of the load is monitored by attaching to it a thermocouple, which is then transported through the oven with it. If the two legs of the thermocouple are sufficiently homogeneous, then the temperature profile in the oven can have no influence when the load has passed through the maximum temperature into a cooler region.

4.1.2.2 Circuits in Regions of Homogeneous Temperature

A thermal emf is not generated in a circuit made up of any number of different materials connected together if all parts of the circuit are maintained at the same temperature.

When applied to the thermocouple this means that inhomogeneous legs of the thermocouple, connecting materials at plugs and connections, etc., have no influence on the measurement, provided that they are maintained at the same temperature [4]. Variations of this temperature with time have no effect provided that the temperatures of all the components remain the same and parts of the measuring circuit do not acquire different temperatures as a result of, for instance, different responses to the temperature change as a result of different thermal inertias.

4.1.2.3 Additivity of the Thermal Electromotive Forces in the Circuit

The thermal emfs generated in a circuit are additive taking into account their signs (+ or −).

Example:

If the connecting lead (compensation lead) of an NiCr-Ni (type K) thermocouple is attached between the connecting head of the thermocouple and the reference junction with the wrong polarity (Figure 4-5), then this produces an error corresponding to twice the temperature difference between the connecting head and the reference junction.

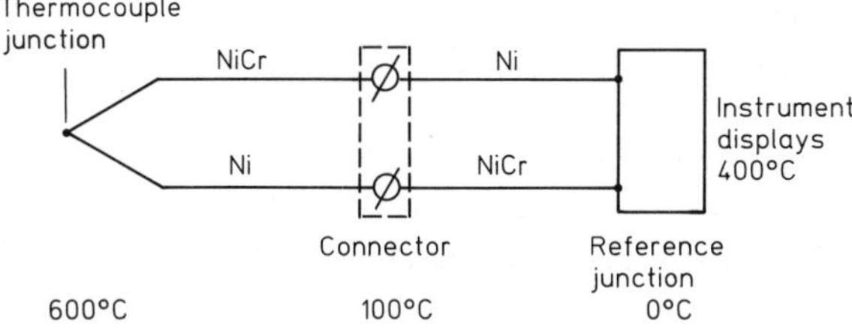

Figure 4-5. Thermocouple with extension lead.

4.2 Materials for Thermocouples

4.2.1 General Requirements

The materials from which thermocouples are constructed must fulfil a number of conditions in order that reliable temperature determinations can be carried out. Their thermal emf should be sufficiently high and the resistance of the thermocouple should not be too high, so that external electrical influences, such as stray pickups, or the effects of the attached measurement instrument, remain small. The characteristic thermal emf curve must be monotonic, as linear as possible, and as steep as possible over the range of application of the thermocouple.

The materials used in the construction of thermocouples should have a high melting point and be capable of being manufactured to a homogeneous, reproducible quality. They must be easily worked into the desired shapes (wires and ribbons). The materials should possess adequate corrosion resistance and be resistant to reducing environments. Their characteristic curves should not exhibit discontinuities or bends within their application range (eg, as a result of allotropic transitions). Scaling and embrittlement during operation must not alter the thermal emf characteristics, nor must the composition of alloys change (eg, as a result of oxidation) within the application range. Mechanical stresses, such as drawing, bending and forming, change the thermoelectric properties of the thermocouple wire and are to be avoided. It is vitally important that the thermocouple wire be annealed after every cold-forming operation.

The temperature duration and atmosphere employed for recrystallizing annealing all depend on the particular material. Thermocouple wires delivered by the manufacturer are annealed and, hence, homogeneous. The method of connection at the junction (twisting, soldering, welding) does not affect the thermal emf unless the junction is situated in a region of steep temperature gradient. Then the temperature of the last point of contact is indicated (Figure 4-6).

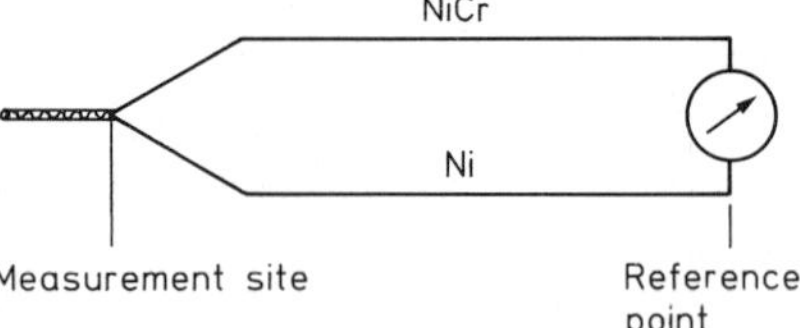

Figure 4-6.
Measuring junction when thermocouple wires are twisted.

4.2.2 Platinum and Platinum-Rhodium System

Platinum and platinum-rhodium alloys were used by Henry le Chatelier for temperature measurement as early as 1886. Both platinum and rhodium can be manufactured in a very pure state (purity >99.99%); they are easily drawn into wires and ribbons without including foreign materials. High oxidation and chemical resistance of platinum and platinum-rhodium alloys are prerequisites for their reliable measurement and low measurement uncertaintly. The thermal emf depends on the rhodium content of the alloyed arm. The following standard pairs listed in IEC 584 [5] are employed in practice.

Type R:	PtRh (87/13) – Pt	up to 1500 (1600) °C
Type S:	PtRh (90/10) – Pt	up to 1500 (1600) °C
Type B:	PtRh (70/30) – PtRh (94/6)	up to 1750 (1800) °C

The influence of the rhodium content on the thermal emf of PtRh alloys is illustrated in Figure 4-7, and the emf characteristics of standard PtRh thermocouples in Figure 4-8. It can be seen that the variation of emf with temperature is not strictly linear.

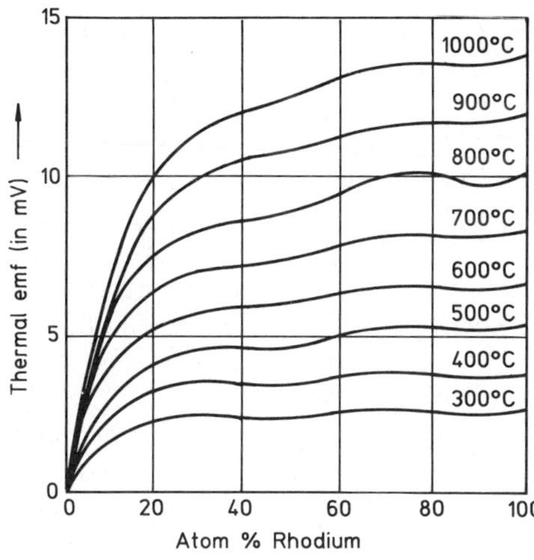

Figure 4-7.
Influence of rhodium content on the thermal emfs of PtRh alloys with respect to platinum.

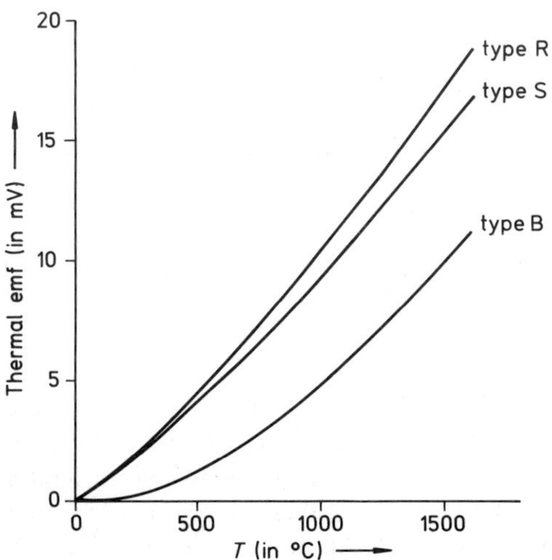

Figure 4-8.
Thermal emfs of PtRh thermocouples of types R, S, and B according to IEC 584-1.

4.2.3 Other Noble Metals

4.2.3.1 Iridium

In order to measure very high temperatures it is necessary to employ materials with very high melting points. Thermocouples consisting of Ir and IrRh (60/40) alloy have proved themselves under laboratory conditions. Short-term measurements are possible up to 2000 °C. In long-term use in air, oxidation of the rhodium and iridium causes the thermal emf to alter. Iridium is such a brittle metal that it is only possible to work it into rolled wire (eg, 0.6 mm diameter). Further the wire is liable to break after prolonged use owing to recrystallization. In spite of these disadvantages it is employed for temperatures over 1800 °C, because it is the only thermocouple that can be used at these temperatures without a protective atmosphere. Such thermocouples are usually calibrated individually. An average characteristic curve is reproduced in Figure 4-9.

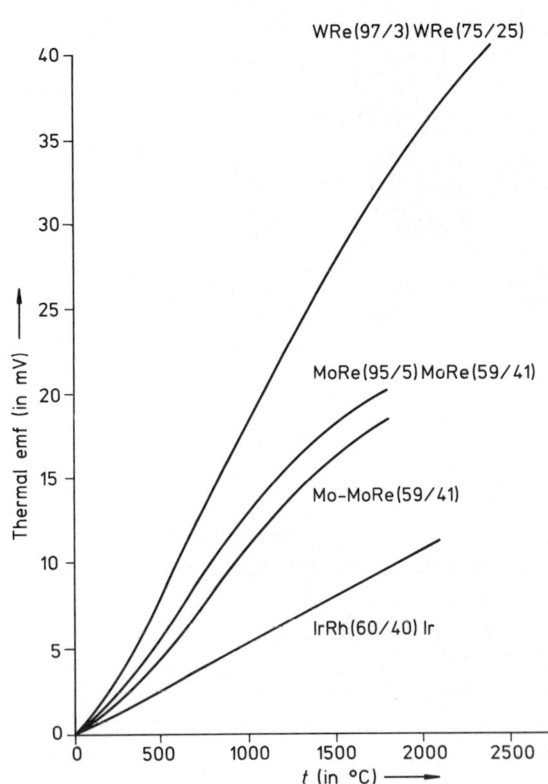

Figure 4-9.
Thermal emfs of W-Re, Mo-Re, and Ir-IrRh.

4.2.3.2 Tungsten, Molybdenum, and Rhenium

Thermocouples constructed from refractory metals and their alloys can be employed at temperatures over 1600 °C. The termocouple WRe (75/25) – WRe (97/3) has been particularly successful in practical use. However, it is necessary to operate it in the absence of air

Table: 4-3 Thermal emf (mV) of WRe (75/25) – WRe (97/3).

°C	0	10	20	30	40	50	60	70	80	90
0	0	0.096	0.197	0.302	0.411	0.525	0.642	0.763	0.888	1.015
100	1.145	1.278	1.414	1.553	1.695	1.840	1.989	2.138	2.290	2.445
200	1.602	2.764	2.928	3.093	3.258	3.426	3.596	3.767	3.940	4.114
300	4.290	4.465	4.642	4.821	5.002	5.186	5.372	5.558	5.747	5.937
400	6.128	6.322	6.518	6.714	6.911	7.108	7.305	7.503	7.701	7.899
500	8.097	8.294	8.492	8.689	8.887	9.088	9.290	9.490	9.691	9.890
600	10.092	10.294	10.496	10.699	10.903	11.107	11.311	11.515	11.720	11.924
700	12.129	12.334	12.540	12.746	12.951	13.157	13.362	13.567	13.772	13.977
800	14.185	14.390	14.595	14.799	15.004	15.208	15.412	14.615	15.820	16.022
900	16.225	16.428	16.631	16.833	17.035	17.237	17.438	17.640	17.841	18.042
1000	18.242	18.441	18.641	18.840	19.039	19.238	19.437	19.635	19.834	20.031
1100	20.232	20.428	20.624	10.821	21.017	21.213	21.409	21.605	21.801	21.997
1200	22.192	22.382	22.573	22.762	22.952	23.142	23.331	23.519	23.708	23.896
1300	24.083	24.272	24.461	24.650	24.838	25.024	25.212	25.396	25.582	25.770
1400	25.956	26.142	26.326	26.510	26.694	26.877	27.060	27.242	27.422	27.601
1500	27.780	27.959	28.138	28.316	28.494	28.671	28.847	29.033	29.199	29.375
1600	29.551	29.726	29.900	30.073	30.246	30.418	30.589	30.760	30.930	31.100
1700	31.269	31.437	31.605	31.772	31.938	32.104	32.269	32.433	32.597	32.760
1800	32.922	33.083	33.244	33.404	35.563	33.722	33.881	34.037	34.193	34.349
1900	34.504	34.658	34.811	34.964	35.115	35.266	35.417	35.567	35.715	35.861
2000	36.007	36.152	36.296	36.440	36.583	36.725	36.864	37.001	37.137	37.273
2100	37.406	37.538	37.669	37.799	37.928	38.055	38.181	38.307	38.431	38.554
2200	38.677	38.799	38.914	39.029	39.143	39.255	39.365	39.471	39.575	39.677
2300	39.777	39.875	39.971	40.065	40.157	40.248	40.338	40.426	40.512	40.596
2400	40.678									

in a neutral or reducing atmosphere. The alloys for both legs allow the manufacture of ductile wires. Since both legs recrystallize and become brittle at high temperatures, it is necessary to use arc welding in a protective atmosphere to join them. After production the welded joint should, if possible, no longer be subjected to mechanical stresses. Table 4-3 lists the basic data for WRe (75/25) – WRe (97/3). In individual cases MoRe (95/5) – MoRe (59/41) and Mo-MoRe (59/41) thermocouples are also employed in the high-temperature range. They have advantages for measurements in nuclear reactors if tungsten reacts with the building materials of the reactor itself. Figure 4-9 illustrates typical characteristics for such thermocouples.

4.2.3.3 *Gold and Silver*

Au-Pt thermocouples have recently been employed for the calibration of other thermal elements. In contrast to the PtRh-Pt thermocouple, both legs are pure noble metals and not alloys. Both metals can be manufactured in a very pure form (>99.999%). Therefore, such

Table 4-4. Thermal emf (mV) of Ag-CuNi.

°C	0	10	20	30	40	50	60	70	80	90
0	0	0.4	0.81	1.23	1.65	2.07	2.49	2.92	3.36	3.81
100	4.26	4.71	5.17	5.64	6.12	6.61	7.11	7.61	8.12	8.64
200	9.16	9.68	10.21	10.74	11.28	11.82	12.37	12.92	13.48	14.04
300	14.61	15.18	15.76	16.34	16.93	17.52	18.12	18.72	19.33	19.94
400	20.56	21.18	21.80	22.43	23.06	23.70	24.34	24.99	25.64	26.30
500	26.96	27.62	28.29	28.96	29.63	30.31	30.99	31.68	32.37	33.06
600	33.76									

Note: since there are no standard values for thermocouples of the type Ag-CuNi, the thermal emf is dependent on the composition of the CuNi alloy (constantan). The values given here are for CuNi according to DIN 43710.

a thermocouple follows a given characteristic with very little variation and can be employed practically without calibration. Another advantage of pure metals is that they yield appreciably more homogeneous wires than do alloys. Stability test are in progress in the main national standards institutes. This thermocouple can be employed at temperature up to 1000 °C.

Silver is mainly employed in thermocouples in conjuction with CuNi (constantan). The Ag-CuNi thermocouple is completely nonmagnetic and yields reliable results even in strong magnetic fields. Its basic thermal emf data are listed in Table 4-4.

AuCo and AuFe for low-temperature measurements are treated in Section 4.2.5.

4.2.4 Thermocouples Employed in Practice

International standards have been set up for the thermocouples employed in technological and manufacturing processes and these are applied in almost every country. In addition, a number of national standards have been set up, which are still in force during a prolonged period of transition. Thus, in addition to the international standard IEC 584, there are also the national standards GOST 3044 and 3045 in the USSR and the national standard DIN 43 710 in the FRG.

Cu-CuNi is primarily employed in the laboratory for the temperature range −250 to +250 °C. The thermocouple wires are ductile and very homogeneous. The standards are type J according to IEC 584 and type U according to DIN 43 710.

Fe-CuNi allows temperature measurements to be made up to 600 °C (800 °C) on account of the greater resistance of iron to oxidation. It is frequently employed in chemical production plants.

NiCr-NiAl is the most frequently employed thermocouple in industrial plants. In addition to type K, which has been in use for many years, the modification type N (Nicrosil-Nisil), with superior stability properties at high temperatures, has been introduced. The thermal emfs of types K and N differ from each other.

PtRh-Pt thermocouples yield very reliable temperature determinations up to 1600 °C (1800 °C). They should be protected from the influence of foreign substances by means of a ceramic tube. Type R is widely employed in the steel industry in the UK, USA and Japan. It offers almost no advantages over the widely employed type S. PtRh-Pt type B yields more stable measurements, particularly at temperatures over 1600 °C and in aggressive atmospheres. The thermocouple yields negative thermal emfs below 40 °C. A reference temperature between 0 and 60 °C has a negligible effect on the result.

A survey of important data concerning the thermocouples described above is given in Table 4-5.

All standard thermocouples are supplied with a given tolerance. This applies only to positive and negative legs that are delivered together. The use of thermocouple wires which are not labelled as pairs can lead to significantly higher deviations. The data in Table 4-5 are nominal values, from which there are variations depending on the manufacturer and the batch. Fe, Ni, CuNi, NiCr and Ni types contain, in addition to the quoted composition, other substances such as Si, P, C and antioxidants which are necessary for their manufacture. These substances also alter the thermoelectrical, mechanical, and chemical properties of the thermocouple wires. Thermocouples from different manufacturers can, therefore, behave differently under different conditions of operation. The characteristics of various commonly used thermocouples are illustrated in Figure 4-10.

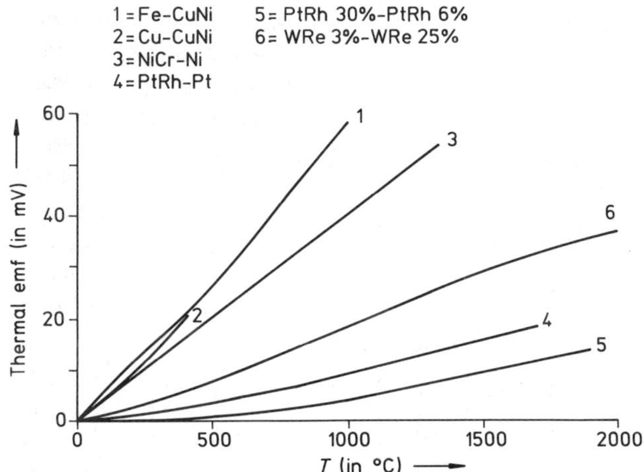

Figure 4-10. Thermal emfs of commonly used thermocouples [6, 9].

4.2.5 Special Thermocouples

The characteristic curves for Cu-CuNi and Fe-CuNi rapidly become less steep, ie, the sensitivity decreases, at very low temperatures (−250 °C). This means that the uncertainty of measurement becomes larger in this region. Thermocouples constructed of AuFe-NiCr do not exhibit this sensitivity decrease so that it is possible to use them for measurements close to absolute zero (0 K = −273.15 °C). The positive leg of the thermocouple consists of pure gold containing small quantities (a few hundredths of a per cent) of Fe. A combination with 0.03 at.% Fe is often employed (Figure 4-11; Table 4-6).

The thermocouple wires are ductile; bending and stretching introduce strain and, thus, alter the thermal emf. Since small quantities of impurities in the gold and in the NiCr (positive leg

Table 4-5. Physical properties of standard thermocouple wires.

Material		Cu	CuNi	Fe	Ni	NiCr	Pt	Pt6Rh	Pt10Rh	Pt13Rh	Pt30Rh
Composition		Cu > 99.5%	55 Cu/ 45 Ni	Fe technically pure	NiMn 3 Al	90 Ni/ 10 Cr	Pt > 99.99	94 Pt/ 6 Rh	90 Pt/ 10 Rh	87 Pt/ 13 Rh	70 Pt/ 30 Rh
Density at 20 °C	$\frac{kg}{dm^3}$	8.9	8.8 to 8.9	7.8 to 7.9	8.6 to 8.8	8.5 to 8.6	21.4	20.6	20.0	19.6	17.7
Specific electrical resistance at 20 °C	$\frac{\Omega mm^2}{m}$	0.017	≈0.49	≈0.12	≈0.27	≈0.72	0.107	0.185	0.193	0.197	0.21
Mean temperature coefficient of the electrical resistance	$\frac{1}{K} \cdot 10^3$	20 to 600 · C			20 to 1000 °C						
		4.3	≈0.05	≈9.5	≈1.2	≈0.27	3.1	1.8	1.4	1.3	1.3
Thermal conductivity	$\frac{W}{mK}$	at 20 °C: 390 at 500 °C: 360	0 to 300 °C 40	at 20 °C: 75 at 800 °C: 35	20 to 700 °C 60	0 to 300 °C 15	70	35	30	27	23
Specific heat	$\frac{J}{kgK}$	at 20 °C: 380 at 500 °C: 440	0 to 300 °C 400	at 20 °C: 460 at 800 °C: 710	20 to 400 °C 550	0 to 300 °C 420	135	142	145	150	168
Mean coefficient of linear expansion from 20 to 600 °C	$\frac{1}{K} \cdot 10^6$	18.0	16.8	14.6	16.0	15.7	9.3	9.1	9.0	9.0	8.8

Table 4-6. Thermal emfs (mV) of Au 0.03 at.% Fe-chromel.

°C	0	−10	−20	−30	−40	−50	−60	−70	−80	−90
0	0	0.196	0.395	0.596	0.799	1.004	1.214	1.424	1.634	1.893
−100	2.039	2.229	2.414	2.594	2.769	2.939	3.104	3.264	3.419	3.569
−200	3.709	3.839	3.973	4.111	4.254	4.402	4.555	4.714		

Table 4-7. Thermal emfs (mV) of Au 2.11 at.% Co – Au 0.37 at.% Ag (reference temperature 0°C).

°C	0	−10	−20	−30	−40	−50	−60	−70	−80	−90
0	0	0.3979	0.7973	1.1975	1.5980	1.9866	2.3991	2.7993	3.1990	3.5978
−100	3.9953	4.3912	4.7850	5.1764	5.5640	5.9470	6.3246	6.6953	7.0573	7.4084
−200	7.7460	8.0674	8.3696	8.6459	8.8904	9.0970				

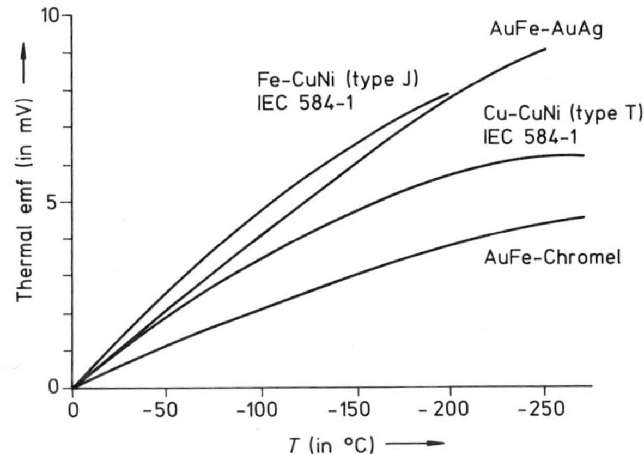

Figure 4-11.
Thermocouples for use at low temperatures.

of type K IEC 584-1) affect the thermal emf characteristic, it is necessary to calibrate each individual thermocouple for accurate measurement.

Thermocouples consisting of Au 2.11 at.% Co – Ag 0.37 at.% Au are occasionally employed at low temperatures. This combination offers the advantage of a lower electrical resistance; however, it is very sensitive to deformation (Table 4-7).

Since metals form carbides on coming into contact with carbon, which causes very large changes in the thermal emf and W, Mo etc., become brittle, nonmetallic thermocouples are employed, for example, to determine high temperatures in graphite furnaces. A thermocouple consisting of B_4C and C (Figures 4-12 and 4-13) is suitable for application in graphite furnaces. The element can be employed up to 2200 °C even in carbon-containing atmospheres. The reference points are at the spring contacts. The temperature here must either be maintained constant (thermostat) or monitored by a temperature sensor (eg, type K thermocouple) and taken into account. After long periods of operation at elevated temperature there can be a lack of insulation between the two legs of the thermocouple begins, and this can result in

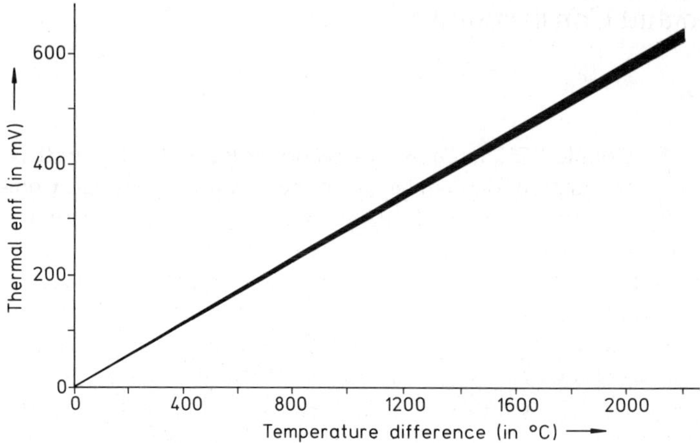

Figure 4-12. Thermal emf of a B_4 C-C thermocouple.

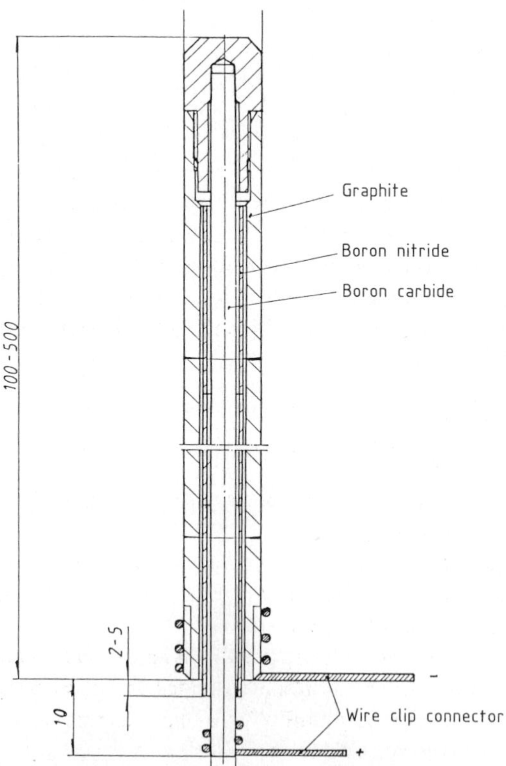

Figure 4-13.
Construction of a B_4 C-C thermocouple.

measurement errors. The fault can be cured by replacing the boron nitride insulating tube. At high temperatures oxidizing gases or nitrogen react with B_4C and C, which considerably reduces to working life of the elements.

4.3 Standardization and Construction

4.3.1 Standardization

IEC publication 584, "Thermocouples" [5], has now been incorporated into virtually all national standards and is a European standard for all members of the European Economic Community. It is intended to replace the DIN 43 710 and Ghost 3044 and 3045 after a period of transition. Only thermocouples according to IEC 584 should be incorporated into new plants.

IEC 584-1, "Thermocouples, Part 1: Reference Tables" [5], includes the basic data concerning types T, J, E, K, N, R, S, and B thermocouples.

IEC 584-2, "Thermocouples, Part 2: Tolerances" [5], provides tolerances for three accuracy grades. They apply to thermocouples and matched thermocouple wires as delivered.

IEC 584-3, "Thermocouples, Part 3: Tolerances for Compensating and Extension Cables; Identification System" [5], contains an identification system (color code) in addition to the data referred to in its title. A compilation of important IEC tables can be found in the Appendix at the end of this chapter (Tables 4-8 to 4-16).

4.3.2 Construction

Thermocouples are contact thermometers. Their measuring function must be in direct thermal contact with the medium whose temperature is being measured. Since it is almost always necessary to protect the thermocouple wires from external influences, such as oxidation or corrosion, the complete sensor unit affects the temperature distribution in the medium being monitored. In order to measure the temperature of the medium being monitored as accurately as possible, the sensor must be matched to the measurement requirements and the installation arrangements. There are many forms of construction, a few of which will be discussed here as examples.

4.3.2.1 Generally Applicable Designs

In the case of industrial measurements an important requirement is the interchangeability of sensors from different manufacturers. For this reason, national standards contain designs that usually differ slightly from the standards applicable in other countries. Figure 4-14 illustrates some designs according to DIN 43 733 that are applicable to firing, electric furnaces, and flue gas ducts.

A metal protective tube of heat-resistant or corrosion-resistant steel (design a) is adequate for thermocouples constructed of non-noble metals (Ni, NiCr, Fe, CuNi, etc.). It is necessary to protect noble metal wires from the uptake of foreign materials by means of a ceramic tube (Figure 4-14 b).

The materials employed here are aluminium silicate or aluminium oxide, depending on the temperature and medium being monitored. If rapid temperature variations (eg, flames) occur at the monitoring site then thermocouples of the design illustrated in Figure 4-14 c are employed. Here are porous, temperature-resistant ceramic outer tube reduces the thermal stress on the gas-tight internal tube.

In the production facilitites for the chemical industry and in power stations it is necessary to be able to use thermocouples with exchangeable sensors, otherwise it would not be possible to replace a thermocouple without shutting down. DIN 43 735 covers sensors and DIN 43 736 protective tubes for various purposes. Figure 4-15 illustrates a design for high pressures and high flow velocities that is frequently employed in power stations.

4.3.2.2 Sheathed Thermocouples

Sheathed thermocouples (Table 4-17) are versatile in application. They are constructed by embedding the thermocouple wires in an insulating powder (eg, MgO, Al_2O_3) and surround by a metal sheath. The insulation is so tightly packed that bending and even deformation do not affect the insulation. Sheathed thermocouple elements are available in the usual thermocouple types (types J, K, R, S) and with different sheath materials of external diameter from 0.25 to 10 mm.

Table 4-17. Commonly used sheathed thermocouples.

Thermocouple	Sheath
Fe-CuNi type J	Stainless steel (18/8 CrNi)
NiCr-Ni type K, N	NiCr 15 Fe (Inconel)
	NiCr 80/20
PtRh-Pt	Stainless steel (18/8 CrNi)
Type R and S	NiCr 15 Fe
	PtRh
WRe-WRe	Molybdenum
	Tungsten

The limits for the use of sheathed thermocouples depend on the thermocouple wires, the insulator, and on the sheath. The upper limit for the application of PtRh thermocouples of this type is 1200–1400 °C because the resistance of the insulators between the thermocouple wires and the sheath decreases greatly above these temperatures (see Section 4.4). The diameter of the thermocouple wire is 10–15% of that of the external diameter and the thickness of the sheath wall 15–20% of it.

Sheathed thermocouples are easily shaped and, thus, readily adapted to the required measuring arrangement. Their properties remain stable as long as the protective sheath retains its integrity and no impurities contaminate the thermocouple wires. Thermocouples with two sheaths are manufactured for use in special circumstances (Figure 4-16 c). The actual thermocouple junction can be insulated from the sheath or connected to it (Figure 4-16 c and d). A junction connected to the sheath has a shorter response time. An insulated junction offers to advantage that it is possible to determine the insulation resistance of the completed thermocouple. Low insulation resistances indicate a defective thermocouple.

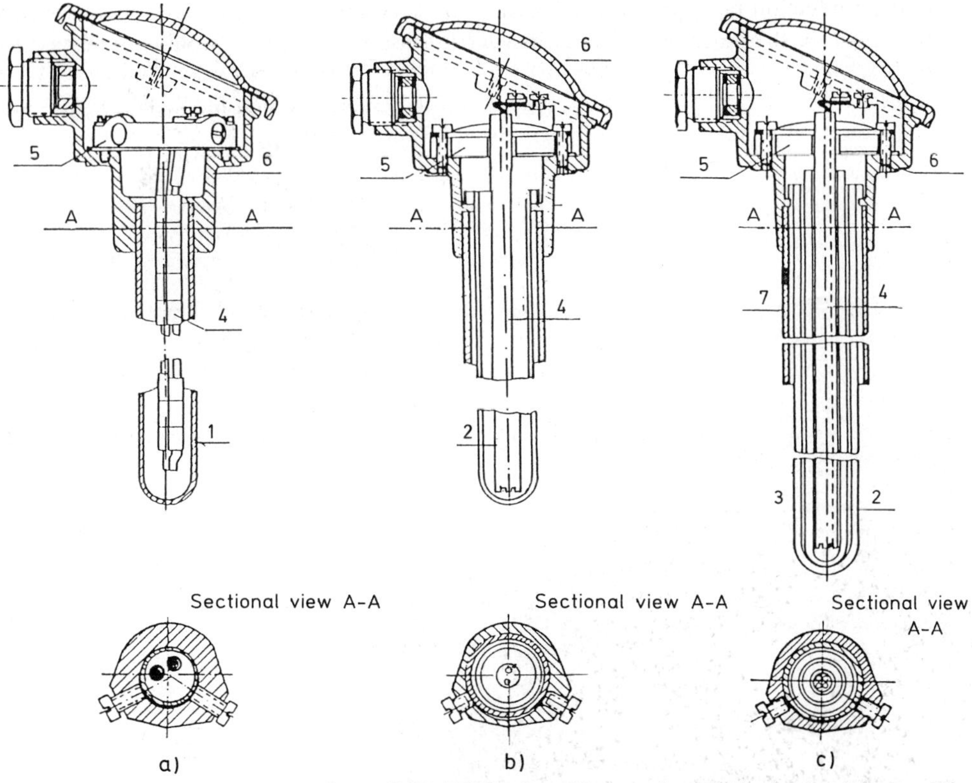

Figure 4-14. Thermocouples according to DIN 43733: a) metal protective tube for base metal wires (Ni, NiCr, Fe, CuNi, etc.); b) noble metal wires protected by a ceramic tube; c) configuration b) protected by a porous, temperature-resistant ceramic outer tube against rapid temperature variations. 1 metal protective tube; 2 ceramic protective tube; 3 ceramic protective tube (porous); 4 insulation; 5 clamp; 6 connecting head; 7 metal fixing tube.

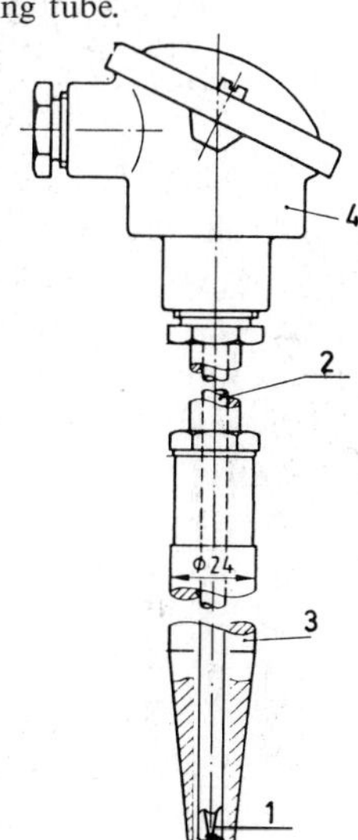

Figure 4-15.
Weld-in thermocouple sensor for superheated steam. 1: Thermocouple junction; 2: measurement insert; 3: protective tube; 4: connecting head.

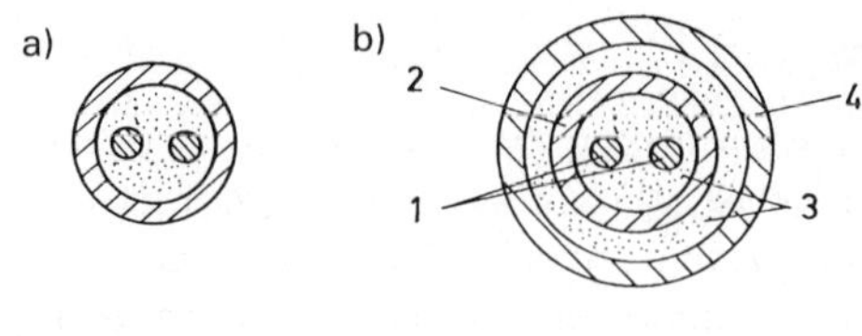

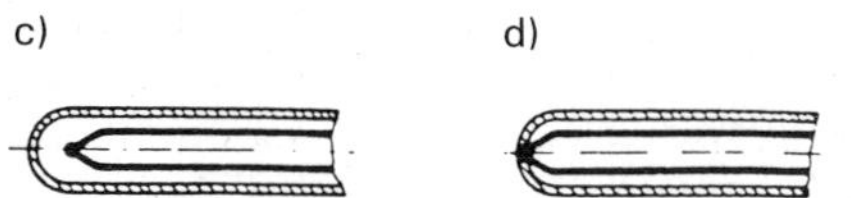

Figure 4-16.
Sheathed thermocouples. a) Cross section through a sheathed thermocouple; b) cross section through a doubly sheathed thermocouple; c) insulated junction; d) junction electrically connected to sheath. 1 Thermocouple wires; 2 insulating powder; 3 and 4 sheaths.

4.3.2.3 Designs for Specific Purposes

The multiplicity of thermocouple types in use is very large. The few examples here are intended to indicate how it is possible to adapt thermocouples to their particular task and environment.

Example 1: Thermocouples for salt hardening baths

Salt hardening baths are often heated directly, that is the heating current is passed through the conducting salt via electrodes. Since the heat is released directly in the salt, such baths heat up very rapidly. The thermocouple employed to control the temperature changes must, therefore, respond to these temperature changes as rapidly as possible. As illustrated in Figure 4-17 this is achieved by employing concentric thermocouple legs. The external iron tube serves as the positive leg of the Fe-CuNi thermocouple. The working life of the thermocouple largely depends on the impurities (eg, carbon) which penetrate the outer tube and can range from 50 to 2000 h.

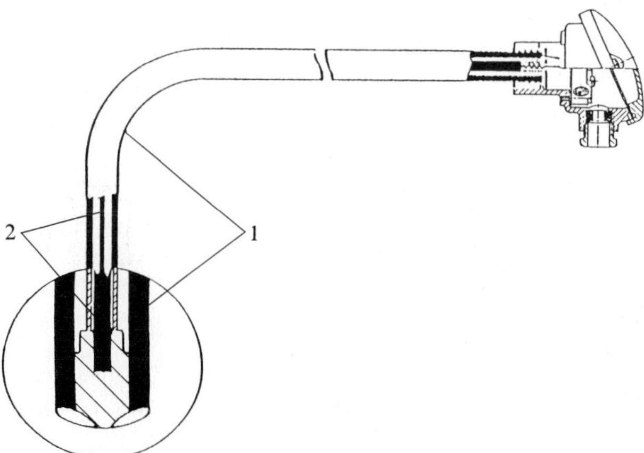

Figure 4-17.
Thermocouple unit for salt heating bath [5]. 1: Protective tube of pure iron (positive leg). 2: thermocouple wire of CuNi (negative leg).

Example 2: Thermocouples for steel smelting

Steel smelting is monitored by means of immersion thermocouple elements. The thermocouple of PtRh-Pt type S, with a wire diameter of 50–100 μm, is stretched through a U-shaped quartz tube and connected to a socket contact as illustrated in Figure 4-18. The socket and the plug body are protected by a multilayered cardboard tube. A steel cap protects the quartz tube with its thermocouple from being damaged on immersion. The measurement head is attached to a lance which is also protected by a multilayered cardboard tube. To make a measurement, the measuring head is thrust down through the slag layer into the liquid steel. The steel protective cap melts and the steel comes into contact with the quartz tube. The measurement is completed in the 8–10 s required for the thermocouple reaches the temperature of the liquid steel. A part of the multilayered cardboard tube burns away during this time. The cushion of gas that this produces prevents the socket contact from being heated exessively. The reproducibility is very good (1–2 °C at 1600 °C) since a new thermocouple is employed for each measurement.

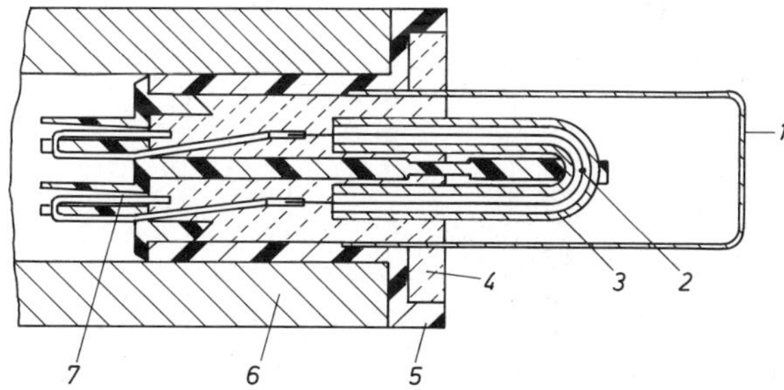

Figure 4-18.
Thermocouple for molten metal [6, 9]. 1: Protective cap; 2: thermocouple; 3: protective quartz tube; 4: thermally resistant ceramic putty; 5: plastic molding; 6: card tube; 7: plug contact [6].

Example 3: Thermocouples for superheated steam

In power stations it is necessary to measure the temperature of superheated steam travelling at high velocities (590 °C, 320 bar, 30 m s^{-1}). To insure that the control system operates properly, it is necessary to limit the delay of the thermocouple display to only 3–5 s. In addition the measurement head must be replaceable while the plant is in operation. Thermocouples according to Figure 4-15 respond too slowly to temperature changes. A modified sensor head and a thermally conducting gold disc at the tip of the sensor (Figure 4-19) provide the desired properties.

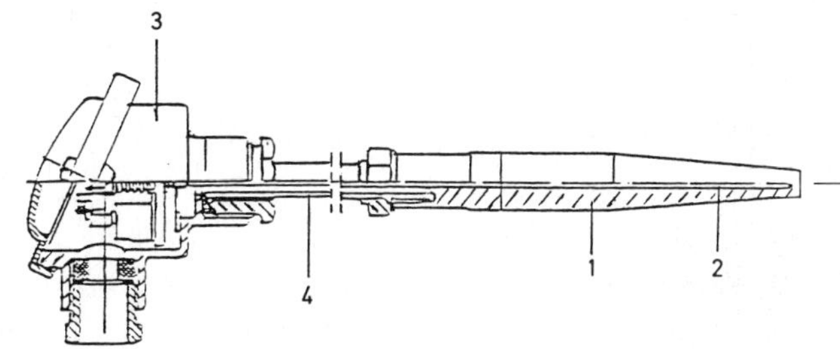

Figure 4-19.
Rapidly responding thermocouple with gold contact. 1: Protective tube; 2: sheathed thermocouple; 3: connecting head; 4: lagging extension.

Example 4: Thermocouples for steam turbine walls

When steam turbines are being run up large temperature gradients are set up in their walls causing thermal tension, which can lead to cracks in the walls if limiting values are exceeded. It is necessary, therefore, to monitor the temperature gradient. Thermocouples according to Figure 4-20 determine the temperature at a precisely defined depth and yield the desired data.

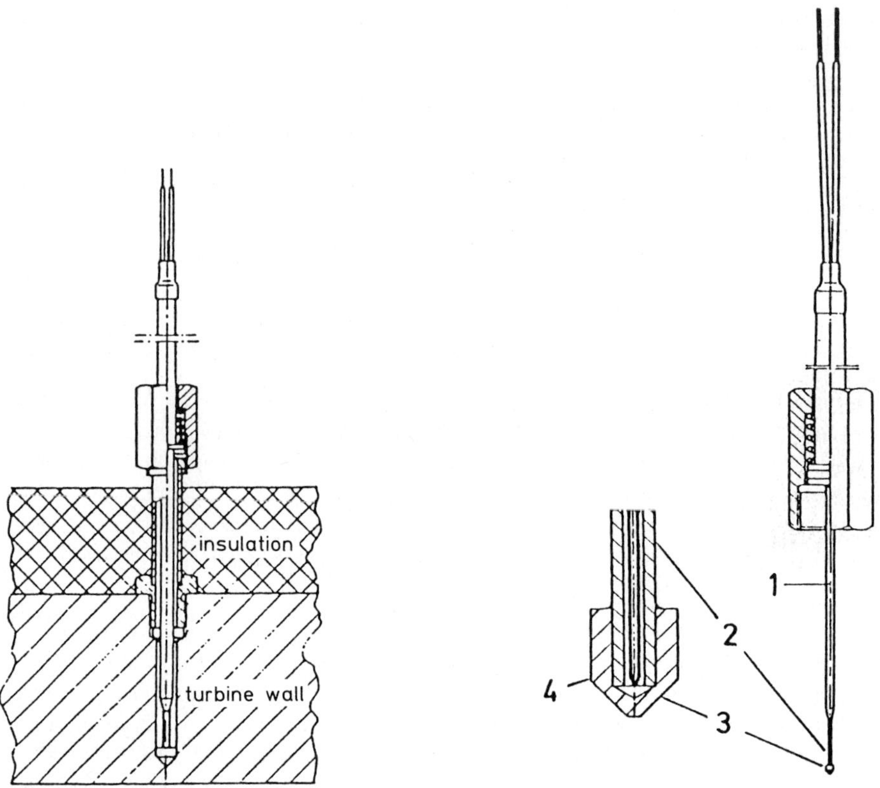

Figure 4-20. Thermocouple for turbine walls. 1: Sheathed thermocouple, 3 mm diameter; 2: reduction to 1.5 mm diameter; 3: sensing tip 3 mm diameter; 4: measuring junction of thermocouple.

Example 5: Thermocouples for gaseous media

In gaseous media there are particularly large errors as a result of conduction losses and, at high temperatures (>800 °C), of radiation losses. In order to increase the heat transfer to the sensor, the gas being measured is sucked away so that is passes the sensor at an enhanced velocity. Figure 4-21 illustrates a particular from the suction thermocouple. An additional heating wire (Pt) heats up the freely suspended thermocouple (PtRh-Pt) electrically and, thus, compensates for heat losses. If the heating current is adjusted to be too low, then the temperature indicated will increase if the suction velocity is increased. The heating current is

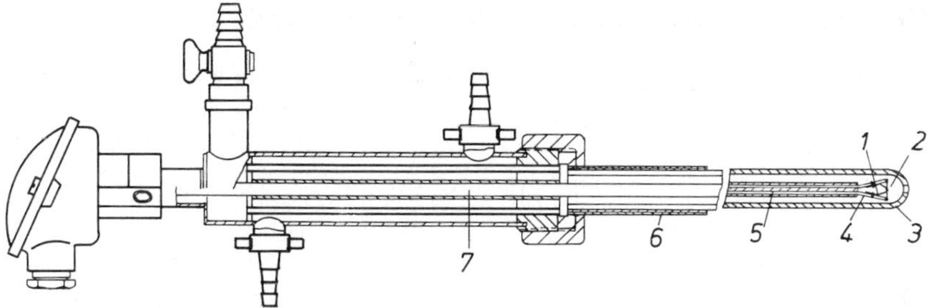

Figure 4-21. Suction thermocouple with supplementary heating [6, 9]. 1: Thermocouple (PtRh-Pt); 2: heating wire (Pt); 3: ceramic protective tube; 4: leads (PtRh and Pt for thermocouple, Pt for heating); 5: 4-bore ceramic insulator; 6: support tube; 7: water cooling [6].

correctly adjusted when the indicated temperature remains constant on alteration of the suction rate.

Example 6: Gaseous media of varying temperature

If the temperature of a gaseous medium changes rapidly, then a measurement error can occur if the thermocouple cannot follow the temperature change rapidly enough. The mass of the sensor must, therefore, be reduced as much as possible. In the case of the sensor illustrated in Figure 4-22, three thermocouples made from wires of different diameters (0.1, 0.06 and 0.03 mm) and, hence, of different thermal intertias, are exposed to the gas stream. The remaining measurement error can be estimated by extrapolation to zero wire diameter.

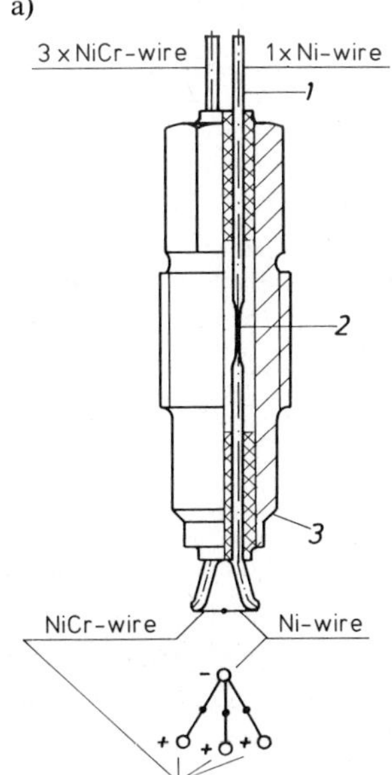

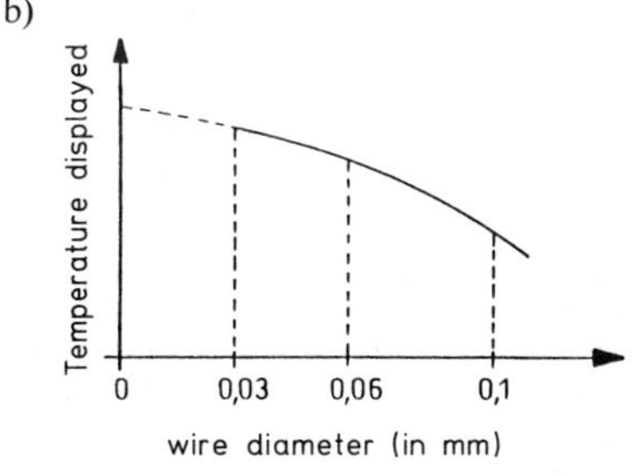

Figure 4-22.
Thermocouple for rapidly changing gas temperatures [6, 9]. 1: Leads (thermocouple wire); 2: pressure-tight insulator; 3. seal [6].

4.4 Factors Affecting the Thermal Electromotive Force

4.4.1 Reference Temperature

Since thermocouples are only able to determine temperature differences the temperature of the reference junction affects the result of the measurement. All tables of thermal emfs are constructed for a reference temperature of 0 °C, so that if the reference temperature deviates from 0 °C it is necessary to add or subtract the thermal emf associated with the reference temperature (see Figure 4-23 a).

The influence of the reference temperature (t_r) depends on the ratio (C) of the gradient of the thermal emf curve at the measurement junction (t) and at the reference junction (t_r):

$$t = t_m + C t_r \tag{4-8}$$

where t_m = temperature displayed (display calibrated to the reference temperature 0 °C).

C values for some thermocouples in general use are reprinted in Figure 4-23 b.

Three methods are employed in industrial plants to account for the true temperature of the reference junction:

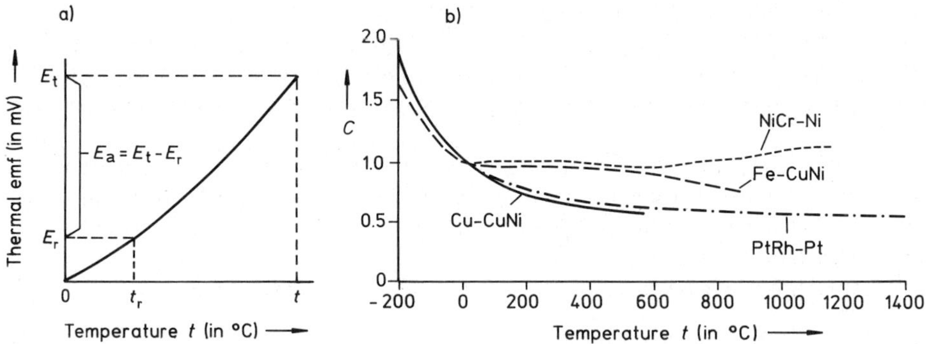

Figure 4-23. Influence of the reference temperature. a) Thermal emf curve; t measurement temperature; t_r reference temperature; E_t thermal emf for measurement temperature with reference temperature 0 °C; E_r thermal emf at the reference temperature; E_a displayed thermal emf b) ratio of gradients of the thermal emf curve at the measurement junction and the reference junction (c) for some generally used thermocouples.

1. Thermostat

The electrically insulated junction is attached to a heated metal block which is maintained at, say, 50 °C (±0.2 °C). The monitoring apparatus (controller, recorder, etc.) is then calibrated for this reference temperature.

2. Reference temperature correction

A temperature-dependent resistor in the thermocouple circuit is employed to create the missing thermal emf. An additional current supply is necessary for this purpose (Figure 4-24).

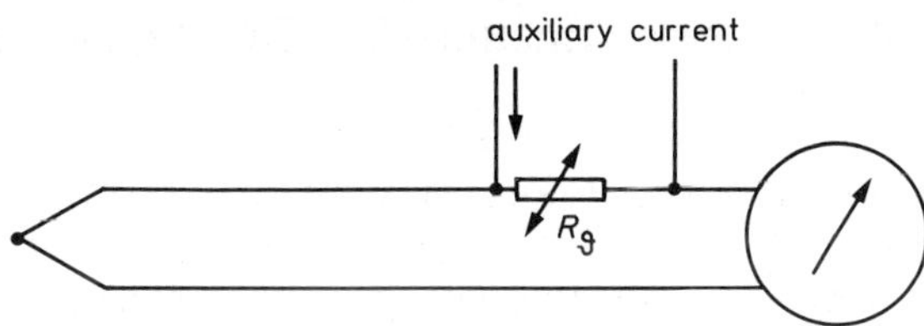

Figure 4-24. Reference temperature compensation by a temperature-dependent resistance R_θ. The supplementary power supply is already built into many electronic display instruments. Such instruments are calibrated for a reference temperature of 0 °C.

3. Computer correction of the reference temperature

Modern process-control units with computers often correct the junction temperature (transition from extension lead to copper conductor) with seperate temperature sensors (eg, Pt 100 resistance thermometers) and then the temperature of the measuring junction (t) is calculated by the computer according to Equation 4-8.

When nonstandard thermocouples are employed, there is often no suitable extension lead for connection to a reference point with constant temperature. Here it is possible to use a circuit as in Figure 4-25, applying a standard extension or compensating lead.

Extension leads are two or more core leads made up of the thermocouple material. Their insulation consists of thermoplastic or heat-resistant plastic. Hygroscopic insulators such as glass wool, rock wool, silk, or cotton should only be employed in exceptional cases (see Section 4.4.3).

Compensating leads are two or more core leads of materials which have the same thermal emf as their associated thermocouples up to 200 °C (or 100 °C). Such leads are primarily employed for extending noble metal, WRe and NiCr thermocouples. The connection between the positive leg and the compensation lead *must* be at the same temperature as the connection to the negative leg otherwise there will be considerable measurement errors.

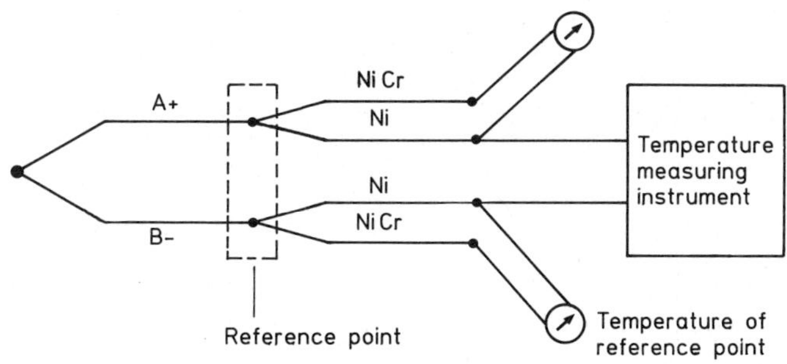

Figure 4-25. Determination of the reference temperature.

4.4.2 Aging

The term aging covers all the changes in the thermal emf of a thermocouple which take place while it is in service. These are caused by oxidation, recrystallization, uptake of impurities, etc. It is only possible to make general comments, since even small changes in the environment of the thermocouple wires (atmosphere, insulation components, etc.) can affect the aging.

Cu-CuNi and Fe-CuNi exhibit only slight aging since they are employed at low temperatures (Cu up to 250 °C, Fe up to 400 °C). Fe readily takes up impurities above 400 °C (see Figure 4-26). CuNi ages by becoming more positive, that is, the negative thermal emf of CuNi with respect to Pt becomes smaller.

The deviation of a thermocouple is given by the deviation of its positive leg minus that of its negative leg. The deviation of the Fe-CuNi thermocouple is higher than the deviations of its individual legs. The initial rapid aging is followed by a period where the rate of change is less steep.

NiCr-NiAl ages as a result of the different oxidation rates of the components of the alloy. Chromium oxidizes more readily than nickel. This selective oxidation leads to a loss of chromium and, hence, to a lower thermal emf. If the atmosphere surrounding the thermocouple wires contains only a small amount of oxygen (eg, as a result of insertion into a narrow protective tube), then chromium is oxidized appreciably more rapidly than nickel. The chromium oxide coats the NiCr wire as a porous green film. This phenomenon is, therefore, often described as "green rot". If sufficient oxygen is present, a dark grey to black oxide layer is formed; this adheres well and provides protection from further attack by oxygen.

The oxide skin must be removed when the thermocouple junction is being made (eg, by welding). The oxide coating is then reformed by annealing in air. NiCr and Ni thermocouple wires are mainly delivered in the preoxidized state.

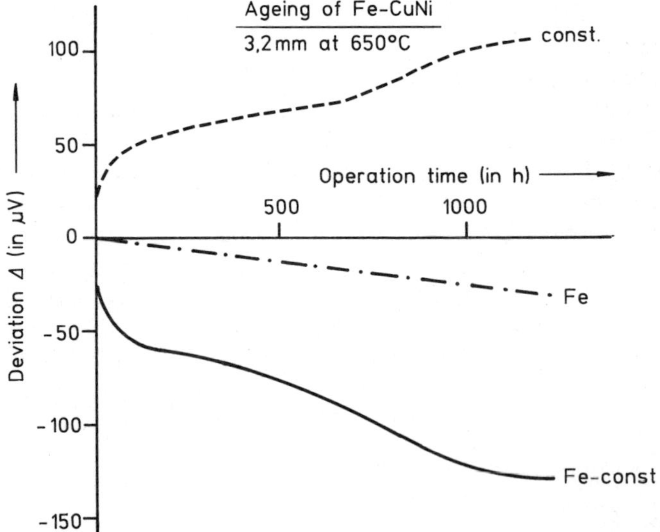

Figure 4-26. Aging of Fe-CuNi thermocouples (3.2 mm diameter) in air at 650 °C.

Sulfur-containing gases (eg, flue gases) cause a great deal of damage to the nickel leg. Inclusions of sulfur at the grain boundaries yield a coarse, friable structure, and the thermal emf is also altered.

PtRh-Pt (types R and S): Thermocouples of this type have high resistance to oxygen. There is, therefore, very little aging in air at temperatures below 1400 °C and its effects only have to be taken into account at temperatures over 1500 °C. In fact, at these temperatures volatile rhodium oxide is produced, leading to a reduction in the rhodium content in the alloy leg. Fine-grained platinum is empoyed to improve mechanical stability at high temperatures. Special measures taken during manufacture ensure that the grain growth is delayed at high temperatures. Platinum poisons such as silicon and phosphorus have a large effect on the thermal emf. Silicon especially is a frequent cause of PtRh thermocouples becoming unusable. Silicon alloys with platinum and forms a coarse, friable structure. Since platinum acts as a catalyst for the reduction of metal oxides, the source of the penetrating Si is usually SiO_2, which is a constituent of the insulating components (ceramic tubes, etc.). In an oxidizing atmosphere and below 1000 °C there is little reduction of SiO_2 so that little Si is taken up. At high temperatures and in reducing atmospheres, however, the damage to platinum commences within a few hours. The oxides of other metals are also reduced and contaminate the thermocouple. Aluminium oxide is much less reduced on account of its high bonding energy. The insulating components of PtRh thermocouples should, therefore, be constructed from very pure Al_2O_3 (>99.5%).

PtRh alloys are less affected by the uptake of foreign substances (eg, Si) than pure platinum. If it is not possible to avoid operating a thermocouple in a neutral or reducing atmosphere at least some of the time, a type B thermocouple should be installed since both legs of these are constructed of PtRh alloys.

Hydrogen penetrates metals at temperatures above 800 °C. Ceramic tubes are often not absolutely impermeable to H_2. Flushing the protective tube with air or oxygen can often be advantageous.

WRe and Mo thermocouple wires are not oxygen-resistant and, therefore, must always be operated in neutral or reducing atmospheres. On contact with carbon at high temperatures they form carbides, which lead to embrittlement and alteration of the thermal emf. The only insulating material compatible with them for operation above 1800 °C is beryllium oxide (BeO).

Statements concerning the aging of thermocouples apply only to the precise operating conditions described. Even small changes in the atmosphere to which the thermocouple wires are exposed can change their behavior decisively. This means that the aging behavior quoted in literature can only rarely be applied directly to other conditions. During practical operation under constant operating conditions, the thermal emfs of thermocouples often change according to an exponential function. A straight line is often obtained on plotting the change with time on a log-log scale (Figure 4-27). Under these conditions it becomes possible, after a short period of operation, to calculate when the deviation caused by aging will reach the maximum allowable deviation.

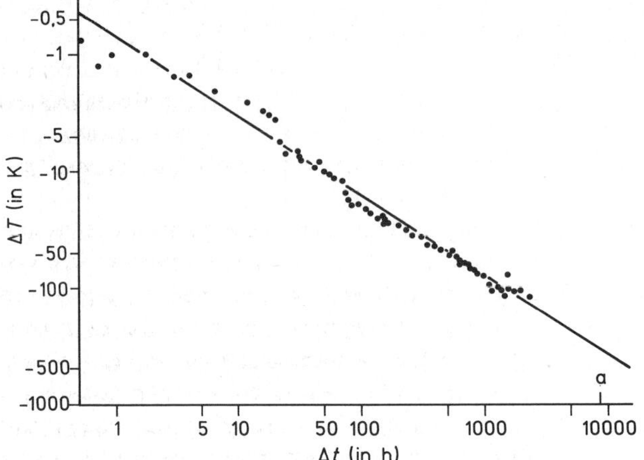

Figure 4-27. Aging of NiCr-Ni (0.8 mm diameter) in a weakly carbonizing atmosphere (95% CO_2 – 5% CO) [7].

4.4.3 Humiditiy and Electrochemical Potentials

The moisture in hygroscopic insulating materials (glass wool, rock wool, asbestos, etc.) and deposits of dust on the thermocouple wires can lead to the formation of electrochemical cells which distort the thermoelectric potential. The error is dependent on the thermocouple, the electrolytes produced, and the distance from the site of interference and the sensor junction, as illustrated in Figure 4-28.

As can be seen from Figure 4-28, a voltage divider is created from R_i and R_l. Since R_i is usually much greater than R_l, only a small portion of the electrochemical potential comes into effect. Electrochemical potentials are generally in the range 0.01–0.5 V. Even with $R_i/R_l \approx 0.1\%$ this effect results in a significant error. For this reason thermocouple elements should only be insulated with moisture-resistant materials.

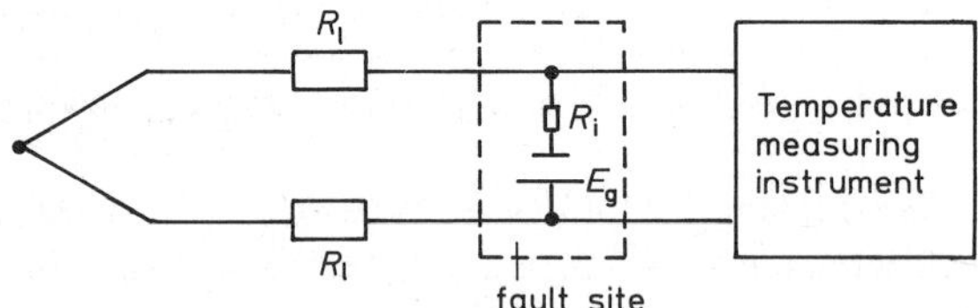

Figure 4-28. Influence of moisture. R_l: conduction resistance of the thermocouple wires between measurement site and site of interference; R_i: internal resistance of the thermoelectric element produced (dependent on electrolytes and on separation of thermocouple wires); E_G: electrochemical interference potential (dependent on composition of thermocouple wires and on nature of electrolytes).

4.4.4 Imperfect Insulation

Imperfect insulation of thermocouples leads to secondary connections and also frequently to secondary measurement sites if the insulation defect is at a region of higher temperatures. Figure 4-29 illustrates the effects of insulation faults.

a)

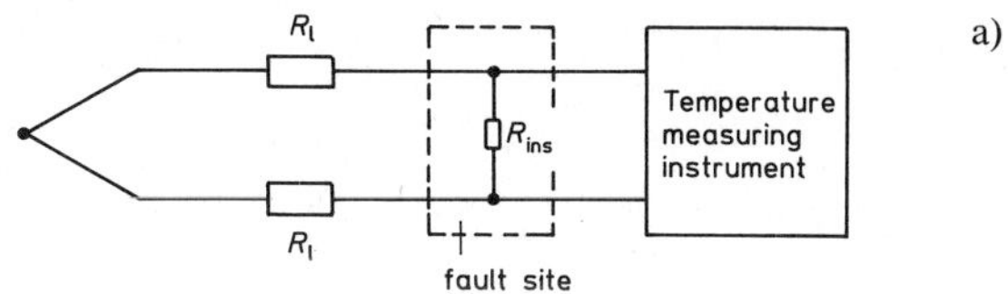

b)

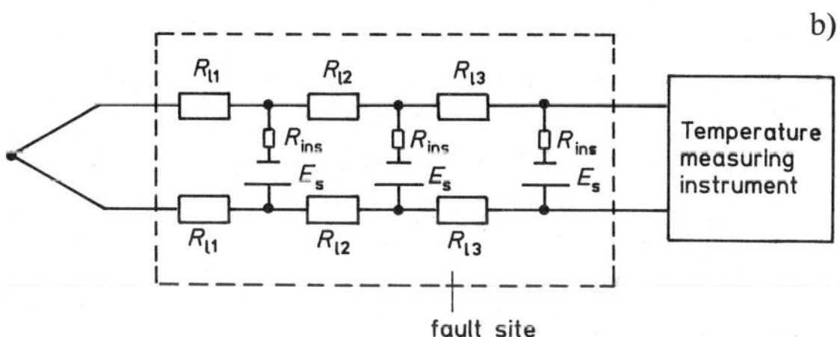

Figure 4-29. Effect of insulation faults. a) Insulation fault at some position in the thermocouple leads; b) insulation fault and secondary thermal emf where part of a lead at elevated temperature suffers insulation fault. R_l: Resistance of the thermocouple lead; R_{ins}: resistance of insulation at fault; E_s: secondary thermal emf; T_1: temperature of measurement junction; T_2: temperature at fault (eg, a part of lead). The situation illustrated in (a) causes the display of a result which is too low because of the potential division R_l/R_{ins}; that in (b) can yield a temperature display higher than T_1 at the temperature sensor if the fault occurs where the temperature T_2 is higher.

4.4.5 Inhomogeneities

Inhomogeneities are produced in the thermocouple legs by cold working (stretching, bending etc.) but also by differing thermal stresses and by the effects of foreign substances on parts of the thermocouple legs (see also Section 4.4.2).

Such inhomogneities only affect the measurement results if they lie in a region of inhomogeneous temperature. No measurement errors are caused in regions of constant temperature (see Section 4.1.2.2).

A check on homogeneity should always be carried out when checking thermocouples, since normally only a length of 30–40 cm can be inserted into the test furnace. The inhomogeneities produced by cold forming can be eliminated by annealing (> 800 °C for Pt and PrRh, and at 700 °C for NiCr and Ni). An annealing time of a few minutes is generally sufficient. Figure 4–30 illustrates a device for testing for inhomogeneity and Figure 4–31 reproduces the record of the inhomogeneities detected by the device in various sheathed thermocouples.

4.4.6 Electric and Magnetic Fields

Electrical fields can be eliminated in a simple manner by using a screen (eg, protective tube, braiding) connected to the ground of the measuring instrument. Strong magnetic fields affect ferromagnetic materials. Screening is usually impossible. The only possibility here is the use of nonferrous materials (Cu, CuNi, Pt etc.).

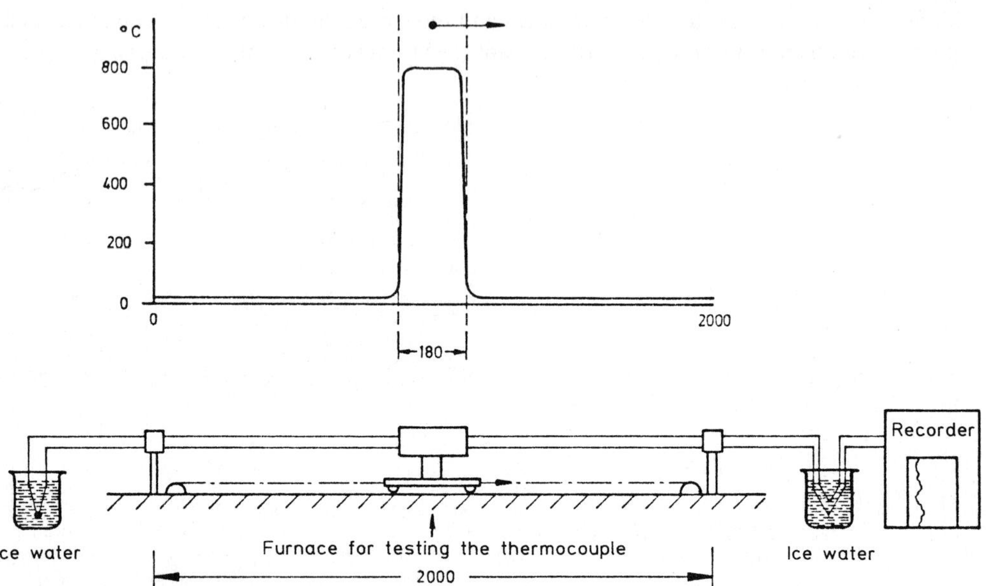

Figure 4-30. Apparatus for determining the inhomogeneity of thermocouple wires [9].

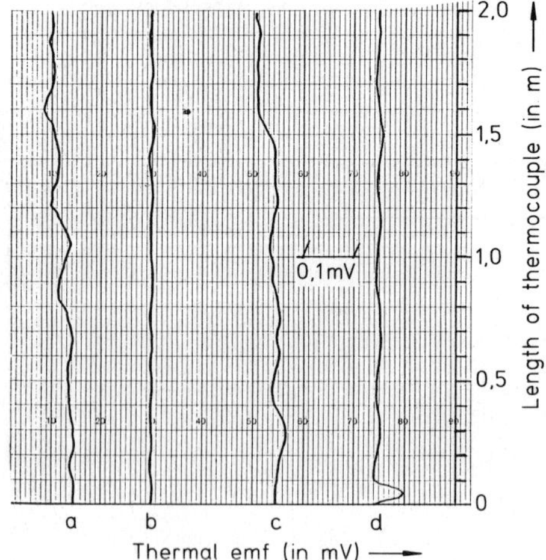

Figure 4-31.
Inhomogeneity of sheathed thermocouples. (a) Fe-CuNi as delivered; (b) Fe-CuNi after annealing at 700 °C; (c) NiCr-Ni as delivered; (d) NiCr-Ni after annealing at 700 °C.

Alternating magnetic fields produce an induction potential in the thermocouple loop. This is dependent on the field strength and the area of the loop that is exposed to the magnetic field. The smaller the area, the smaller is the potential induced. The alternating potential that is induced has no effect on pure DC meters such as moving coil instruments. Semiconductors are employed in many instruments (eg, in the input circuits or to guard against the effects of element breakage). These rectify a part of the induced alternating current which leads to an erroneous display. Twisting the leads causes a great reduction in the induced potential. The higher the frequency of the alternating field and the more inhomogeneous the magnetic field the more tightly it is necessary to twist the leads (1–5 twists per cm). AC pickup can be reduced by a low-pass filter (Figure 4-32).

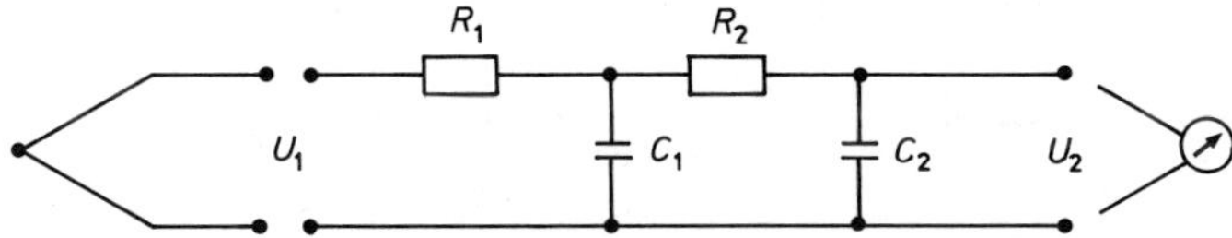

Figure 4-32. RC interference filter.

For a filter consisting of R_1 and C_1 the inhibition of the AC potential (Figure 4-32) is given by

$$U_1/U_2 \approx \omega RC \tag{4-9}$$

and in the case of the L_1C_1 filter (Figure 4-33) by

$$U_1/U_2 \approx \omega^2 LC \tag{4-10}$$

The DC component is only affected by the ratio R_1/R_i or $R_\mathrm{L}/R_\mathrm{i}$, where R_i is the internal resistance of the measurement instrument R_L is the DC resistance of the inductivity. The inhibition is multiplied when several filters are combined in series.

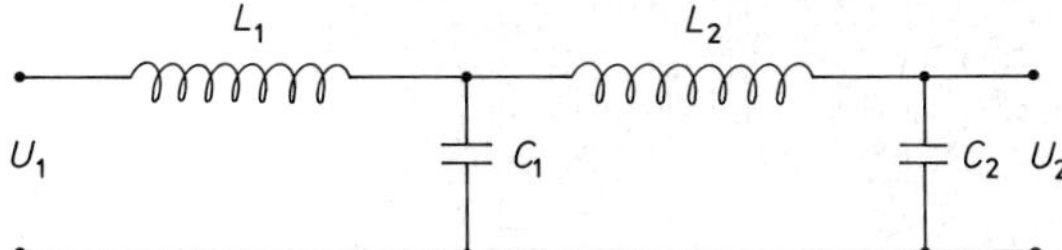

Figure 4-33. LC interference filter.

4.4.7 Alpha, Beta, and Gamma Radiation and Neutrons

Alpha and beta radiation have almost no effect on the determination of temperature by thermocouples. Gamma radiation causes heating, which depends on the intensity of the radiation and the volume irradiated. The thermocouple is not permanently affected.

Neutrons can cause changes in the thermocouple material; these depend on the intensity and type (slow or fast neutrons) of the radiation, time of exposure, and the nature of the material (neutron capture cross section). An overview is provided in Tables 4-18 and 4-19.

Table 4-18. Influence of thermal neutrons (10^{14} n cm^{-2}) on thermocouples (BICC 2852).

Material	Composition (%)										
	Al	Si	Mn	V	Cr	Fe	Co	Ni	Cu	Zr	Zn
Chromel	–	0.36	–	–	9.58	–	–	89,87	–	–	–
(after 10 years)	–	0.36	–	0.14	9.43	–	–	89,76	0.10	–	–
Alumel	1.6	0.74	3.35	–	–	0.02	0.33	95.20	0.04	–	–
(after 10 years)	1.59	0.75	2.21	–	–	1.16	0.22	95.19	0.14	–	–
Iron	–	–	–	–	–	100	–	–	–	–	–
(after 10 years)	–	–	0.22	–	–	99.77	0.01	–	–	–	–
Copper	–	–	–	–	–	–	–	–	100	–	–
(after 10 years)	–	–	–	–	–	–	–	5.04	89.92	–	5.02
CuNi	–	0.05	0.96	–	–	0.20	–	45.01	55.10	–	–
(after 10 years)	–	0.05	0.63	–	–	0.53	–	47.73	49.59	2.77	–

Table 4-19. Influence of thermal neutrons (10^{14} n cm^{-2}) on thermocouples (BICC 2852).

Material	Composition (%)							
	Pt	Rh	Pd	Au	Hg	W	Re	Os
Pt	100	–	–	–	–	–	–	–
(after 10 years)	98.15	–	–	0.32	0.92	–	–	–
PtRh	90	10	–	–	–	–	–	–
(after 10 years)	88,33	0.12	9.88	0.29	0.83	–	–	–
Tungsten	–	–	–	–	–	100	–	–
(after 10 years)	–	–	–	–	–	81,02	6.51	12.37
W Re/26%	–	–	–	–	–	74	26	–
(after 10 years)	–	–	–	–	–	59.95	7.12	32.93

4.5 References

[1] *The Theory and Properties of Thermocouple Elements.* ASTM Special Technical Publication 492; Philadelphia: American Society for Testing and Materials, 1971.

[2] Blatt, F. G., Schroeder, P. H., Foiles, C. W., Greig, D., *Thermoelectric Power of Metals:* New York: Plenum Press, 1976.

[3] Mott, N. F., Jones, H., The Theory and the Properties of Metals and Alloys, New York: Dover, 1936.

[4] *Manual of the Use of Thermocouples in Temperature Measurement,* ASTM Special Technical Publication 470; Philadelphia: American Society for Testing and Materials, 1970.

[5] *IEC Publication 584-1 to 584-3,* Bureau Central de la Commission Electrotechnique International, 3. Rue Varambe, Geneve, Suisse.

[6] Lieneweg, F., *Handbuch der Technischen Temperaturmessung;* Braunschweig: Vieweg, 1976.

[7] Von Körtveleyssy, L., *Thermoelement Praxis;* Essen: Vulkan, 1987.

[8] *Measurement and Control, Electrical Sensors, Reference Table Type U and Type L for Thermocouples, DIN 43 710;* Berlin: Beuth, Dec. 1985.

[9] Weichert, L., *Temperaturmessung in der Technik, Grundlagen u. Praxis;* Greifenau: Expert Verlag, 3. ed. 1981.

4.6 Appendix IEC Tables

Table 4-8: Thermal emfs (mV) of Cu-CuNi type T according to IEC 584-1 [4].

°C	0	−10	−20	−30	−40	−50	−60	−70	−80	−90
−200	−5.603	−5.753	−5.889	−6.007	−6.105	−6.181	−6.232	−6.258		
−100	−3.378	−3.656	−3.923	−4.177	−4.419	−4.648	−4.865	−5.069	−5.261	5.439
0	0	−0.383	−0.757	−1.121	−1.475	−1.819	−2.152	−2.475	−2.788	−3.089
°C	0	10	20	30	40	50	60	70	80	90
0	0	0.391	0.789	1.196	1.611	2.035	2.467	2.908	3.357	3.813
100	4.277	4.749	5.227	5.712	6.204	6.702	7.207	7.718	8.235	8.757
200	9.286	9.820	10.360	10.905	11.456	12.011	12.572	13.137	13.707	14.281
300	14.860	15.443	16.030	16.621	17.217	17.816	18.420	19.027	19.638	20.252
400	20.869									

The Copper/Copper-Nickel reference tables were developed from the following polynomials:

temperature range −270 °C to 0 °C

$$E = \sum_{i=0}^{14} a_i t_{68}^i \ \mu\text{V}$$

$a_0 = 0$
$a_1 = 3.8740773840 \times 10^{+1}$
$a_2 = 4.4123932482 \times 10^{-2}$
$a_3 = 1.1405238498 \times 10^{-4}$
$a_4 = 1.9974406568 \times 10^{-5}$
$a_5 = 9.0445401187 \times 10^{-7}$
$a_6 = 2.2766018504 \times 10^{-8}$
$a_7 = 3.6247409380 \times 10^{-10}$
$a_8 = 3.8648924201 \times 10^{-12}$
$a_9 = 2.8298678519 \times 10^{-14}$
$a_{10} = 1.4281383349 \times 10^{-16}$
$a_{11} = 4.8833254364 \times 10^{-19}$
$a_{12} = 1.0803474683 \times 10^{-21}$
$a_{13} = 1.3949291026 \times 10^{-24}$
$a_{14} = 7.9795893156 \times 10^{-28}$

temperature range 0 °C to 400 °C

$$E = \sum_{i=0}^{8} b_i t_{68}^i \ \mu\text{V}$$

$b_0 = 0$
$b_1 = 3.8740773840 \times 10^{+1}$
$b_2 = 3.3190198092 \times 10^{-2}$
$b_3 = 2.0714183645 \times 10^{-4}$
$b_4 = -2.1945834823 \times 10^{-6}$
$b_5 = 1.1031900550 \times 10^{-8}$
$b_6 = -3.0927581898 \times 10^{-11}$
$b_7 = 4.5653337165 \times 10^{-14}$
$b_8 = -2.7616878040 \times 10^{-17}$

Table 4-9: Thermal emfs of Fe-CuNi type J according to IEC 584-1 [4].

°C	0	−10	−20	−30	−40	−50	−60	−70	−80	−90
−200	−7.890	−8.096								
−100	−4.632	−5.036	−5.426	−5.801	−6.159	−6.499	−6.821	−7.122	−7.402	−7.659
0	0	−0.501	−0.995	−1.481	−1.960	−2.431	−2.892	−3.344	−3.785	−4.215
°C	0	10	20	30	40	50	60	70	80	90
0	0	0.507	1.019	1.536	2.058	2.585	3.115	3.649	4.186	4.725
100	5.268	5.812	6.359	6.907	7.457	8.008	8.560	9.113	9.667	10.222
200	10.777	11.332	11.887	12.442	12.998	13.553	14.108	14.663	15.217	15.771
300	16.325	16.879	17.432	17.984	18.537	19.089	19.640	20.192	20.743	21.295
400	21.846	22.397	22.949	23.501	24.054	24.607	25.161	25.716	26.272	26.829
500	27.388	27.949	28.511	29.075	29.642	30.210	30.782	31.356	31.933	32.513
600	33.096	33.683	34.273	34.867	35.464	36.066	36.671	37.280	37.893	38.510
700	39.130	39.754	40.382	41.013	41.647	42.283	42.922	43.563	44.207	44.852
800	45.498	46.144	46.790	47.434	48.076	48.716	49.354	49.989	50.621	51.249
900	51.875	52.496	53.115	53.729	54.341	54.948	55.553	56.155	56.753	57.349
1000	57,942	58.533	59.121	59.708	60.293	60.876	61.459	62.039	62.619	63.199
1100	63.777	64.355	64.933	65.510	66.087	66.664	67.240	67.815	68.390	68.964
1200	69.536									

The Iron/Copper-Nickel reference tables were developed from the following polynomials:

temperature range −210 °C to 760 °C

$$E = \sum_{i=0}^{7} a_i t_{68}^i \ \mu V$$

$a_0 = 0$
$a_1 = 5.0372753027 \times 10^{+1}$
$a_2 = 3.0425491284 \times 10^{-2}$
$a_3 = -8.5669750464 \times 10^{-5}$
$a_4 = 1.3348825735 \times 10^{-7}$
$a_5 = -1.7022405966 \times 10^{-10}$
$a_6 = 1.9416091001 \times 10^{-13}$
$a_7 = -9.6391844859 \times 10^{-17}$

temperature range 760 °C to 1200 °C

$$E = \sum_{i=0}^{5} b_i t_{68}^i \ \mu V$$

$b_0 = 2.9721751778 \times 10^{+5}$
$b_1 = -1.5059632873 \times 10^{+3}$
$b_2 = 3.2051064215$
$b_3 = -3.2210174230 \times 10^{-3}$
$b_4 = 1.5949968788 \times 10^{-6}$
$b_5 = -3.1239801752 \times 10^{-10}$

Table 4-10: Thermal emfs (mV) of NiCr-CuNi type E according to IEC 584-1 [4].

°C	0	−10	−20	−30	−40	−50	−60	−70	−80	−90
−200	−8.824	−9.063	−9.274	−9.455	−9.604	−9.719	−9.797	−9.835		
−100	−5.237	−5.680	−6.107	−6.516	−6.907	−7.279	−7.631	−7.963	−8.273	−8.561
0	0	−0.581	−1.151	−1.709	−2.254	−2.787	−3.306	−3.811	−4.301	−4.777
°C	0	10	20	30	40	50	60	70	80	90
0	0	0.591	1.192	1.801	2.419	3.047	3.683	4.329	4.983	5.646
100	6.317	6.996	7.683	8.377	9.078	9.787	10.501	11.222	11.949	12.681
200	13.419	14.161	14.909	15.661	16.417	17.178	17.942	18.710	19.481	20.256
300	21.033	21.814	22.597	23.383	24.171	24.961	25.754	26.549	27.345	28.143
400	28.943	29.744	30.546	31.350	32.155	32.960	33.767	34.547	35.382	36.190
500	36.999	37.808	38.617	39.426	40.236	41.045	41.853	42.662	43.470	44.278
600	45.085	45.891	46.697	47.502	48.306	49.109	49.911	50.713	51.513	52.312
700	53.110	53.907	54.703	55.498	56.291	57.083	57.873	58.663	59.451	60.237
800	61.022	61.806	62.588	63.368	64.147	64.924	65.700	66.473	67.245	68.015
900	68.783	69.549	70.313	71.075	71.835	72.593	73.350	74.104	74.857	75.608
1000	76.358									

The Nickel-Chromium/Copper-Nickel reference tables were developed from the following polynomials:

temperature range −270 °C to 0 °C

$$E = \sum_{i=0}^{13} a_i t_{68}^i \ \mu V$$

$a_0 = 0$
$a_1 = 5.8695857799 \times 10^{+1}$
$a_2 = 5.1667517705 \times 10^{-2}$
$a_3 = -4.4652683347 \times 10^{-4}$
$a_4 = -1.7346270905 \times 10^{-5}$
$a_5 = -4.8719368427 \times 10^{-7}$
$a_6 = -8.8896550447 \times 10^{-9}$
$a_7 = -1.0930767375 \times 10^{-10}$
$a_8 = -9.1784535039 \times 10^{-13}$
$a_9 = -5.2575158521 \times 10^{-15}$
$a_{10} = -2.0169601996 \times 10^{-17}$
$a_{11} = -4.9502138782 \times 10^{-20}$
$a_{12} = -7.0177980633 \times 10^{-23}$
$a_{13} = -4.3671808488 \times 10^{-26}$

temperature range 0 °C to 1000 °C

$$E = \sum_{i=0}^{9} b_i t_{68}^i \ \mu V$$

$b_0 = 0$
$b_1 = 5.8695857799 \times 10^{+1}$
$b_2 = 4.3110945462 \times 10^{-2}$
$b_3 = 5.7220358202 \times 10^{-5}$
$b_4 = -5.4020668085 \times 10^{-7}$
$b_5 = 1.5425922111 \times 10^{-9}$
$b_6 = -2.4850089136 \times 10^{-12}$
$b_7 = 2.3389721459 \times 10^{-15}$
$b_8 = -1.1946296815 \times 10^{-18}$
$b_9 = 2.5561127497 \times 10^{-22}$

Table 4-11: Thermal emfs (mV) of NiCr-Ni type K according to IEC 584-1 [4].

°C	0	−10	−20	−30	−40	−50	−60	−70	−80	−90
−200	−5.891	−6.035	−6.158	−6.262	−6.344	−6.404	−6.441	−6.458		
−100	−3.553	−3.852	−4.138	−4.410	−4.669	−4.912	−5.141	−5.354	−5.550	−5.730
0	0	−0.392	−0.777	−1.156	−1.527	−1.889	−2.243	−2.586	−2.920	−3.242
°C	0	10	20	30	40	50	60	70	80	90
0	0	0.397	0.798	1.203	1.611	2.022	2.436	2.850	3.266	3.681
100	4.095	4.508	4.919	5.327	5.733	6.137	6.539	6.939	7.338	7.737
200	8.137	8.537	8.938	9.341	9.745	10.151	10.560	10.969	11.381	11.793
300	12.207	12.623	13.039	13.456	13.874	14.292	14.712	15.132	15.552	15.974
400	16.395	16.818	17.241	17.664	18.088	18.513	18.938	19.363	19.788	20.214
500	20.640	21.066	21.493	21.919	22.346	22.772	23.198	23.624	24.050	24.476
600	24.902	25.327	25.751	26.176	26.599	27.022	27.445	27.867	28.288	28.709
700	29.128	29.547	29.965	30.383	30.799	31.214	31.629	32.042	32.455	32.866
800	33.277	33.686	34.095	34.502	34.909	35.314	35.718	36.121	36.524	36.925
900	37.325	37.724	38.122	38.519	38.915	39.310	39.703	40.096	40.488	40.879
1000	41.269	41.657	42.045	42.432	42.817	43.202	43.585	43.968	44.349	44.729
1100	45.108	45.486	45.863	46.238	46.612	46.985	47.356	47.726	48.095	48.462
1200	48.828	49.192	49.555	49.916	50.276	50.633	50.990	51.344	51.697	52.049
1300	52.398	52.747	53.093	53.439	53.782	54.125	54.466	54.807		

The Nickel-Chromium/Nickel reference tables were developed from the following polynomials:

temperature range −270°C to 0°C

$$E = \sum_{i=0}^{10} a_i t_{68}^i \ \mu V$$

$a_0 = 0$
$a_1 = 3.9475433139 \times 10^{+1}$
$a_2 = 2.7465251138 \times 10^{-2}$
$a_3 = -1.6565406716 \times 10^{-4}$
$a_4 = -1.5190912392 \times 10^{-6}$
$a_5 = -2.4581670924 \times 10^{-8}$
$a_6 = -2.4757917816 \times 10^{-10}$
$a_7 = -1.5585276173 \times 10^{-12}$
$a_8 = -5.9729921255 \times 10^{-15}$
$a_9 = -1.2688801216 \times 10^{-17}$
$a_{10} = -1.1382797374 \times 10^{-20}$

temperature range 0°C to 1372°C

$$E = \sum_{i=0}^{8} b_i t_{68}^i + 125 \exp\left[-1/2\left(\frac{t_{68} - 127}{65}\right)^2\right] \mu V$$

$b_0 = -1.8533063273 \times 10^{+1}$
$b_1 = 3.8918344612 \times 10^{+1}$
$b_2 = 1.6645154356 \times 10^{-2}$
$b_3 = -7.8702374448 \times 10^{-5}$
$b_4 = 2.2835785557 \times 10^{-7}$
$b_5 = -3.5700231258 \times 10^{-10}$
$b_6 = 2.9932909136 \times 10^{-13}$
$b_7 = -1.2849848798 \times 10^{-16}$
$b_8 = 2.2239974336 \times 10^{-20}$

Table 4-12: Thermal emfs (mV) of NiCr-Ni type N according to IEC 584-1 [4].

°C	0	−10	−20	−30	−40	−50	−60	−70	−80	−90
−200	−3.990	−4.083	−4.162	−4.227	−4.227	−4.313	−4.336	−4.345		
−100	−2.407	−2.612	−2.807	−2.994	−3.170	−3.336	−3.491	−3.634	−3.766	−3.884
0	0	−0.260	−0.518	−0.772	−1.023	−1.268	−1.509	−1.744	−1.972	−2.193
°C	0	10	20	30	40	50	60	70	80	90
0	0	0.261	0.525	0.793	1.064	1.339	1.691	1.902	0.188	2.479
100	2.774	3.072	3.374	3.679	3.988	4.301	4.617	4.936	5.258	5.584
200	5.912	6.243	6.577	6.914	7.254	7.596	7.940	8.287	8.636	8.987
300	9.340	9.695	10.053	10.412	10.772	11.135	11.499	11.865	12.233	12.602
400	12.972	13.344	13.717	14.091	14.467	14.844	15.222	15.601	15.981	16.362
500	16.744	17.127	17.511	17.896	18.282	18.668	19.055	19.443	19.831	20.220
600	20.609	20.999	21.390	21.781	22.172	22.564	22.956	23.348	23.740	24.133
700	24.526	24.919	25.312	25.705	26.098	26.491	26.885	27.278	27.671	28.063
800	28.456	28.849	29.241	29.633	30.025	30.417	30.800	31.199	31.590	31.980
900	32.370	32.760	33.149	33.538	33.926	34.315	34.702	35.089	34.476	35.062
1000	36.248	36.633	37.018	37.402	37.786	38.169	38.552	38.934	39.315	39.696
1100	40.076	40.456	40.835	41.213	41.590	41.966	42.342	42.717	43.091	43.464
1200	43.836	44.207	44.577	44.947	45.315	45.682	46.048	46.413	46.777	47.140
1300	47.502									

The Nickel-Chromium/Nickel reference tables were developed from the following polynomials:

temperature range −270 °C to 0 °C

$$E = \sum_{i=0}^{8} a_i t_{68}^i \ \mu\text{V}$$

$a_1 = 2.6153540164 \times 10^{1}$
$a_2 = 1.0933114132 \times 10^{-2}$
$a_3 = -9.3917128470 \times 10^{-5}$
$a_4 = -5.3592739285 \times 10^{-8}$
$a_5 = -2.7406835184 \times 10^{-9}$
$a_6 = -2.3370710645 \times 10^{-11}$
$a_7 = -7.8250681060 \times 10^{-14}$
$a_8 = -9.5885491371 \times 10^{-17}$

temperature range 0 °C to 1300 °C

$$E = \sum_{i=0}^{9} b_i t_{68}^i$$

$b_1 = 2.5897798582 \times 10^{1}$
$b_2 = 1.6656127713 \times 10^{-2}$
$b_3 = -3.1234962101 \times 10^{-5}$
$b_4 = -1.7248130773 \times 10^{-7}$
$b_5 = 3.6526665920 \times 10^{-10}$
$b_6 = -4.4390833504 \times 10^{-13}$
$b_7 = 3.1553382729 \times 10^{-16}$
$b_8 = -1.2150879468 \times 10^{-19}$
$b_9 = 1.9557197559 \times 10^{-23}$

Table 4-13: Thermal emfs (mV) of PtRh-Pt type R according to IEC 584-1 [4].

°C	0	−10	−20	−30	−40	−50	−60	−70	−80	−90
0	0	−0.051	−0.1	−0.145	−0.188	−0.226				
°C	0	10	20	30	40	50	60	70	80	90
0	0	0.054	0.111	0.171	0.232	0.296	0.363	0.431	0.501	0.573
100	0.647	0.723	0.800	0.879	0.959	1.041	1.124	1.208	1.294	1.380
200	1.468	1.557	1.647	1.738	1.830	1.923	2.017	2.111	2.207	2.303
300	2.400	2.498	2.596	2.695	2.795	2.896	2.997	3.099	3.201	3.304
400	3.407	3.511	3.616	3.721	3.826	3.933	4.039	4.146	4.254	4.362
500	4.471	4.580	4.689	4.799	4.910	5.021	5.132	5.244	5.356	5.469
600	5.582	5.696	5.810	5.925	6.040	6.155	6.272	6.388	6.505	6.623
700	6.741	6.860	6.979	7.098	7.218	7.339	7.460	7.582	7.703	7.826
800	7.949	8.072	8.196	8.320	8.445	8.570	8.696	8.822	8.949	9.076
900	9.203	9.331	9.460	9.589	9.718	9.848	9.978	10.109	10.240	10.371
1000	10.503	10.636	10.768	10.902	11.035	11.170	11.304	11.439	11.574	11.710
1100	11.846	11.983	12.119	12.257	12.394	12.532	12.669	12.808	12.946	13.085
1200	13.224	13.363	13.502	13.642	13.782	13.922	14.062	14.202	14.343	14.483
1300	14.624	14.765	14.906	15.047	15.188	15.329	15.470	15.611	15.752	15.893
1400	16.035	16.176	16.317	16.458	16.599	16.741	16.882	17.022	17.163	17.304
1500	17.445	17.585	17.726	17.866	18.006	18.146	18.286	18.425	18.564	18.703
1600	18.842	18.981	19.119	19.257	19.395	19.533	19.670	19.807	19.944	20.080
1700	20.215	20.350	20.483	20.616	20.748	20.878	21.006			

The Platinum-13% Rhodium/Platinum reference tables were developed from the following polynomials:

temperature range −50°C to 630.74°C

$$E = \sum_{i=0}^{7} a_i t_{68}^i \ \mu V$$

$a_0 = 0$
$a_1 = 5.289139$
$a_2 = 1.391111 \times 10^{-2}$
$a_3 = -2.400524 \times 10^{-5}$
$a_4 = 3.620141 \times 10^{-8}$
$a_5 = -4.464502 \times 10^{-11}$
$a_6 = 3.849769 \times 10^{-14}$
$a_7 = -1.537264 \times 10^{-17}$

temperature range 630.74°C to 1064.43°C

$$E = \sum_{i=0}^{3} b_i t_{68}^i \ \mu V$$

$b_0 = -2.641801 \times 10^{+2}$
$b_1 = 8.046868$
$b_2 = 2.989229 \times 10^{-3}$
$b_3 = -2.687606 \times 10^{-7}$

temperature range 1064.43 °C to 1665 °C

$$E = \sum_{i=0}^{3} c_i (t^*)^i \ \mu V$$

$$t^* = (t_{68} - 1365)/300$$

$$c_0 = 1.5540414 \times 10^{+4}$$
$$c_1 = 4.2357773 \times 10^{+3}$$
$$c_2 = 1.4693087 \times 10^{+1}$$
$$c_3 = -5.2213890 \times 10^{+1}$$

temperature range 1665 °C to 1767.6 °C

$$E = \sum_{i=0}^{3} d_i (t^*)^i \ \mu V$$

$$t^* = (t_{68} - 1715)/50$$

$$d_0 = 2.0416695 \times 10^{+4}$$
$$d_1 = 6.6850914 \times 10^{+2}$$
$$d_2 = -1.2301472 \times 10^{+1}$$
$$d_3 = -2.7861521$$

Table 4-14: Thermal emfs (mV) of PtRh-Pt type S according to IEC 584-1 [4].

°C	0	−10	−20	−30	−40	−50	−60	−70	−80	−90
0	0	−0.053	−0.103	−0.150	−0.194	−0.236				
°C	0	10	20	30	40	50	60	70	80	90
0	0	0.055	0.113	0.173	0.235	0.299	0.365	0.432	0.502	0.573
100	0.645	0.719	0.795	0.872	0.950	1.029	1.109	1.190	1.273	1.356
200	1.440	1.525	1.611	1.698	1.785	1.873	1.962	2.051	2.141	2.232
300	2.323	2.414	2.506	2.599	2.692	2.786	2.880	2.974	3.069	3.164
400	3.260	3.356	3.452	3.549	3.645	3.743	3.840	3.938	4.036	4.135
500	4.234	4.333	4.432	4.532	4.632	4.732	4.832	4.933	5.034	5.136
600	5.237	5.339	5.442	5.544	5.648	5.751	5.855	5.960	6.064	6.169
700	6.274	6.380	6.486	6.592	6.699	6.805	6.913	7.020	7.128	7.236
800	7.345	7.454	7.563	7.672	7.782	7.892	8.003	8.114	8.225	8.336
900	8.448	8,560	8.673	8.786	8.899	9.012	9.126	9.240	9.355	9.470
1000	9.585	9.700	9.816	9.932	10.048	10.165	10.282	10.400	10.517	10.635
110	10.754	10.872	10.991	11.110	11.229	11.348	11.467	11.587	11.707	11.827
1200	11.947	12.067	12.188	12.308	12.429	12.550	12.671	12.792	12.913	13.034
1300	13.155	13.276	13.397	13.519	13.640	13.761	13.883	14.004	14.125	14.247
1400	14.368	14.489	14.610	14.731	14.852	14.973	15.094	15.215	15.336	15.456
1500	15.576	15.697	15.817	15.937	16.057	16.176	16.296	16.415	16.534	16.653
1600	16.771	16.890	17.008	17.125	17.243	17.360	17.477	17.594	17.711	17.826
1700	17.942	18.056	18.170	18.282	18.394	18.504	18.612			

(For polynomials see next page)

The Platinum-10% Rhodium/Platinum reference tables were developed from the following polynomials:

temperature range $-50\,°C$ to $630.74\,°C$

$$E = \sum_{i=0}^{6} a_i t_{68}^i \ \mu V$$

$a_0 = 0$
$a_1 = 5.399578$
$a_2 = 1.251977 \times 10^{-2}$
$a_3 = -2.244822 \times 10^{-5}$
$a_4 = 2.845216 \times 10^{-8}$
$a_5 = -2.244058 \times 10^{-11}$
$a_6 = 8.505417 \times 10^{-15}$

temperature range $630.74\,°C$ to $1064.43\,°C$

$$E = \sum_{i=0}^{2} b_i t_{68}^i \ \mu V$$

$b_0 = -2.982448 \times 10^{+2}$
$b_1 = 8.237553$
$b_2 = 1.645391 \times 10^{-3}$

temperature range $1064.43\,°C$ to $1665\,°C$

$$E = \sum_{i=0}^{3} c_i (t^*)^i \ \mu V$$

$$t^* = (t_{68} - 1365)/300$$

$c_0 = 1.3943439 \times 10^{+4}$
$c_1 = 3.6398687 \times 10^{+3}$
$c_2 = -5.0281206$
$c_3 = -4.2450546 \times 10^{+1}$

temperature range $1665\,°C$ to $1767.6\,°C$

$$E = \sum_{i=0}^{3} d_i (t^*)^i \ \mu V$$

$$t^* = (t_{68} - 1715)/50$$

$d_0 = 1.8113083 \times 10^{+4}$
$d_1 = 5.6795375 \times 10^{+2}$
$d_2 = -1.2112492 \times 10^{+1}$
$d_3 = -2.8117589$

Table 4-15: Thermal emfs (mV) of PtRh-PtRh type B according to IEC 584-1 [4].

°C	0	10	20	30	40	50	60	70	80	90
0	0	−0.002	−0.003	−0.002	0	0.002	0.006	0.011	0.017	0.025
100	0.033	0.043	0.053	0.065	0.078	0.092	0.107	0.123	0.140	0.159
200	0.178	0.199	0.220	0.243	0.266	0.291	0.317	0.344	0.372	0.401
300	0.431	0.462	0.494	0.527	0.561	0.596	0.632	0.669	0.707	0.746
400	0.786	0.827	0.870	0.913	0.957	1.002	1.048	1.095	1.143	1.192
500	1.241	1.292	1.344	1.397	1.450	1.505	1.560	1.617	1.674	1.732
600	1.791	1.851	1.912	1.974	2.036	2.100	2.164	2.230	2.296	2.363
700	2.430	2.499	2.569	2.639	2.710	2.782	2.855	2.928	3.003	3.078
800	3.154	3.231	3.308	3.387	3.466	3.546	3.626	3.708	3.790	3.873
900	3.957	4.041	4.126	4.212	4.298	4.386	4.474	4.562	4.652	4.742
1000	4.833	4.924	5.016	5.109	5.202	5.297	5.391	5.487	5.583	5.680
1100	5.777	5.875	5.973	6.073	6.172	6.273	6.374	6.475	6.577	6.680
1200	6.783	6.887	6.991	7.096	7.202	7.308	7.414	7.521	7.628	7.736
1300	7.845	7.953	8.063	8.172	8.283	8.393	8.504	8.616	8.727	8.839
1400	8.952	9.065	9.178	9.291	9.405	9.519	9.634	9.748	9.863	9.979
1500	10.094	10.210	10.325	10.441	10.558	10.674	10.790	10.907	11.024	11.141
1600	11.257	11.374	11.491	11.608	11.725	11.842	11.959	12.076	12.193	12.310
1700	12.426	12.543	12.659	12.776	12.892	13.008	13.124	13.239	13.354	13.470
1800	13.585	13.699	13.814							

The Platinum-30% Rhodium/Platinum-6% Rhodium reference tables were developed from the following polynomial (temperature range 0°C to 1820°C):

$$E = \sum_{i=0}^{8} a_i t_{68}^i \ \mu\text{V}$$

$$\begin{aligned}
a_0 &= 0 \\
a_1 &= -2.4674601620 \times 10^{-1} \\
a_2 &= 5.9102111169 \times 10^{-3} \\
a_3 &= -1.4307123430 \times 10^{-6} \\
a_4 &= 2.1509149750 \times 10^{-9} \\
a_5 &= -3.1757800720 \times 10^{-12} \\
a_6 &= 2.4010367459 \times 10^{-15} \\
a_7 &= -9.0928148159 \times 10^{-19} \\
a_8 &= 1.3299505137 \times 10^{-22}
\end{aligned}$$

Table 4-16. Tolerance classes for thermocouples according to IEC 584-1 [4] (reference junction at 0 °C).

Types	Tolerance class 1	Tolerance class 2	Tolerance class 3 [1)]
Type T			
Temperature range	−40 °C to +125 °C	−40 °C to + 133 °C	−67 °C to +40 °C
Tolerance value	±0.5 °C	±1 °C	±1 °C
Temperature range	125 °C to 350 °C	133 °C to 350 °C	−200 °C to −67 °C
Tolerance value	$\pm 0.004 \cdot \lvert t \rvert$	$\pm 0.0075 \cdot \lvert t \rvert$	$\pm 0.015 \cdot \lvert t \rvert$
Type E			
Temperature range	−40 °C to +375 °C	−40 °C to +333 °C	−167 °C to +40 °C
Tolerance value	±1.5 °C	±2.5 °C	±2.5 °C
Temperature range	375 °C to 800 °C	333 °C to 900 °C	−200 °C to −167 °C
Tolerance value	$\pm 0.004 \cdot \lvert t \rvert$	$\pm 0.0075 \cdot \lvert t \rvert$	$\pm 0.015 \cdot \lvert t \rvert$
Type J			
Temperature range	−40 °C to +375 °C	−40 °C to +333 °C	–
Tolerance value	±1.5 °C	±2.5 °C	–
Temperature range	375 °C to 750 °C	333 °C to 750 °C	–
Tolerance value	$\pm 0.004 \cdot \lvert t \rvert$	$\pm 0.0075 \cdot \lvert t \rvert$	–
Type K, type N			
Temperature range	−40 °C to 375 °C	−40 °C to +333 °C	−167 °C to +40 °C
Tolerance value	±1.5 °C	±2.5 °C	±2.5 °C
Temperature range	375 °C to 1000 °C	333 °C to 1200 °C	−200 °C to −167 °C
Tolerance value	$\pm 0.004 \cdot \lvert t \rvert$	$\pm 0.0075 \cdot \lvert t \rvert$	$\pm 0.015 \cdot \lvert t \rvert$
Type R, type S			
Temperature range	0 °C to 1100 °C	0 °C to +600 °C	–
Tolerance value	±1 °C	±1.5 °C	–
Temperature range	1100 °C to 1600 °C	600 °C to 1600 °C	–
Tolerance value	$\pm[1 + 0.003\ (t - 1100)]$ °C	$\pm 0.0025 \cdot \lvert t \rvert$	–
Type B			
Temperature range	–	–	600 °C to 800 °C
Tolerance value	–	–	+4 °C
Temperature range	–	600 °C to 1700 °C	800 °C to 1700 °C
Tolerance value	–	$\pm 0.0025 \cdot \lvert t \rvert$	$\pm 0.005 \cdot \lvert t \rvert$

[1)] Thermocouple materials are normally supplied to meet the manufacturing tolerances specified in the table for temperatures above −40 °C. These materials, however, may not fall within the manufacturing tolerances for low temperatures given under glass 3 for types T, E, K and N. If thermocouples are required to meet limits of class 3, as well as those of class 1 or 2 the purchaser shall state this, as selection of materials is usually required.

5 Radiation Thermometers

Teresio Ricolfi, CNR-Istituto di Metrologia "G. Colonnetti", Turin, Italy,
Roy Barber, Land Instruments International Ltd., Dronfield (Sheffield), UK

Contents

5.1 Introduction

Radiation thermometers can provide very accurate and reliable measurements of temperature under very difficult conditions. However, if they are incorrectly specified, installed, or maintained, they can be unsatisfactory.

The correct selection and use of radiation thermometers are certainly more difficult than for other thermometers, owing to their relatively complex principles of measurement, the large choice of models, the wide variety of application problems, and the prevalence of operational errors over instrumental errors. These factors have been taken into account in preparing this chapter.

To assist the reader who does not wish initially to understand the details of theory and design, Section 5.2 explains briefly the principles of the various types of thermometers that are available. Full analyses and details are then given in Sections 5.3–5.7. To account for the importance of operational conditions, a large portion of the chapter is devoted to applications. Measurement problems and related solutions are discussed in Section 5.8 with reference to major industrial applications.

5.2 Overview

All radiation thermometers measure some part (or parts) of the thermal radiation leaving a surface, and the temperature of the surface is deduced from this measurement. Thermal radiation occurs as part of the electromagnetic spectrum in the region from, and including, the visible spectrum (0.38–0.78 μm) to the short-wavelength end of microwaves (1 mm). The part utilized in radiation thermometers usually lies between 0.5 and 20 μm.

Well defined laws describe the temperature and wavelength dependence of the radiation emitted by a surface and suggest different measurement approaches that are adopted in thermometers of different types.

The use of radiation thermometers is simple when dealing with surfaces showing black-body conditions (see Section 5.3.2), because in this instance the application of radiation laws is straightforward. Unfortunately, in most applications departures from black-body conditions occur as a result of emissivity, absorption, or reflection effects (see Section 5.4).

Many measurement problems arising from these effects can be overcome by a careful choice of the type of thermometer and its signal-processing facilities. In some instances, a knowledge of the surface emissivity may also be required, and this may not be a serious problem in those industrial applications where the product shows reproducible surface conditions. The most difficult problems arise with unknown or variable emissivity (or absorption or reflection) when special measurement techniques are required.

5.2.1 Types of Thermometers

Radiation thermometers can be subdivided into four groups, as follows.

1. Total-Radiation Thermometers

These measure radiant energy in a very broad waveband (several micrometers). Since a broad waveband has the advantage of containing a large amount of energy at any temperature, they are normally used in general-purpose, low-temperature thermometers. These instruments can, in some instances, have limitations such as a dependence on atmospheric conditions and sighting distance, and are therefore becoming less common as better detectors have become available.

2. Single-Waveband Thermometers

These thermometers are the most common and, together with the total-radiation thermometers, are the simplest and cheapest instruments. They measure radiant energy in a narrow waveband, which may be at a short or long wavelength, or at a specially selected position in the spectrum.

The very common thermometers using silicon or germanium detectors fall into the category of short-wavelength thermometers.

Short wavelengths have the major advantage that, in this spectral region, the rate of change of radiant energy with temperature is high (up to 2–3% $°C^{-1}$ at ordinary temperatures; see Figure 5-1) and, consequently, the temperature readings are less influenced by changes in the energy collected due, for example, to changes in the surface emissivity or atmospheric absorption.

Short wavelengths generally imply visible or near-infrared bands, eg, situated between 0.5 and 2 μm, but at very low temperatures longer wavelengths will still exhibit a high rate of change of energy with temperature.

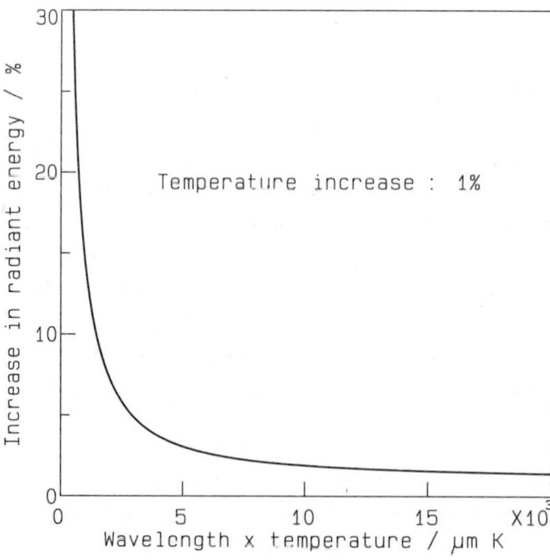

Figure 5-1.
Relative increase in radiant energy produced by a 1% increase in the absolute temperature of a black-body radiator. It is shown that the rate of change of radiant energy is high at short wavelengths and/or low temperatures.

Thermometers operating in selected wavebands limited by filters are often necessary to enable measurements to be made on semitransparent materials. For example, most glasses are transparent up to 2 μm but become opaque and exhibit a high and stable emissivity above 4.5 μm. Similarly, many thin plastic materials are transparent over wide parts of the spectrum, but have high opacity and low reflectance in known narrow spectral bands. Thermometers can be specially designed to measure the temperature of such plastics.

Finally, many low-temperature thermometers have their waveband limited by filters to a band of 8–14 μm. This fairly wide band contains sufficient energy to allow temperature measurements to be made down to, say, −50 °C while at the same time avoiding the region where atmospheric water vapor can give a false reading by absorbing energy between the surface and the thermometer.

3. Ratio Thermometers

These instruments work on the principle of measuring the ratio of the energy in two (usually narrow) wavebands. Their obvious advantage is that their readings will be independent of target size, obscuration of optical path and emissivity, provided that the obscuration and emissivity are the same at the two wavelengths. Unfortunately, this does not generally occur and it is very important to assess the effect of wavelength variations before choosing between a ratio and a single-waveband model. Ratio thermometers can be very effective but can also be a great disappointment. A good rule is *not* to choose a ratio thermometer if a single-waveband thermometer is satisfactory (see Section 5.4).

4. Multi-Waveband Thermometers

A ratio thermometer is independent of emissivity values only if the emissivities are the same at the two wavelengths chosen. Multi-waveband thermometers permit account to be taken of more complex emissivity-wavelength relationships (see Section 5.4.4). However, the added complexity and the increasing precision of measurement required result in multi-waveband thermometers being useful only in special situations.

5.2.2 Special Thermometers and Methods

The search for a truly emissivity-independent thermometer has much in common with the Philosopher's Stone. However, several methods have been proposed which are worthy of attention and some of them are described in Section 5.4.3. These methods are:

- auxiliary-reflector methods (see Figure 5-2);
- hot-source methods;
- polaradiometer method;
- reflectance methods;
- temperature-invariant method.

None of these special methods is likely to supersede the simple and less expensive single-waveband or ratio thermometers for most applications. However, they can provide measurements in applications where simple methods are inadequate. It is usually better to try to improve the measurement conditions so that a simple thermometer can be used than to use a more complex system to accommodate the difficult conditions. However, this is not always possible.

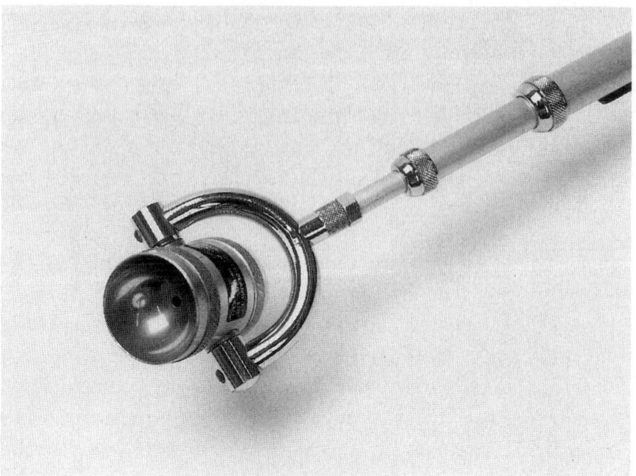

Figure 5-2. Land surface pyrometer based on the auxiliary-reflector method. The reflector is a gold-plated hemisphere. When it is placed over the target of measurement, multiple reflections occur between the target and reflector, thus enhancing the apparent emissivity of the target (courtesy of Land Infrared).

5.2.3 Thermometer Output Signals

Early thermometers gave low-level signals (microamperes or millivolts) which were proportional to the energy being detected and hence very nonlinear with temperature. Such thermometers are still used, but most modern thermometer systems incorporate electronic amplification and signal processing that enable high-level signals (milliamperes or volts) to be produced that are usually linear functions of temperature. It is important to select a thermometer with the necessary signal-processing capability to suit the application needs.

Typical processing capabilities include preamplification, various analog and digital outputs linear with temperature, peak picking, valley picking, sample and hold, emissivity corrector, memory circuits, and time constant adjustment (see Section 5.6.3). Modern thermometers also provide communication links for computer processing.

5.3 Principles of Measurement

The basic physical laws of radiation thermometry have already been discussed in Chapter 2. Supplementary information is first given in this section. A more detailed analysis is then made of the measuring principles of the various types of thermometer.

5.3.1 Quantities and Properties

In any discussion on thermal radiation phenomena, use is made of terms defining radiometric quantities and radiative properties of materials. Additional terms and symbols (descriptors) are used to specify their dependence on temperature, wavelength, geometric conditions, and, sometimes, state of polarization.

A detailed treatment of the question of nomenclature and interrelationships of quantities and properties was given by Touloukian and DeWitt [1] and, more recently, by DeWitt and Incropera [2]. In this section only those terms, symbols, and interrelationships are reported that are basic tools for understanding any discussion on radiation thermometers.

5.3.1.1 Definitions and Symbols

Definitions and symbols for quantities and properties are reported in Tables 5-1 and 5-2, respectively.

Concerning properties, it must be noted that the use of the two endings *−ance* and *−ivity* is often a source of confusion as they are used interchangeably in the literature. Although the term emissivity has been used in the past mainly in discussions concerning black-body cavities, it has now come into common usage, whichever the nature of the radiation source. For this reason, in accordance with [2], it will be generally used here.

Table 5-1. Radiometric quantities

Quantity	Symbol	Definition*	Unit symbol
Radiant energy	Q	Energy propagated as electromagnetic waves or photons	J
Radiant flux	Φ	Rate of flow of radiant energy $\Phi = \mathrm{d}Q/\mathrm{d}t$.	W
Radiant intensity	I	Flux per unit solid angle from a source $I = \mathrm{d}\Phi/\mathrm{d}\omega$	W sr^{-1}
Radiant exitance	M	Flux per unit area leaving a surface $M = \mathrm{d}\Phi/\mathrm{d}A$	W m^{-2}
Radiance	L	Flux propagated in a given direction, per unit solid angle about that direction and per unit area projected normal to the direction $L = \mathrm{d}^2\Phi/(\mathrm{d}A \cos\theta \,\mathrm{d}\omega)$	W m^{-2} sr^{-1}
Irradiance	E	Flux per unit area incident on a surface	W m^{-2}

* $\mathrm{d}A$ = element of surface; $\mathrm{d}t$ = element of time; $\mathrm{d}\omega$ = element of solid angle; θ = angle between direction of propagation and normal to emitting surface. Note: The symbol λ will be used to indicate reference to a given wavelength (spectral quantities).

Table 5-2. Radiation properties

Property*	Symbol	Definition
Absorptance, absorptivity	α	Ratio of absorbed flux to incident flux
Reflectance, reflectivity	ρ	Ratio of reflected flux to incident flux
Transmittance	τ	Ratio of transmitted flux to incident flux
Emittance, emissivity	ε	Ratio of the radiant exitance of a body at a given temperature to that of a black body at the same temperature and under the same viewing conditions

* The ending -ivity should be used when referring to an ideal specimen having an optically smooth and uncontaminated surface and thick enough to be opaque. Refer to text for the use of the term "emissivity" in the present treatment. The symbol λ will be used to indicate reference to a given wavelength (spectral properties).

5.3.1.2 *Interrelationships of Properties*

On the basis of the conservation of energy, the radiative properties integrated over the whole spectrum of black-body radiation (total properties) of a specimen of any kind are related by the equation

$$\alpha + \rho + \tau = 1, \tag{5-1}$$

where α is the absorptance, ρ the reflectance, and τ the transmittance.

For opaque materials, ie, for $\tau = 0$, Equation (5-1) becomes

$$\alpha + \rho = 1. \tag{5-2}$$

Emissivity is introduced into the above relationships by referring to Kirchhoff's law. This states in simple terms that any body which is a good absorber of radiation is also a good emitter, and just in the same proportion.

Kirchhoff's law leads to

$$\alpha = \varepsilon, \tag{5-3}$$

where ε is the emissivity, and for opaque materials to

$$\varepsilon + \rho = 1. \tag{5-4}$$

A deeper insight into Kirchhoff's law leads to some qualifications of Equation (5-3) and, consequently, Equation (5-4):

1. When referring to broad spectral intervals, Equation (5-3) applies only if the spectral composition of the incident radiant flux to which α is referred is that of black-body radiation at the temperature of the specimen.

2. Equation (5-3) applies to any particular wavelength λ in addition to the whole spectrum of black-body radiation. Equations (5-3) and (5-4) can then be written as

$$\alpha_\lambda = \varepsilon_\lambda \tag{5-5}$$

and

$$\varepsilon_\lambda + \rho_\lambda = 1. \tag{5-6}$$

3. Equation (5-3) also holds for the properties α and ε (or α_λ and ε_λ) being referred to a particular direction (specified by elevation and azimuthal angles θ and ϕ, respectively) over a particular solid angle (ω), ie,

$$\alpha(\theta, \phi, \omega) = \varepsilon(\theta, \phi, \omega) \tag{5-7}$$

From Equation (5-7) and making use of the principle of conservation of energy, Equation (5-4) can be converted to

$$\varepsilon(\theta, \phi, \omega) = 1 - \rho(\theta, \phi, \omega; 2\pi) = 1 - \rho(2\pi; \theta, \phi, \omega) \tag{5-8}$$

where the angles in ρ specify the conditions of irradiation and viewing during its measurement. More precisely, $\rho(\theta, \phi, \omega; 2\pi)$ is the reflectance for directional irradiation and hemispherical viewing, while the reverse condition holds for $\rho(2\pi; \theta, \phi, \omega)$.

5.3.2 Black-Body Laws

A black-body radiator is an ideal body having the following properties:

i. It absorbs all radiant energy incident on it, irrespective of wavelength, direction, and state of polarization.

ii. It emits at any temperature the maximum amount of radiant flux per unit area, and this applies to both integrated and spectral fluxes. This, and the preceding statement, lead to

$$\alpha = \varepsilon = \alpha_\lambda = \varepsilon_\lambda = 1. \tag{5-9}$$

iii. Its emission is independent of direction. This property is described by calling it a *perfectly diffuse* radiator.

The mathematical expressions for black-body laws, if the refractive index of the medium (see Chapter 2) is neglected, as is usually done in practical thermometry, are as follows:

– Stefan-Boltzmann law:

$$M = \sigma T^4 \tag{5-10}$$

where M is the total radiant exitance in W m^{-2}, T is the absolute temperature, and $\sigma \approx 5.670 \cdot 10^{-8}$ W m^{-2} K^{-4} is the Stefan-Boltzmann constant.

– Planck's law:

$$M_\lambda = c_1 \lambda^{-5} (\exp (c_2/\lambda T) - 1)^{-1} \tag{5-11}$$

where M_λ is the spectral radiant exitance in W m^{-2} per unit wavelength interval, λ is the wavelength in vacuum, and $c_1 \approx 3.742 \cdot 10^{-16}$ W m^2 and $c_2 \approx 1.439 \cdot 10^{-2}$ mK (the International Temperature Scale, ITS-90, value is $1.4388 \cdot 10^{-2}$ mK) are the so-called first and second radiation constants, respectively.

To express Planck's law in terms of spectral radiance, the term on the right-hand side of Equation (5-11) must be divided by π.

– Wien's approximation of Planck's law (valid for low values of λT):

$$M_\lambda = c_1 \lambda^{-5} \exp (-c_2/\lambda T) \tag{5-12}$$

– Wien's displacement law:

$$\lambda_M T = 2898 \ \mu\text{m K} \tag{5-13}$$

where λ_M is the wavelength at which the maximum occurs of the Planck's curve at temperature T.

– Lambert's cosine law:

$$I_\theta = I_0 \cos \theta \tag{5-14}$$

where I_θ is the directional radiant intensity in W sr^{-1} of a plane source in the direction θ and I_0 is the intensity in the direction normal to the source.

5.3.3 Real Bodies

It has been said that a black-body radiator is the maximum emitter at any temperature. Consequently, the emission of any real body is lower than that of a black body at the same temperature. However, the radiation laws derived for black bodies still apply to real bodies provided that a proportionality factor is introduced. This factor is the emissivity, and is always < 1. The spectral radiance, which is the quantity of main interest in radiation thermometry, is then given by

$$L_\lambda = \varepsilon_\lambda (\theta, T) c_1 \lambda^{-5} \pi^{-1} (\exp (c_2/\lambda T) - 1)^{-1} \tag{5-15}$$

where $\varepsilon_\lambda (\theta, T)$ is the directional spectral emissivity at the observation angle θ and temperature T.

In most thermometric applications, the normal spectral emissivity is assumed and the meaning of normal is that of angles $\leq 15°$.

The main features of the emissivity of materials are described in Section 5.4.1.

5.3.4 Measuring Principles of Thermometers

5.3.4.1 Total-Radiation Thermometers

The output $S(T)$ of a total-radiation thermometer is related to the temperature of a black-body source by

$$S(T) = F \int_0^\infty L_\lambda(T)\, s_\lambda\, d\lambda \tag{5-16}$$

where F is a constant geometrical factor depending on the optical system, $L_\lambda(T)$ is the black-body spectral radiance, and s_λ is the spectral responsivity of the thermometer (for a more precise expression of Equation (5-16), see Chapter 2). The responsivity is determined by the transmittance of the optical components and the responsivity of the detector. If s_λ is independent of wavelength, Equation (5-16) becomes

$$S(T) = F' \int_0^\infty L_\lambda(T)\, d\lambda = F' \pi^{-1} \sigma T^4, \tag{5-17}$$

where F' includes all constant factors.

Equation (5-17) is often referred to as the interpolating equation for total-radiation thermometers.

The use of aperture optics (see Section 5.6.1.1) and a black-body detector is the only way to realize a true total-radiation thermometer. This was done by Quinn and Martin [3] for high-precision thermodynamic temperature measurements (See Chapter 2). In thermometers for industrial measurements, the condition of constant responsivity over an infinite spectral interval is never fulfilled because of the use of lenses or mirrors and non-black-body detectors. Thus, in principle, the use of Stefan-Boltzmann law is incorrect, and Equation (5-17) must be replaced by some empirical relationship between thermometer output and target temperature.

Since total-radiation thermometers operate over a limited, although wide, spectral interval, they are sometimes included in the following class of single-waveband thermometers and are called *broad-waveband* thermometers.

5.3.4.2 Single-Waveband Thermometers

Narrow-waveband or *monochromatic* thermometers are alternative names to designate these thermometers, which are the most widely used in both precision and industrial measurements.

The interpolating equation for single-waveband thermometers is obtained from Equation (5-16) by setting the integration limits at two wavelength values λ_1 and λ_2. The width of the interval between λ_1 and λ_2 ranges from about 10 nm for high-precision thermometers to a maximum of a few micrometers for industrial types.

In discussions concerning single-waveband thermometers, the concept of *effective wavelength* is frequently used. The reasoning leading to the definition of effective wavelength and methods for calculating it have been widely discussed in the literature [4, 5] and will not be reported here. For the present purposes it is sufficient to say that the working waveband of a thermometer can be represented by a single wavelength, which is the effective wavelength. In this way, integration in Equation (5-16) can be avoided and it can be simply written as

$$S(T) = GL_{\lambda}(T) \tag{5-18}$$

where λ is the effective wavelength and G includes geometrical and responsivity factors.

Equation (5-18), although holding over a limited temperature span for many thermometers, is useful for the discussion of their basic properties.

An important property that must be accounted for in designing or selecting a thermometer is its *sensitivity,* which describes the change in the thermometer output produced by a change in the target temperature. The choice of the working waveband largely determines the thermometer sensitivity, although it depends also on the efficiency of the optical system and the detectivity of the detector.

Sensitivity can be obtained by differentiating Equation (5-18) with respect to T. If Wien's approximation is used for $L_{\lambda}(T)$, one obtains

$$\frac{\Delta L_{\lambda}}{L_{\lambda}} = \frac{c_2}{\lambda T}\,\frac{\Delta T}{T}. \tag{5-19}$$

This equation indicates that the sensitivity of a single-waveband thermometer decreases as the wavelength increases.

From a different point of view, it is also indicated that equal errors in measuring the target radiance produce larger temperature errors at longer wavelengths. This means that the accuracy of single-waveband thermometers also decreases at longer wavelengths.

It is concluded that, in order to obtain the maximum sensitivity and accuracy, it is convenient to work at the *shortest possible wavelength.* For this reason, the effective wavelength of precision thermometers is always shorter than 1 µm. However, in some applications operational errors due to the measurement conditions (eg, emissivity of the target) can be much more effective than the sensitivity and accuracy of the thermometer in determining the measurement accuracy. Consequently, the selection of the working waveband can be based on different criteria, as shown in Section 5.4.2.

5.3.4.3 Ratio Thermometers

In ratio thermometers, temperature is derived by measuring the ratio of the detector signals obtained in two different wavebands of the incoming radiation. If the wavebands are represented by the two wavelengths λ_1 and λ_2, the measuring principle of ratio thermometers is described by

$$R(T) = G'\,\frac{L_{\lambda_1}(T)}{L_{\lambda_2}(T)} \tag{5-20}$$

where $R(T)$ is the measured ratio and G' is a constant representing the ratio of the spectral responses at the two wavelengths.

By applying the same calculation procedure as for single-waveband thermometers, the following expression is obtained for the sensitivity of ratio thermometers:

$$\frac{\Delta R}{R} = -\frac{c_2}{T}\left(\frac{1}{\lambda_1} - \frac{1}{\lambda_2}\right)\frac{\Delta T}{T}. \tag{5-21}$$

If Equations (5-19) and (5-21) are compared, it is seen that, with respect to sensitivity, a ratio thermometer is equivalent to a single-waveband thermometer whose effective wavelength λ is given by

$$\frac{1}{\lambda} = \frac{1}{\lambda_1} - \frac{1}{\lambda_2}. \tag{5-22}$$

This equation shows that λ decreases as the separation between λ_1 and λ_2 increases. Hence, referring to the conclusion derived for single-wavelength thermometers, the sensitivity of a ratio thermometer increases with increasing separation of the working wavelengths. However, it must be noted that, if the separation of wavelengths increases, the probability of having different emissivities at the same wavelengths will also increase. Therefore, the measurement error can be larger, as will be shown in Section 5.4.3.

The sensitivity and accuracy of ratio thermometers are lower than those of single-waveband thermometers operating in approximately the same spectral region. This stems from the effective wavelength of ratio thermometers as given by Equation (5-22) being longer than each of the wavelengths λ_1 and λ_2.

Owing to their lower sensitivity and accuracy, ratio thermometers are not used for precision laboratory measurements where single-waveband thermometers are preferable. However, they may be advantageous in some practical applications where emissivity, absorption, or reflection is a major concern, as will be shown in Section 5.4.

5.3.4.4 *Multi-Waveband Thermometers*

Various techniques can be devised where radiant fluxes in three or more wavebands are measured in order to obtain more information on the emitted radiation. However, following the same reasoning as for single-waveband and ratio thermometers, the conclusion is drawn that the addition of any further waveband entails a decrease in sensitivity and accuracy. Therefore, recourse to a technique of this kind may be justified only when it allows an emissivity problem to be solved, as shown in the following section.

5.4 The Problem of Emissivity, Absorption, and Reflection

Radiation thermometers are always calibrated in terms of black-body temperatures, so measurement errors can be made whenever departures from black-body conditions occur. These may be due to a low emissivity of the source, absorption of radiation along the optical path, or reflection of extraneous radiation at the source.

The choice of the type of thermometer and the use of its emissivity corrector are essential in order to minimize errors of this kind. However, in some instances conventional thermometers are insufficient to obtain reliable measurements, so special methods must be adopted.

5.4.1 Emissivity of Materials

The emissivity of materials depends on the following parameters:

a. Wavelength: The spectral emissivity of polished metals generally decreases as the wavelength increases, whereas the opposite applies with glasses, ceramic materials, and paints. The spectral emissivity of semitransparent materials such as plastics, rather than showing a definite trend, is characterized by strong irregularities and peaks and valleys can be found in narrow spectral bands. In contrast, the emissivity of oxidized metals and of some categories of materials such as carbides has a very regular behavior and does not show a marked dependence on wavelength.

b. Temperature: The dependence of spectral emissivity on temperature is not very strong, although it cannot be neglected when deriving data from handbooks. This dependence is generally stronger for total emissivity. Generally, total emissivity increases with increase in temperature for materials that show a decrease in spectral emissivity at longer wavelengths.

c. Surface conditions: Surface roughness, oxide layers, and structural defects can greatly affect the emissivity of metallic materials. Emissivity increases as the roughness or degree of oxidation increase. However, the effect of thin oxide layers (thickness $\approx \lambda$) can be selective with wavelength. This effect must be accounted for mainly when operating with ratio thermometers in order to avoid large measurement errors. Dielectric materials are generally less affected by surface conditions.

d. Angle of emission: The dependence of emissivity on the angle of emission is stronger for metallic than for dielectric materials. The directional emissivity of smooth metals increases with increase in the angle from normal to a maximum near 90°, whereas that of dielectrics is nearly constant up to about 70°. In the inmediate proximity of 90° for metals and above 70° for dielectrics the emissivity drops rapidly to zero.

e. State of polarization: For some materials, mainly polished metals, the emitted radiation is partially polarized and the emissivity depends on the state of polarization. However, this phenomenon generally is not a primary concern in practical thermometry, although it is utilized in the polaradiometer method (see Section 5.4.5).

5.4.2 Influence of Emissivity, Absorption and Reflection on Thermometer Readings

5.4.2.1 *Single-Waveband Thermometers*

The temperature T_λ measured by a single-waveband thermometer aimed at a source at temperature T and having a spectral emissivity ε_λ is given by

$$\frac{1}{T_\lambda} = \frac{1}{T} - \frac{\lambda}{c_2} \ln \varepsilon_\lambda \tag{5-23}$$

as can be derived by Wien's approximation of Planck's law.

The temperature T_λ, which is called the *radiance temperature*, is always lower than the true temperature T, except in the case where $\varepsilon_\lambda = 1$, ie, the source is a black-body radiator.

A relationship of the same kind between T_λ and T holds in the presence of absorption. In this case, ε_λ in Equation (5-23) must be multiplied by the spectral transmittance of the medium between the source and the thermometers.

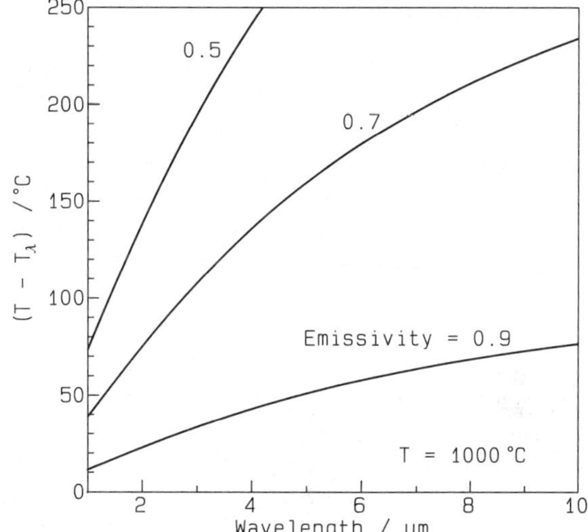

Figure 5-3. Difference between true (T) and radiance (T_λ) temperature as a function of wavelength at a temperature $T = 1000\,°C$.

The effect of emissivity on the readings of a single-waveband thermometer is shown in Figure 5-3. The wavelength behavior of the curves shown and the general features of the spectral emissivity of materials (see Section 5.4.1) suggest the following criteria for the selection of the working wavelengths:

i. The effect of emissivity (and the same applies for absorption) increases with increase in wavelength. Therefore short wavelengths are more adequate for general-purpose thermometers.

ii. For metals, whose emissivity decreases as the wavelength increases, there is a further advantage in using short wavelengths.

iii. Most glasses and some ceramics are partially transparent at short wavelengths. Longer wavelengths are therefore necessary for the accurate measurement of these materials.

iv. The emissivity of plastic materials shows peaks at definite wavelengths in the infrared region. Hence operation at these wavelengths is essential.

v. The same reasoning as in the preceding point applies to atmospheric absorption, which shows minima in definite wavebands that are called *atmospheric windows* (see Figure 5.4).

Whereas emissivity and absorption lead to underestimates of the temperature of the source, the contrary happens with reflection, as it gives rise to an increase in the apparent radiance of the source. Owing to the opposite effects of emissivity and reflection, a measured radiance temperature can result that is closer to the true temperature than in the absence of reflection.

An application of this principle is found in the auxiliary-reflector methods described in Section 5.4.5. Generally, however, reflection effects are difficult to account for. Hence they must be avoided, as they can lead to large measurements errors (see examples in Table 5.3).

Proper countermeasures to minimize the effect of reflections, when they cannot be removed by shielding, are described in Section 5.8.1.

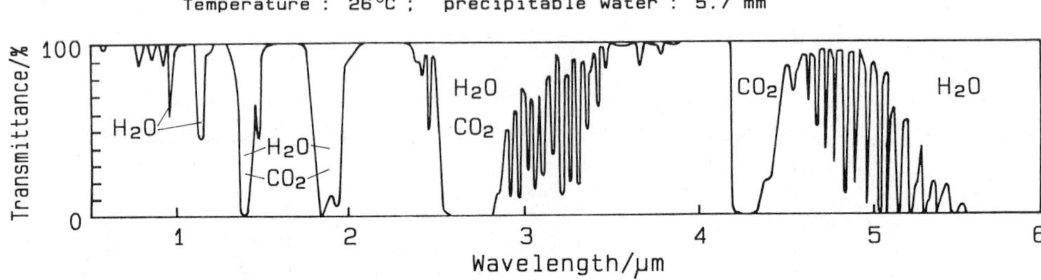

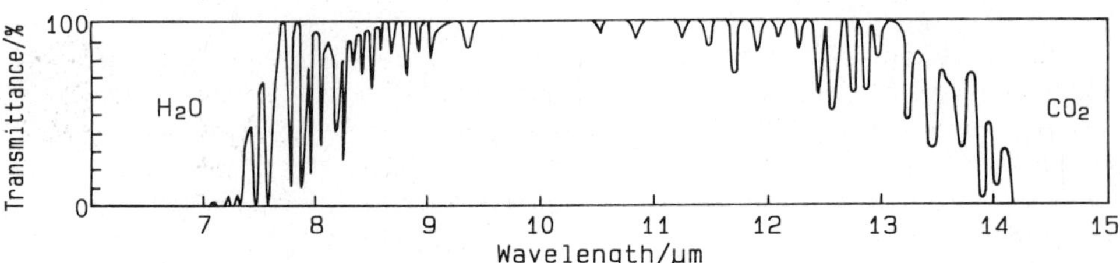

Figure 5-4. Atmospheric transmittance at sea level over a 300-m path at the indicated temperature and humidity (adapted from [6]). Spectral regions of high transmittance (atmospheric windows) are suitable for radiation thermometry since they neither absorb nor emit radiation.

Table 5-3. Errors (in °C) of an 8–14-µm thermometer due to stray radiation from the surroundings

	ε								
	0.3			0.6			0.9		
T_s^*/°C	50°C	100°C	150°C	50°C	100°C	150°C	50°C	100°C	150°C
0	−38	−25	−19	−10	− 7	− 5	−1.5	−1	−1
10	−18	−13	−10	− 5	− 3.5	− 3	−1	−0.5	−0.5
20	0	0	0	0	0	0	0	0	0
30	+18	+13	+10	+ 5	+ 4	+ 3	+1	+0.5	+0.5
40	+35	+26	+22	+11	+ 8	+ 6	+2	+1	+1
50	+51	+40	+33	+16	+12	+10	+3	+2	+1.5

* T_s is the temperature of the surroundings. It is assumed that the reference temperature in the thermometer is 20°C.

5.4.2.2 Ratio Thermometers

If the black-body radiances in Equation (5-20) are replaced by the corresponding radiances for a real body, one obtains

$$R(T) = G' \frac{\varepsilon_{\lambda_1} L_{\lambda_1}(T)}{\varepsilon_{\lambda_2} L_{\lambda_2}(T)} . \tag{5-24}$$

Whereas the temperature measured by a single-waveband thermometer equals the true temperature only in the case of unit emissivity, Equation (5-24) shows that the less restrictive condition for which

$$\varepsilon_{\lambda_1} = \varepsilon_{\lambda_2} \tag{5-25}$$

is sufficient for a ratio thermometer to give correct readings. In fact, when the condition of equal emissivity at both working wavelengths exists (*grey-body* condition), Equation (5-24) is identical with Equation (5-20) for black-body conditions. The same reasoning can be applied to absorption.

Although the capability of ratio thermometers of tolerating departures from black-body conditions turns out to be useful in some instances, nevertheless this capability is often overestimated. In fact, equality of emissivity or transmittance at the two working wavelengths is not commonly found.

The departure of measured from true temperatures, when the condition of equal emissivities is not fulfilled is given by

$$\frac{1}{T_{\mathrm{R}}} = \frac{1}{T} + \frac{\lambda_1 \lambda_2}{c_2 (\lambda_1 - \lambda_2)} \ln \frac{\varepsilon_{\lambda_1}}{\varepsilon_{\lambda_2}} \tag{5-26}$$

where T_{R} and T are the measured (by ratio thermometer) and true temperature, respectively. Equation (5-26) shows that the measured temperature can either be lower or higher than the true value, depending on the slope of the spectral emissivity. The extent of the departure between the two temperatures increases as the slope increases.

Criteria for comparing the errors with single-wavelength and ratio thermometers were given by Pyatt [7]. Nevertheless, the best choice can only be dictated by some preliminary knowledge of the behavior of the emissivity of the source material.

The situations where ratio thermometers certainly perform better than single-waveband types are when mechanical obstructions interfere with the sighting cone or the target size is insufficient to fill the field of view, since in these instances the attenuation of radiation is identical at the two working wavelengths.

5.4.2.3 Multi-Waveband Thermometers

It has been shown that ratio thermometers, as compared with the single-waveband type, permit broader assumptions about emissivity. This suggests that a further increase in the number of wavebands might allow the assumption of more complex and perhaps more valid relationships between emissivity and wavelength. In fact, a three-wavelength thermometer can give a true temperature when the emissivity varies linearly with wavelength.

A three-wavelength thermometer can operate according to the approach suggested by Hornbeck [8], where the ratio

$$R = \frac{\varepsilon_{\lambda_1} \varepsilon_{\lambda_3}}{\varepsilon^2_{\lambda_2}} \cdot \frac{L_{\lambda_1} L_{\lambda_3}}{L^2_{\lambda_2}} \tag{5-27}$$

can be obtained from the measurement of radiant fluxes in three wavebands.

If the emissivity is linear with wavelength and the wavelengths are chosen such that

$$\lambda_2 = (\lambda_1 + \lambda_3)/2 \tag{5-28}$$

then

$$\varepsilon_{\lambda_1} \varepsilon_{\lambda_3} \approx \varepsilon^2_{\lambda_2} \tag{5-29}$$

and the measured ratio R becomes independent of emissivity.

The major limitation in this approach is that the emissivity must be known to be linear with wavelength to a high degree of accuracy. The same limitation exists for thermometers operating in more than three wavebands and thus allowing the assumption of more complex models for the dependence of the emissivity on wavelength. In addition, as noted by Coates [9] who analyzed a least-squares approach to multi-waveband thermometry, the validity of the mathematical models assumed cannot be determined from the experimental data, and the magnitude of the resulting systematic error cannot be predicted.

These, and the further limitation that the sensitivity decreases as the number of wavebands increases, as previously noted, seem to justify the conclusion by Coates that a single-waveband thermometer, together with a crude estimate of the emissivity, will in general yield a more reliable estimate of the temperature.

5.4.3 Special Thermometers and Methods

Several methods have been proposed to compensate for departures from black-body conditions and some of them are described in Section 5.8 with reference to particular applications. In this section are separately described the methods which have led to the realization of emissivity-independent thermometer systems. These methods are as follows:

1. Auxiliary-Reflector Methods

These rely on the principle that the apparent emissivity of a surface can be enhanced by placing in close proximity over it a highly reflecting surface (see Figure 5-2). The emissivity increase results from multiple reflections between the two surfaces.

The most common instrument based on this principle is the Land surface pyrometer [10]. It incorporates a gold-plated hemispherical reflector with a small hole at the top through which the instrument sensor can view the surface. Gold was chosen as it has the highest reflectance for near-infrared wavelengths.

Owing to the difficulty of keeping the reflector cool and clean, the Land surface pyrometers have been designed only as portable spot-reading instruments. They can allow a measurement of emissivity to be made by taking a second reading with a black hemisphere covering the gold reflector, and provide a useful reference instrument for checking the calibration of fixed systems.

A different application of the auxiliary-reflector principle has been proposed by Iuchi [11] for the continuous measurement on bright steel strips. His device has been well tested but is not commercially available, possibly owing to the difficulty of keeping the specular reflecting surfaces clean.

2. Hot-Source Methods

Several methods have been proposed and evaluated to enable the emissivity and true temperature of a surface to be measured by introducing a black-body source of known temperature into the equipment [11–13]. As an example, let us consider the TERM method proposed by Iuchi [11]. The principle of the method is illustrated in Figure 5-5. Iuchi showed that from the measurement of two thermometer signals corresponding to two temperatures of the black-body source, it is possible to derive the true temperature and emissivity of the hot surface.

The accuracy of the method relies on the assumption that all the energy received by the thermometer has been either emitted by the surface or reflected from the black-body source. This is true only when the surface is specularly reflecting. In fact, if the reflection is partially diffuse, the reflected radiation will originate from sources other than the black body (eg, the furnace walls).

Iuchi found that the assumption held very well for cold-rolled steel sheet and bright annealed stainless steel when a long-wavelength (8 μm) thermometer was used. The errors that could result from partially diffuse reflection were estimated to be within ±10°C. In contrast, the assumption did not hold for roughened (shot-blasted) surfaces or at short wavelengths (1–2 μm).

Iuchi's main finding was that the degree of specularity of a surface depended only on its finish and not on its degree of oxidation. Consequently, measurements could be taken in a furnace in which the material was being progressively oxidized.

Although an equipment based on the TERM method is commercially available, it is not widely used.

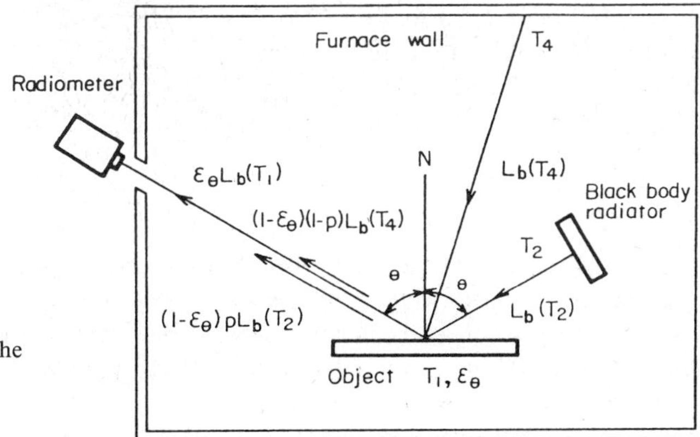

Figure 5-5.
Principle of the TERM method. L_b is the black-body radiance at the indicated temperatures and p ($0 \leqslant p \leqslant 1$) the fraction of specular reflectance.

3. Polaradiometer Method

Murray [14] first developed a variant of the hot-source method which consists in examining for polarization the sum of the radiation emitted by a target and the radiation originated from a black-body source and reflected by the target itself.

The radiation emitted by a metal sample is such that the component of polarization in the plane of incidence is larger than that in the plane perpendicular to the plane of incidence. Since Kirchhoff's law holds for each component, the opposite occurs to the originally unpolarized radiation from a black-body source after reflection at the sample. Consequently, the resultant of the emitted and reflected radiation will be unpolarized when the temperature of the sample equals that of the black body.

The experimental equipment includes a black-body source, a detector, and a rotating polarizer (analyzer) in front of it. The detector signal consists of a steady and an alternating component. Vanishing of the alternating component indicates equal temperatures for the sample and black body, regardless of the sample emissivity.

The accuracy of this method is largely independent of the surface finish. The only requirement is that whatever radiation is scattered must be unpolarized, which means that the surface texture must not show preferential directions.

4. Reflectance Methods

Two types of instrument have been produced which incorporate a light source and which make a measurement of the reflectance (or of some factors related to reflectance) of the target in order to establish a value for the emissivity.

The first portable instruments, developed by Exxon and manufactured under license by The Pyrometer Instrument Co., are short-wavelength single-waveband thermometers that incorporate a visible laser source. The laser beam is focused on the target and the reflected component is measured in addition to the emitted component. The major limitation is that the reading will be very dependent on the angle at which the target is viewed, except when the target has a perfectly diffuse surface.

The second instrument, developed by Minolta Camera Co., uses a light source focused on the target and largely overcomes the problem of surface finish by measuring both the emitted and reflected energy at three wavelengths (0.8, 0.9 and 1.0 μm). It is possible to solve the resulting six simultaneous equations and to determine both the emissivity and true temperature of the target. The only assumption made is that the fraction of energy reflected at any angle is the same at each wavelength. Although this is not true over a wide spectral range, it has been shown that for the wavelengths used the deviation does not cause significant errors.

5. Temperature-Invariant Method

It was pointed out by Svet [15] that ratio thermometers and multi-waveband thermometers using narrow spectral bands can give more information than is normally used. For example, it can be shown that if the radiance temperatures T_1 and T_2 (in K) are measured at two wavelengths λ_1 and λ_2 then the value $(1/T_1 - 1/T_2)$ is independent of the temperature being measured and is a function only of the emissivities at the two wavelengths and the wavelength values. It is often possible with some prior knowledge of the conditions to correlate the values of

$(1/T_1 - 1/T_2)$ with the emissivity. Alternatively, it can be shown that the true temperature T can be calculated from the equation

$$\frac{1}{T} = \frac{1}{T_1} + A\left(\frac{1}{T_1} - \frac{1}{T_2}\right) \tag{5-30}$$

where the value of A can be determined by prior experimentation for the particular target conditions being encountered.

5.5 Factors Affecting the Choice of the Best Thermometer System

The importance of selecting the optimum thermometer system for any application cannot be overstressed. Unfortunately, there is no simple set of rules that is applicable to all situations. However, the following check list can provide a very useful guide in most applications. Further examples are given in Section 5.8.

A. The Target

A1. Target material(s), condition and stability.
A2. Temperature span.
A3. Size range and positional stability.
A4. Distance range from thermometers.
A5. Rate of temperature change.

B. The Surroundings

B1. Much colder than target.
B2. Same temperature as target.
B3. Hotter than target.
B4. Temperature range.
B5. Atmosphere: can it absorb or emit radiation that affects the readings?
B6. Moving parts: is the line of sight periodically obscured?
B7. Electrical fields.

C. Thermometer

C1. Are there any size restraints?
C2. Ambient temperature range.
C3. Cleanliness of environment.
C4. Is it necessary to sight through a window?

D. Processing

D1. Type of signal required (current, voltage, digital).

Taking these factors in turn, the following major considerations and solutions can be noted:

A1.
- If the target condition is stable and the emissivity is above 0.5: use single-waveband short-wavelength thermometer (see B1, also).
- If the target material has low and/or variable emissivity: study use of ratio thermometer; alternatively, study special methods.

  Note: A single-waveband short-wavelength thermometer may still be the best solution.

A2. If a very wide span is essential, it will be necessary to compromise on the short-wavelength requirement: choose the shortest wavelength model that meets the minimum temperature requirement.

A3. If the target is small and/or not in a constant position: study the use of a rectangular field of view system (see Section 5.8.1) or ratio thermometers.

A4. If the target distance is large or very variable:
  i. ensure that a model is chosen that is free from atmospheric absorption effects;
  ii. ensure that the field of view is filled at all distances.

A5. If the target can heat or cool very rapidly: ensure that the speed of the thermometer system can give the required measurement (very high-speed systems are sometimes available as special options).

B1. If the surroundings are much colder than the target: use the shortest-wavelength thermometer which meets the temperature span requirement.

B2. If the target is in thermal equilibrium with its surroundings: any thermometer can be used with the emissivity set at 1. If, however, the temperature of the surroundings is close to but not the same as that of the target (eg, measurements near room temperature) take great care to assess the correct emissivity setting.

B3. If the surroundings are much hotter than the target: use the compensation method (two-sensor system; see Section 5.8.1), or, if the target is moving, shield the measurement point (see Section 5.8.1).

B5. If the atmosphere between the thermometer and the target can contain significant quantities of steam, water droplets, or particles:
  i. if the obscuration is intermittent: use peak-picking signal processing (see Section 5.6.3);
  ii. if the obscuration is permanent: consider a ratio thermometer (take care that the obscuring particles are sufficiently large that they do not cause highly wavelength-dependent scattering at the wavebands of the ratio thermometer; see Section 5.8.3).

B6. If the line of sight is periodically obscured (eg, by moving machinery) or if the target is intermittent (eg, glass bottles on a conveyor): study the most suitable signal processing method, eg, sample and hold if an electrical signal can be synchronized with the circuit, or peak picking if not.

B7. If there are strong electrical fields near the thermometer: first consider use of the fiber-optic thermometer; if a suitable model cannot be found, take advice on the optimum screening and earthing methods for the thermometer system.

C1. If there are size limitations in positioning the thermometer: first consider the fiber-optic model then special miniature thermometers.

C2. If the ambient conditions at the thermometer position may exceed the specified thermometer range: use good air- or water-cooled jackets round the thermometer. Take care of radiation that may heat the back of the thermometer.

C3. If the environment is likely to dirty the optical system: use an effective lens purging system.

C4. If it is necessary to sight through a window (eg, vacuum or controlled-atmosphere furnaces): use glass or silica windows for short-wavelength thermometers. Allow for reflection by adjusting the emissivity setting. At low temperatures, where long-wavelength thermometers are necessary, take professional advice on the windows intended to be used.

D1. Ensure that you consider carefully the type or types of electrical outputs that will be compatible with your measurement or control system. A large range of options are available.

5.6 Construction of Radiation Thermometers

Whatever their measuring principle, all radiation thermometers contain the following basic elements:

- optical system;
- detector;
- signal-processing facilities.

The general features of these elements are first examined in the following sections. Practical implementations are then shown and discussed with reference to the various classes of thermometers.

5.6.1 Optical System

The optical systems actually used in radiation thermometers can be classified into one of the following classes:

- aperture optics;
- lens or mirror optics;
- fiber optics.

In many thermometers the optical system includes filters that are used to define the working waveband or to lower the irradiance at the detector.

5.6.1.1 Aperture Optics

The simplest form of optical system results from placing a diaphragm at a fixed distance in front of the detector. The aperture of this diaphragm, together with the active area of the detector, determine the angular field of view of the thermometer and, therefore, the useful size of the target at any working distance.

This arrangement is actually used in some simple thermometers for spot measurements. Its main disadvantage is that limiting the cone of view in order to limit the target size means lowering the irradiance at the detector. Consequently, poor sensitivity can result at middle-low temperatures.

5.6.1.2 Lens and Mirror Optics

The way to increase the efficiency in collecting radiation while keeping the target size within reasonable limits is to use transfer optics. To do this, the diaphragm in the previous example may be replaced with a lens (or a mirror) which forms an image of the source in the plane of the detector. The resulting setup is a *single-lens* system as shown schematically in Figure 5-6. The size of the cone of radiation accepted from an axial point on the target and, therefore, the irradiance at the detector, is determined by the clear aperture of the lens, while the active area of the detector determines the field of view. The lens and detector then act as the *aperture* and *field stops* of the system, respectively.

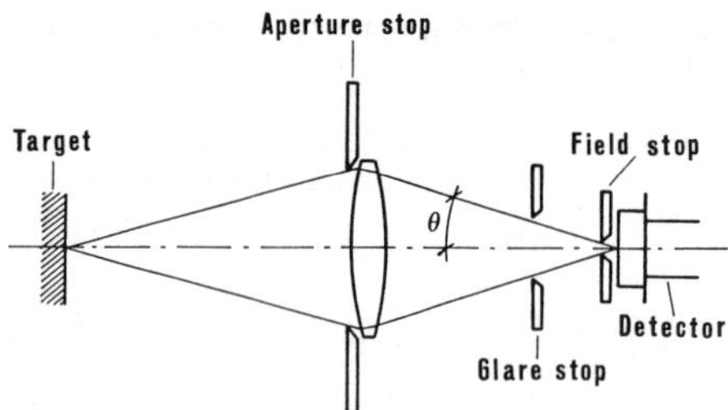

Figure 5-6. Schematic representation of a single-lens system.

In a single-lens system the irradiance E at the detector produced by an axial point source emitting diffusely is given by

$$E = \pi \tau L \sin^2 \theta \tag{5-31}$$

where τ is the transmittance of the lens, L is the radiance of the source, and θ is the angle shown in the figure.

Equation (5-31) shows that at a given source temperature the irradiance is constant provided that θ is constant. A fixed-focus system can therefore be used in thermometers for both permanent installation and portable use, since their reading is independent of the distance from the target and focusing conditions provided that the target is large enough to fill the field of view.

Nevertheless, a need to limit the target size or to give a clearly focused view for the operator at all working distances can suggest a variable-focus system. In this case the condition of constant reading cannot be further maintained unless the aperture stop is placed at a fixed position away from the lens.

This can be done simply by placing the aperture stop somewhere between the lens and detector. The size and position of the aperture stop must be such that the outer diameter of the lens never interferes with the cone of radiation.

The principle of using a fixed-aperture stop also holds in *multi-lens* systems. Owing to their flexibility, which permits the optimization of the position of stops and filters and the condition of incidence on filters and detector, multi-lens (two or more) system are generally used in high-precision thermometers.

An important point to consider when designing an optical system is that of reflection effects. Stray radiation from outside the target can propagate through the system by reflection or scattering from the mechanical or optical elements. Baffles and groovings are effective in suppressing unwanted radiation. However, the most effective way is to use a *glare stop,* which is a diaphragm located at a suitable position between the aperture and field stops. The glare stop passes all the rays passing directly through the system from the aperture stop, but suppresses the reflected or scattered rays.

For radiation thermometers working in the visible or near-infrared region up to about 2 μm, the choice of the lens material is easy since borosilicate glass is a suitable material owing to its low cost and fairly good optical, mechanical, and chemical properties. In critical applications where the objective lens is exposed to strong radiation fluxes, borosilicate can be replaced with fused-silica glass, which has a coefficient of thermal expansion that is more than ten times lower. More problems are encountered when working at longer wavelengths.

The first requirement for a candidate material is to transmit in the spectral region of interest. Many materials having wide transmission ranges in the infrared region are found among oxides, halides, semiconductors, and chalcogenides. The transmission ranges of some of them that are commonly used in radiation thermometers are shown in Figure 5-7.

Other than on the transmission range, the optimum choice for any application depends on several additional parameters which can be classified as follows:

- *Optical:* refractive index, dispersion, homogeneity, scattering properties.
- *Mechanical:* density, hardness, strength, workability.
- *Thermal:* temperature coefficient of the refractive index, coefficient of thermal expansion, conductivity, softening or melting point.
- *Chemical:* solubility, resistance to attack, ability to be coated.
- *Cost.*

All these parameters must be considered simultaneously, and in most cases the choice is a compromise solution.

Although lens optics are mostly used in radiation thermometers, mirror optics can be convenient in some instances. One of these is when the cost of an infrared lens would be too high compared with the overall cost of the thermometer. In some other situations the advantage in using mirrors stems from their being free from chromatic aberrations. This turns out to be

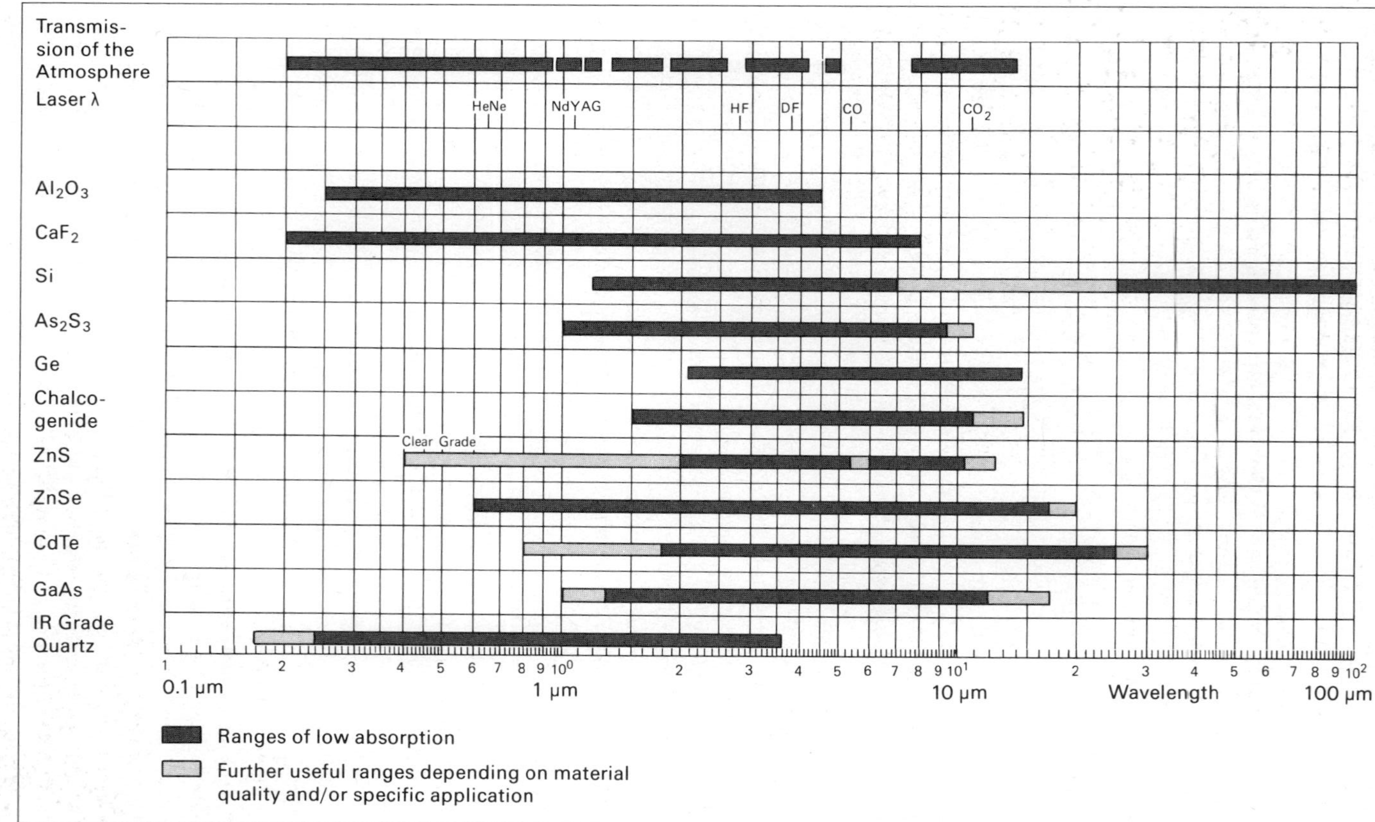

Figure 5-7. Transmission ranges of commonly used infrared optical materials.

useful in thermometers working in more than one waveband and facilitates the design of the visual sighting channel in long-wavelength thermometers of any kind.

A disadvantage of mirrors compared with lenses is that in order to protect the interior of the thermometer from dust, moisture, smoke, temperature transients, and drafts, it is necessary to provide a front window, which can be almost as expensive as a lens of the same material. This inconvenience is sometimes removed by using a window made of a thin plastic sheet, although care must be exercized to avoid deleterious effects from scattering and distortion in the radiation beam.

A further disadvantage with mirror systems is that they are much more difficult to provide with adequate glare protection than lens systems.

5.6.1.3 Fiber Optics

In many practical circumstances, lenses and mirrors can be conveniently replaced in part or fully by fiber-optic components. In fact, the flexibility of fiber-optic cables and their capability of transmitting radiation over long distances allow one to circumvent the difficulties arising when:

- direct sighting of the target is impossible owing to obstructions;
- operating in an environment where large amounts of fumes, smoke, or water vapor are present;
- electromagnetic interferences or nuclear radiation require the electronics to be located at a safe distance from the target;
- a high ambient temperature exists;
- the target is kept in a vacuum vessel and sighting through the window is difficult or impossible;
- a small-sized optical head is necessary, as with induction heating (see Section 5.8.1).

Fiber-optic components can be used in two ways:

- as a means of taking the radiation from an optical head to a distant detector; and
- as an aperture-optics system without lenses or mirrors.

A fiber-optic thermometer of the first type is shown in Figure 5-8. A small optical head containing a lens and field stop focuses the target onto one end of a flexible fiber-optic light guide which conveys the radiation to the detector located at the other end.

Glass or silica fiber bundles are used in most cases. They are characterized by good transmittance in the spectral interval from 0.5 to about 2 μm. The maximum permissible length is set by the amount of radiation reaching the detector, and therefore it depends on the temperature of the target. Whereas for a temperature of 50 °C the length limit is of the order of 0.5 m, it may be increased to 20 m as the temperature increases up to 1000 °C. Silica fibers are particularly useful when the optical head has to be operated in high-temperature environments, since they can withstand temperatures up to 500 °C.

The second approach to the use of fiber optics in radiation thermometry was proposed by Dils [16] and is shown in Figure 5-9. Although being a radiation thermometer, this thermometer is operated just like a thermocouple as it is plunged into the body (furnace, mass of fluid,

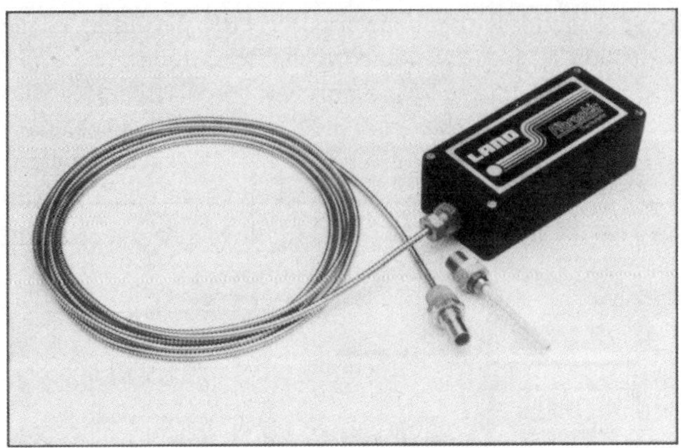

Figure 5-8. Fiber-optic thermometer (courtesy of Land Infrared).

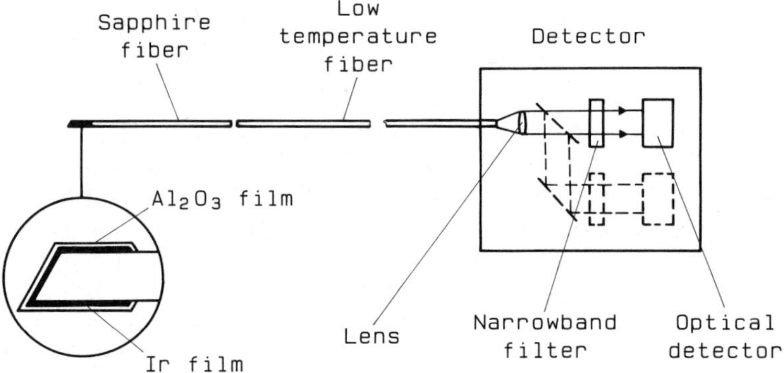

Figure 5-9. Fiber-optic thermometer after Dils [16]. The rigid single-crystal sapphire fiber can withstand temperatures up to about 2000 °C.

etc.) whose temperature is to be measured. To improve the measurement accuracy when dealing with low-emissivity materials, a small black-body cavity can be created at the end of the fiber by sputtering on it a thin metallic coating of platinum or iridium. The actual radiation source as seen by the detector is thus a black-body cavity at the temperature of the target of measurement.

5.6.1.4 Filters

The main purpose of filters is to select the desired operating waveband of the thermometers. The nature of the filters depends on the type of thermometer and the application for which it is intended. Interference filters are generally used in precision single-waveband and ratio thermometers.

Two classes of interference filters are available, namely *bandpass* filters, which transmit within a restricted band, and *edge* filters, which transmit only above (high pass) or below (low pass) a given wavelength (cut-off wavelength).

In most of the above thermometers, bandpass filters are used. Their bandwidth, ie, the width of their transmission band measured between the points where the transmittance is half of the peak transmittance, ranges from about 10 nm for primary-standard thermometers to a maximum of 100 nm for single-waveband thermometers that can be used as secondary standards. Single-waveband thermometers for industrial measurements generally operate in wider bands. For this reason, a bandpass filter is not strictly required since the operating band can be more easily limited at one side by an edge filter (generally a high pass) and at the other hand by the response curve of the detector. In some instances, particularly at short wavelengths, interference edge filters can be conveniently replaced by less expensive color filter glasses.

A different function of filters is that of attenuating the radiation flux reaching the detector. This may be required simply to protect the detector from an excessive illumination which could degrade its performance, or to provide more than one temperature span as in the case of the disappearing-filament pyrometer. For these purposes, neutral density filters are used which show nearly constant transmittance within the working band.

5.6.2 Detectors

The principal factors that must be considered in selecting a detector are the following.

1. Spectral Responsivity: This factor describes in relative terms the behavior of the detector output with the wavelength of the incident radiation. The spectral responsivity for a number of detectors commonly used in radiation thermometers may be derived from the spectral detectivity curves shown in Figure 5-10.

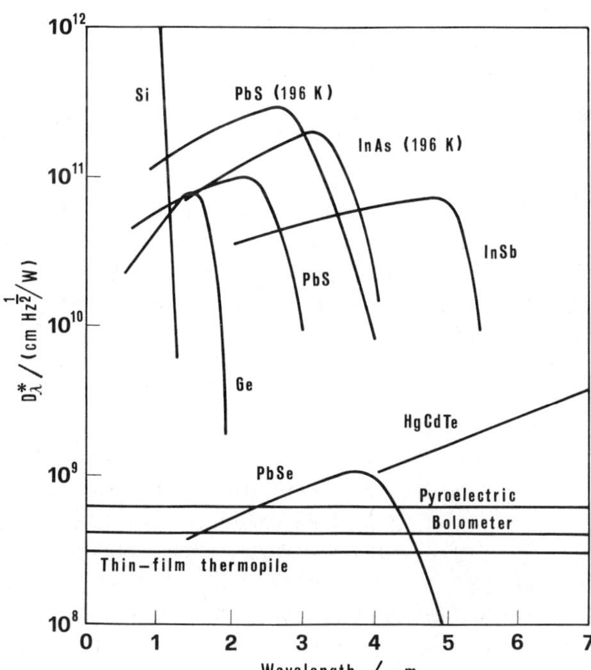

Figure 5-10.
Spectral detectivity curves of detectors commonly used in radiation thermometers.

2. Detectivity: The temperature resolution of a thermometer depends, among other factors, on the detectivity, a figure-of-merit describing in quantitative terms the resolving power of a detector. The detectivity, denoted by the symbol D_λ^*, is defined by

$$D_\lambda^* = (A\,\Delta f)^{1/2}\,[\mathrm{NEP}]^{-1} \text{ in cm Hz}^{1/2}\,\mathrm{W}^{-1} \tag{5-32}$$

where A is the effective sensitive area of the detector, Δf is the electric frequency bandwidth of the post-detector electronics, and NEP (noise equivalent power) is a figure-of-merit describing the level of rms incident radiation required to produce a signal-to-noise ratio of unity.

From the standpoint of resolution, the higher the D_λ^*, the better is the detector. Figure 5-10 shows that the detectivity of photon detectors is generally higher than that of thermal detectors.

3. Stability of Response: This is a most important factor since it determines to a large extent the reproducibility of measurements. Data on stability may be of two kinds according to whether they refer to short- or long-term stability.

Short-term instabilities are mostly due to the dependence of the responsivity on ambient temperature. Among the various detectors, silicon is known for having a relatively low temperature coefficient. Roughly, a variation of 1 °C in ambient temperature produces an output variation which corresponds to less than 0.1 °C in terms of target temperature. In contrast, the temperature coefficient of lead sulphide is large as it can be more than ten times larger than for silicon.

Compensation for the ambient temperature effect can be made according to one of the following approaches:

- control of the temperature of the detector. This is the more straightforward way, but may be too expensive for some thermometers.
- correction of the output. This requires a knowledge of the temperature coefficient and measurement of the temperature of the detector. The correction can be made automatically by using the output of a thermistor as a feedback signal.
- use of a stable reference source in the thermometer. The source can be a lamp (for example, a gallium arsenide lamp is used with lead sulphide detectors) or a small black-body source.

Long-term stability is essential to avoid the need for frequent calibrations. As for short-term stability, also for long-term stability silicon detectors are the best choice. For this reason, they are generally used in secondary-standard thermometers. Stability within 0.2 °C during a year has been obtained in thermometers of this type [17].

4. Linearity: A linear relationship between detector output and incident radiant flux, although not essential, is a useful property as it simplifies the thermometer calibration. In fact, a linear response permits accurate interpolation and extrapolation of the calibration between reference points, thus minimizing the number of points needed to cover a temperature interval. The best linearity is shown by silicon detectors. Good linearity is also shown by germanium and indium antimonide detectors.

5. Speed of response: This is not a major requirement except in specialized high-speed thermometers with a measurement capability of the order of 1 μs. The response time of various detectors ranges from a few nanoseconds for silicon to a few milliseconds for thin-film thermopiles and thermistor bolometers. Lead sulphide, which is the slowest among photon detectors, has a response time of the order of 100–300 μs.

6. Operating Mode: This mainly refers to the use of continuous (DC) or chopped (AC) radiation. Some detectors, such as silicon and germanium, can be operated in both DC and AC modes. For pyroelectric detectors the AC mode is mandatory as they are sensitive only to temperature variations. Generally, the AC mode is convenient when operating at relatively long wavelengths to measure low target temperatures. In fact, one can provide by chopping a periodic reference signal proportional to the temperature of the thermometer itself. In this way it is possible to compensate for variations in this temperature.

7. Operating Temperature: For most photon detectors the detectivity increases as the temperature decreases. Hence cooling can be a means of increasing the temperature resolution. An exception is found with silicon, which has a positive temperature coefficient at wavelengths longer than about 700 nm. For this reason, silicon detectors are usually operated near room temperature. A further effect of cooling is a shift of the responsivity curves towards the short

Table 5-4. Detectors used in radiation thermometers

Detector	Types of thermometer	Remarks
Photomultiplier	Primary standards	DC or photon-counting operation at room temperature. High cost
Si	Primary and secondary standards. Industrial single-waveband and ratio	Photoconductor or photovoltaic. DC or AC operation at room temperature. Best among IR detectors for detectivity, linearity and stability. Minimum target temperature: 400 °C. Low cost
Ge	Industrial single-waveband and ratio	Photoconductor or photovoltaic. DC or AC operation at room temperature. Good linearity and stability. Minimum target temperature: 200 °C. Low cost
PbS	Industrial single-waveband and ratio	Photoconductor. AC operation at room temperature or cooled. High temperature coefficient of responsivity. Minimum target temperature: 100 °C. Low, medium, or high cost depending on operating temperature
PbSe	Industrial single-waveband and ratio	Same as PbS but lower detectivity. Minimum target temperature: 50 °C
InAs	Special single-waveband	Photovoltaic. AC operation and cooled. Minimum target temperature: 0 °C. High cost
InSb	Special single-waveband	Photovoltaic. AC operation and cooled. Minimum target temperature: 0 °C. High cost
HgCdTe	Special single-waveband	Photoconductor. AC operation and cooled. Possibility of tailoring the peak wavelength within a wide range. Minimum target temperature: −50 °C. High cost
Pyroelectric	Industrial single-waveband and total-radiation	Thermal detector. AC operation at room temperature. Minimum target temperature: −50 °C. Low-medium cost
Thin-film thermopile	Industrial single-waveband and total-radiation	Thermal detector. AC operation at room temperature. Minimum target temperature: 0 °C. Low-medium cost
Thermistor	Industrial single-waveband and total-radiation	Same as thermopile

(Ge, InAs, InSb) or the long (PbS, PbSe) wavelength side. This effect can be used to maximize the detectivity within the desired working band. An adverse effect of cooling can be a decrease in the speed of response as with lead sulphide and lead selenide.

8. Cost: The cost of detectors of the same kind may vary according to a number of parameters such as noise performance, active area, speed of response, and inclusion of amplifiers or filters. However, in most instances high costs arise from the facilities for cooling. For this reason, when cost is a major consideration, uncooled detectors are to be prefered. The consequent loss in temperature resolution can sometimes be compensated for by enlarging the working band of the thermometer.

A resumé of the main features of detectors used in radiation thermometers is given in Table 5-4.

5.6.3 Signal-Processing Facilities

In modern thermometers, the low-level signal from the detector is first amplified and then processed to suit a number of application needs. The principal signal-processing facilities are as follows:

i. Linear Analog or Digital Indicator: Because the target radiance is markedly nonlinear with temperature, so is the detector signal. In order to obtain temperature readings on an indicator, variable signal increments must be processed to give each a unit temperature increment. Linearization can be implemented by approximating the signal versus temperature curve with linear segments and using for this linear-amplification circuits. In more recent and sophisticated thermometers, linearization is performed numerically by microprocessors, so residual errors due to an insufficient approximation are avoided.

ii. Linear Analog Output: A linear current or voltage output (4–20 mA, 0–20 mA, 0–10 V, etc.) compatible with process control instrumentation is common in thermometers for fixed installation. In hand-held thermometers the same facility is sometimes provided by a separate A/D converter.

iii. Time Functions: When a radiation thermometer is employed in an on-line system, its signal may be subjected to large fluctuations. Time-function facilities are useful to retain the signal levels that are more likely to represent the true target temperature. The principles of the main time functions are shown in Figure 5-11.

The *peak picker* function holds a peak signal and allows it to decay slowly until the arrival of the next peak. It is used in such applications as rolling mills where the signal is occasionally lowered by steam, smoke, water droplets, metal scale, etc. Just the reverse are the principle and utilization of the *valley picker.*

The *track (sample) and hold* function allows smoothed signals to be obtained from intermittent events. These may originate from objects on a conveyor belt or targets where the view is periodically obscured by rotating machinery. The output sampling is activated by a command signal received by an external switch that can be actuated by the belt or rotating machinery itself. The output is held when the switch is operated until the next sample command is received.

Rapid temperature fluctuations about a true mean value can make the thermometer output unsuitable for recording or control. In these cases the *averager* function can be used to provide a smoothed signal.

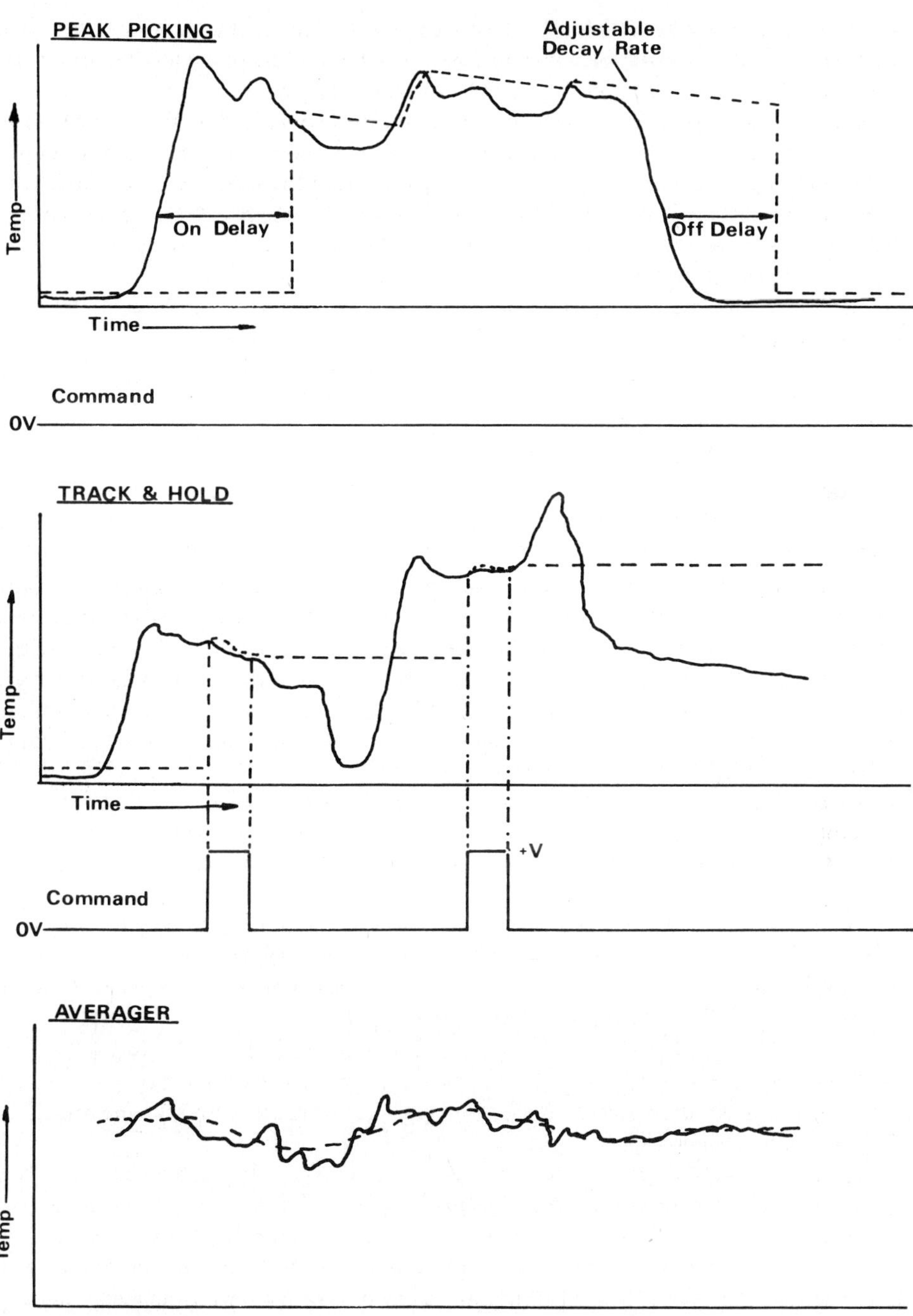

Figure 5-11. Illustration of the principles of time functions.

iv. Emissivity Corrector: As indicated in Section 5.4.2, the temperatures measured by a single-waveband thermometer (and also for a total-radiation thermometer) are lower than the true values when the target emissivity is lower than unity. An emissivity corrector compensates for this effect by increasing the amplifier gain by a multiplicative factor equal to the inverse of the emissivity value. A similar principle applies to the emissivity correctors that are provided in ratio thermometers to compensate for the slope of emissivity with wavelength. Emissivity correctors can also be used to account for absorption losses.

v. Memory Circuits: These allow retrospective analysis of minimum, mean, and maximum values observed over a preset period.

vi. Communication Links: These allow the thermometer to be interrogated or reset by a process computer, or to be linked to a digital indicator or printer.

5.6.4 Single-Waveband Thermometers

This class of thermometers includes the popular disappearing-filament (or optical) pyrometer, precision thermometers that are used as primary or secondary standards, and the large family of infrared thermometers for practical applications.

5.6.4.1 Disappearing-Filament Pyrometer

The basic scheme of this thermometer is illustrated in Figure 5-12. The minimum operating temperatures of the disappearing-filament pyrometer is about 700 °C and the temperature resolution is of the order of 1 °C at 1000 °C. The main limitation in this thermometer is the impossibility of using it as a sensor for automatic control. For this reason, it has been superseded by thermometers with electrical sensors in many applications.

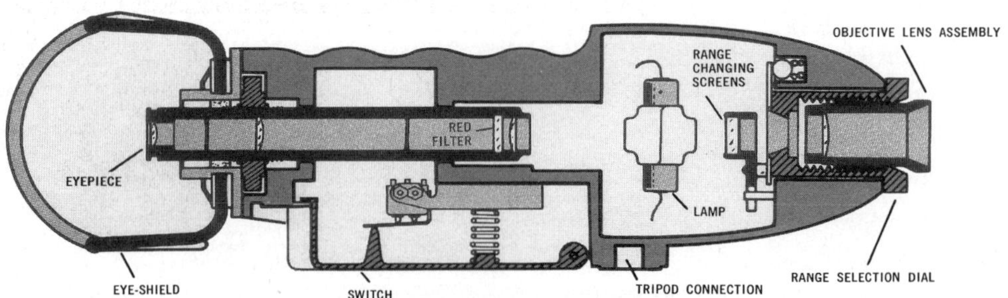

Figure 5-12. Disappearing-filament pyrometer (courtesy of Leeds and Northrup). The objective lens allows the image of the target to be superimposed on the filament of the miniature lamp. The red filter is used to restrict the useful spectral interval to a narrow band around 650 nm.

5.6.4.2 Standard Thermometers

This category includes:

a. *primary-standard* thermometers that are used to realize the International Temperature Scale (ITS);
b. *secondary-standard* thermometers that are used to disseminate the International Temperature Scale.

a. Primary-Standard Thermometers

The optical layout of a primary-standard thermometer is shown in Figure 5-13. Common features of thermometers of this type are the following:

1. They do not carry a calibration but are simply used as photoelectric comparators of the spectral radiances of two external sources. These can be a fixed-point black body and a tungsten ribbon lamp.
2. The effective wavelength is usually about 660 nm. In recent versions a working wavelength between 900 and 1000 nm is also provided.
3. Narrow-band interference filters are used to define the effective wavelength. Their bandwidth is typically of the order of 10 nm. For a good definition of the working wavelength (to an accuracy of at least 0.1 nm) the transmittance outside the working band must be less than 10^{-5} times that of the peak.
4. A photomultiplier with S-20 response (trialkali cathode) and a silicon photodetector (the use of this detector in precision thermometry was first suggested by Ruffino [18]) are generally used to work at 660 and 900–1000 nm, respectively. The two detectors can be included in the same instrument, as in the example in Figure 5-13.
5. The temperature resolution is a few mk at the gold-point temperature (1064 °C).
6. Typical minimum target sizes for these thermometers are about 1 mm diameter.

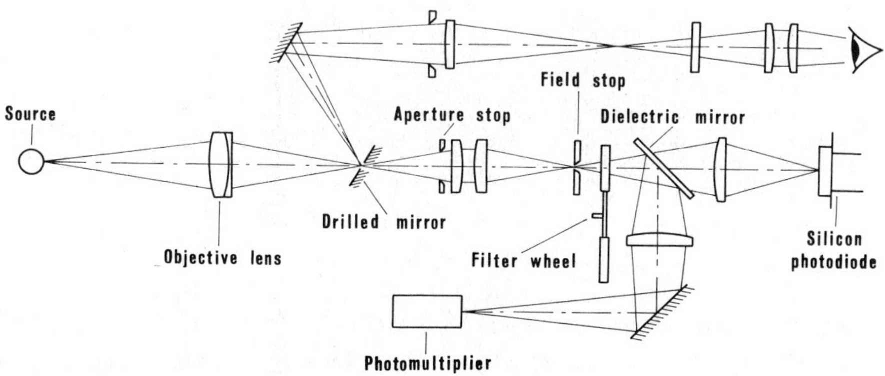

Figure 5-13. Optical layout of the primary-standard thermometer used at IMGC (Italy).

b. Secondary-Standard Thermometers

These are precision instruments that are used by industrial calibration laboratories as local references for calibrating field thermometers. They can also be used as an alternative to tungsten ribbon lamps to transfer the ITS from primary to secondary laboratories. For this reason, they are sometimes called *transfer standards* [19].

The main features of these thermometers and the differences with respect to primary standards are as follows:

1. They are simpler in construction and more compact than primary standards (Figure 5-14).

2. They are calibrated in terms of detector signal versus target temperature. Long-term stability is then an important requirement for the detector. For this reason, silicon detectors are generally used.

3. The minimum working temperature can be as low as 400 °C, whereas that of primary standards is about 700 °C, which is the minimum temperature where tungsten ribbon lamps are sufficiently stable. The possibility of operating down to 400 °C enables them to be used in place of resistance thermometers or thermocouples to measure the temperature of blackbody cavities that are used to calibrate infrared thermometers. The consequent advantages are outlined in Section 5.7.2.

Figure 5-14.
A commercial transfer-standard thermometer (courtesy of Eurotron srl).

5.6.4.3 Infrared Thermometers

Specific features of commercial single-waveband thermometers working in the infrared are:

1. They operate in wavebands within the spectral interval 0.5–14 μm. The criteria for selecting the working wavebands have been discussed in Sections 5.3.4 and 5.4.2. The main applications for thermometers working in different wavebands are shown in Table 5-5.

2. In most infrared thermometers refractive (lens) optics are used. The use of reflective (mirror) optics is generally restricted to thermometers operating at the long-wavelength side, eg, in the atmospheric window 8–14 μm. Optical layouts of thermometers with lens and mirror optics are shown in Figures 5-15 and 5-16, respectively.

Table 5-5. Applications of infrared single-waveband thermometers

Working wavelength/μm	Applications
0.7–1.1	General-purpose, high-precision, and industrial measurements
1.1–1.7	Medium temperature range general-purpose measurements. Metal and glass industries
2.0–2.5	General-purpose industrial measurements
3.43	Thin-film plastics (hydrocarbon polymers), oils, waxes, paints, paper
3.9	Gas-and oil-fired furnaces
4.4–4.6	Flame temperature measurement
4.8–5.2	Glass and ceramics surface temperature
7.9	Thin-film plastics (polyesters, fluoro carbons, polyamides). Thin glass
8 –14	General-purpose low-temperature measurements. Long-range measurements

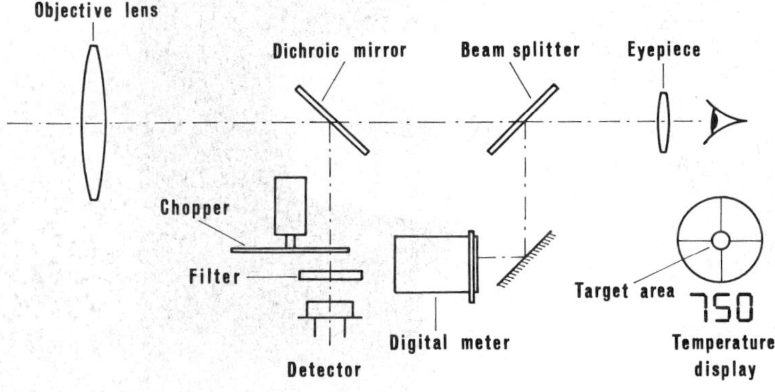

Figure 5-15. Infrared thermometer with lens optics.

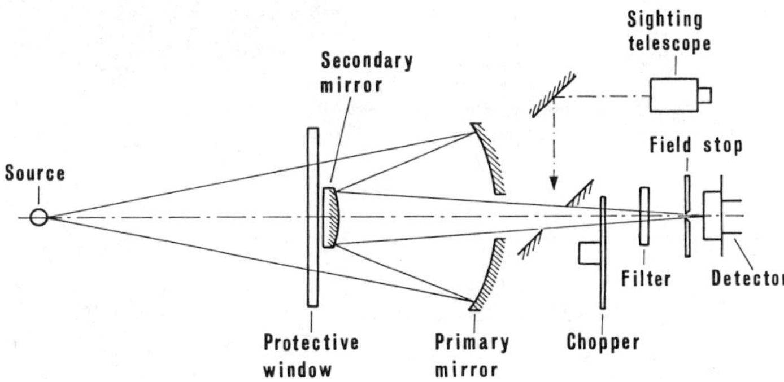

Figure 5-16. Infrared thermometer with mirror optics.

3. Thermometers for either portable use (hand-held) or permanent and semipermanent installation are available. The optical system of the former is generally more complex, which leads to a higher cost. Thermometers for permanent installation are more rugged and can be supplied with cooled (air or water) mounting jackets or screens for operation in adverse environments.

4. The values of the parameters characterizing infrared thermometers (accuracy, resolution, etc.) may vary over a wide range owing to the large variety of available instruments. Moreover, these parameters are often interdependent, so that one of them has to be traded with another (eg, a smaller target entails a higher low-temperature limit). Hence it is impossible to give them absolute limits. However, to give a potential user an idea of what can be achieved with commercially available industrial thermometers, indicative values for a number of parameters are given in Table 5-6.

5. Various signal-processing facilities as described in Section 5.6.3 are usually provided.

Table 5-6. Parameters characterizing commercial single-waveband infrared thermometers

Parameter	Typical values	Best values
Accuracy	±0.5–4% of reading	±0.1% of reading
Resolution	0.1–1 °C	0.01 °C
Repeatability	±0.5–2 °C	±0.1 °C
Distance factor (measuring distance/target size)	20–300	500
Response time	0.1–2 s	0.1–1 ms

5.6.5 Ratio Thermometers

A typical scheme for ratio thermometers is shown in Figure 5-17. In this scheme two bandpass filters mounted on a rotating wheel alternately cross the radiation beam. The detector thus receives in sequence radiation fluxes concentrated in two different spectral bands. From the ratio of the resulting signals it is possible to derive the target temperature according to Equation (5-20).

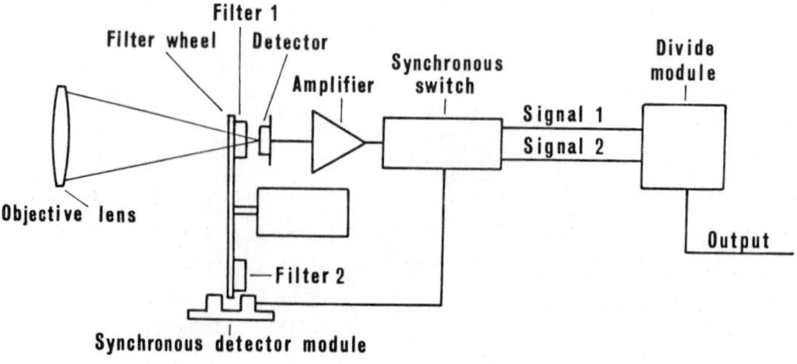

Figure 5-17. Schematic diagram of a ratio thermometer.

Alternative ways to provide a double spectral channel include the use of beam splitters or bifurcated optical fibers. A solution which avoids the use of moving parts or of beam-splitting components relies on the use of dual detectors where two sensing elements of the same or different type (eg, Si and Si or Si and PbS) are stacked together in the same case. In this way, bandpass filters are also not needed, since the two operating bands are delimited by the responsivity curves of the sensing elements.

As outlined in Section 5.3.4, ratio thermometers are inherently less sensitive and accurate than the single-waveband type. In the best cases commercially available instruments have a temperature resolution of 1 °C and an accuracy of ±1% of the reading.

5.6.6 Total-Radiation Thermometers

Mirror components and thermal detectors such as thermopiles, thermistor bolometers or pyroelectrics are generally used in total-radiation thermometers, since they provide a wide operating spectral range.

The main advantages of total-radiation thermometers are their ruggedness and relatively low cost. However, these may be largely offset by the disadvantages stemming from operating in a wide spectral band, ie, a high dependence on emissivity, absorption, and reflection effects.

5.7 Calibration

The characteristics of radiation thermometers are so different among the various models that no internationally agreed standards exist for their calibration. The calibration procedures thus largely depend on the experience and care of manufactures and calibration laboratories.

The purpose of the calibration is to relate the thermometer output to target temperatures, the target being assumed to be a perfect black body.

The calibration procedures depend on both the type of thermometer and the desired accuracy. The latter in turn is related to the place taken by the calibration laboratory within a calibration chain. In any case, an important requirement for a calibration procedure is to ensure the *traceability* to national standards.

Calibration procedures may be classified into two main groups: absolute and comparison calibrations.

5.7.1 Absolute Calibrations

An absolute procedure is one that allows a thermometer to be calibrated by making recourse to fixed-point apparatus and interpolating equations. A knowledge of some properties of the thermometer is required, but there is no need for any reference thermometer.

The two main types of absolute procedure refer to:

1. realization of the ITS;
2. fixed-point calibration.

1. Realization of the ITS

The realization of the ITS-90, which means the calibration of a primary-standard thermometer, requires the following operations: calibration at a fixed-point temperature (any of the silver, gold, or copper freezing points), determination of the effective wavelength of the thermometer, and realization of a scale of radiance ratios or, alternatively, measurement of the nonlinearity of the detector.

Detailed descriptions of procedures for realizing the ITS have been reported in various papers, the most recent being that by Jones and Tapping [20]. The uncertainties in the realization range from ±0.02–0.05 °C at the gold point (1064 °C) to ±0.2–0.5 °C at 2000 °C.

2. Fixed-Point Calibration

An alternative way to perform an absolute calibration relies on the use of more than one fixed point to derive the coefficients of an interpolating equation for the thermometer output versus temperature. According to Sakuma and Hattori [21], the equation

$$V(T) = C \exp(-c_2/(AT + B)) \tag{5-33}$$

where $V(T)$ is the output voltage, T the absolute target temperature, and A, B, and C are constant coefficients, is adequate for silicon-detector thermometers. The coefficients A, B, and C can be determined by measuring $V(T)$ at three fixed points which can be chosen among the Zn, Al, Ag, and Cu points, according to the temperature range of interest. A procedure of this kind, which does not require any measurement of the spectral characteristics of the thermometer, is suitable for the calibration of precision thermometers to be used as secondary standards at temperatures below the gold point. Using this technique, calibration uncertainties of the order of ±0.05 °C were obtained by Battuello et al. [22] in the temperature interval delimited by the Zn and Cu points.

5.7.2 Comparison Calibrations

In a comparison procedure a radiation thermometer is calibrated against another thermometer which is assumed to be a standard. This can be either an immersion thermometer (resistance thermometer or thermocouple) or a radiation thermometer.

The use of a radiation thermometer as a standard is recommended:

a. When a pyrometric lamp can be used as a calibration source. This occurs in the temperature range between 700 and 2300 °C with narrow-waveband thermometers operating near 0.65 μm or in the infrared region below 1 μm [23]. A further requirement is that the minimum target size does not exceed 3 mm, which is the maximum width of the ribbon of pyrometric lamps. The use of a magnifying lens in front of a lamp must be considered with care as it could give reproducibility problems owing to difficult alignment and focusing.

b. When temperature gradients exist inside the black-body cavity to be used as calibration source. This mainly occurs in the high-temperature region above 1100 °C. It has been shown [24] that large temperature gradients can be tolerated when a radiation thermometer is used as a standard, provided that the black-body cavity is properly designed.

A standard immersion thermometer can be convenient:

c. When operating in the low-middle temperature range where precision radiation thermometers are not available. This is the case of temperatures below 400 °C, which is the low operation limit of silicon-detector thermometers. On the other hand, it should be noted that below 400 °C the attainment of isothermal conditions inside a black-body cavity is not a difficult task.

d. When cost is a major consideration. Generally, an immersion thermometer is less expensive than a radiation type, although moderately priced silicon-detector thermometers suitable for use as secondary standards have recently become commercially available. The temperature region where cost is more likely to determine the choice of the standard is from 400–1100 °C.

The calibration source can be either a pyrometric lamp or a black-body cavity. Pyrometric lamps have been used for so many years that their properties are now well established [25–27]. Much more varied are the possibilities of the realization, choice, and use of black-body cavities and related equipment (baths, furnaces). Common solutions are listed in Table 5-7 and some of them shown in Figure 5-18.

Table 5-7. Equipment for realizing black-body calibration sources

Type of equipment	Typical temperature range of use/ °C
Liquid baths (water, silicone oils)	0–250
Salt baths (nitrate salts)	150–600
Wire-wound resistor furnaces (nichrome, Kanthal resistors)	200–1100
Silicon carbide furnaces (SiC rods)	500–1500
Tubular-heater furnaces (graphite, molybdenum, tantalum, tungsten)	1000–3000
Heat-pipe devices (hydrocarbons, water, sodium)	–50–1100

5.8 Applications

It would be impossible to list all the applications of radiation thermometers and this section therefore concentrates on two major application areas, the iron and steel industry and the glass industry. Most of the special instruments and methods that have been developed to overcome difficult measurement problems are found in these areas. It is assumed that the reader is familiar with the general principles of the processes being described.

5.8.1 Iron and Steel Industry

5.8.1.1 Blast Furnace Stove Dome Temperatures

Measurement Objectives and Problems

The optimum cycle time for the use of stoves can best be determined by measuring the refractory temperature either looking upwards at the dome or downwards into the checker brickwork. Problems are due to the high pressure in the stove and difficulty of access.

a

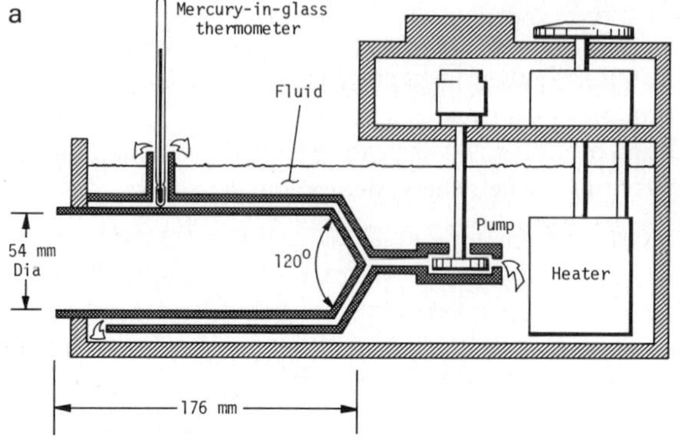

b

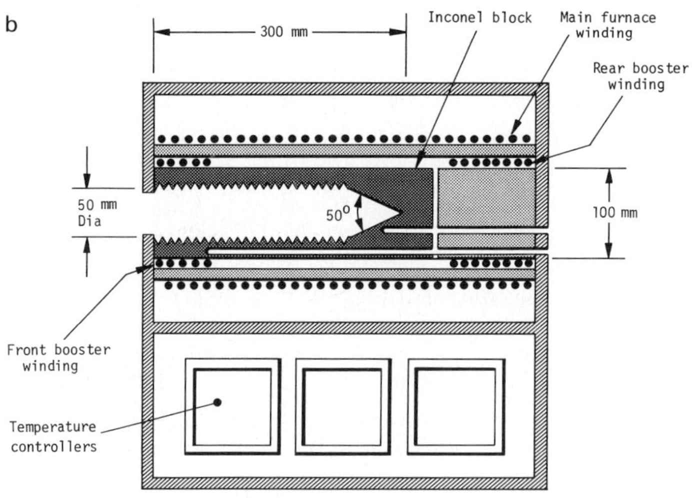

c

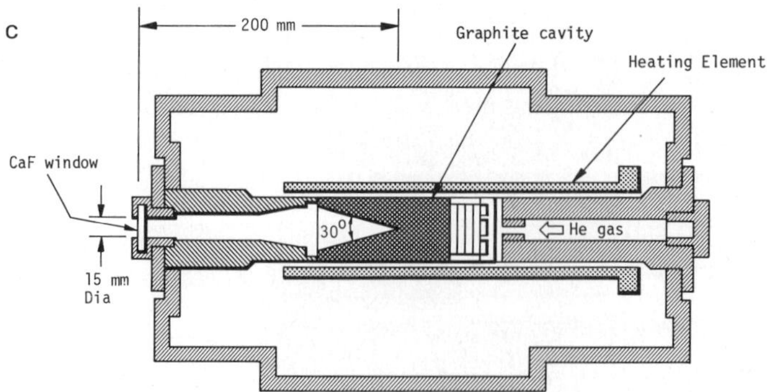

Figure 5-18. Black-body transfer sources (from Barber and Brown [28]). (a) liquid-bath source, 0–200 °C; (b) metal-cavity source, 150–950 °C; (c) high-temperature source, up to 2600 °C.

Solutions

The measurement is straightforward using a single-waveband short-wavelength thermometer. However, specially constructed mounting accessories are essential and provide a high-pressure sealed window with either a manual or motorized valve to seal off the system to facilitate repairs and to protect the window from dirt when the system is not being used (see Figure 5-19). Two or three systems per stove are sometimes used to provide back-up confirmation.

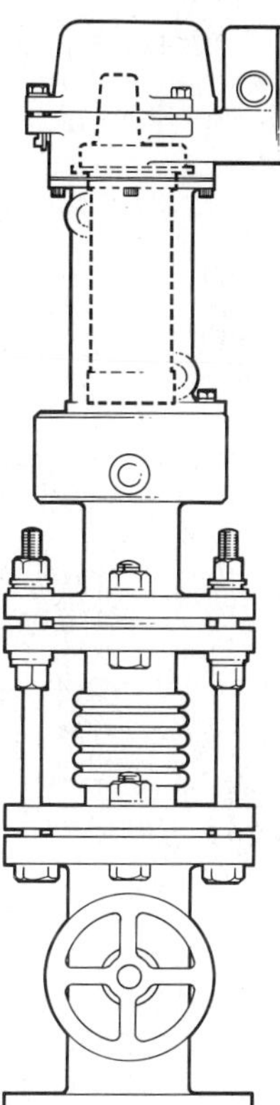

Figure 5-19.
Thermometer mounting for stove-dome temperature measurement in blast furnaces.

5.8.1.2 Coke Ovens

Measurement objectives and problems

The objective is to provide uniform temperatures across the coke-oven batteries. Problems are due to the large number of measurement points involved if flue temperatures are measured and the hazardous and unpleasant nature of the environment.

Solutions

a. Portable Instruments: For many years measurements were made by sighting a disappearing-filament pyrometer (DFP) down each flue and measuring the temperature of a refractory tile at the base. The DFP has now been superseded by a portable short-wavelength thermometer. The thermometer can provide digital communication with a portable data logger which can be programmed to assist the operator in entering the required data in the correct order, and remove the necessity for normal data entry. The data can later be downloaded and analyzed on a computer.

b. Fixed Systems: British Steel Corporation [29] have shown that a profile of each coke oven can be obtained by analyzing the output of three thermometers fixed to the mobile coke car in a position to measure the coke as it is pushed from the ovens into the car. This system requires the use of small-target short-wavelength thermometers, with peak-picking processing. This combination enables the true bulk temperature of the coke to be recorded as the thermometer periodically views the cavities created by the surface "cracking". The information of oven number, pusher position, etc., together with temperature are transmitted by radio link to a nearby data logger/computer which provides the desired control and management data.

An alternative solution has been evaluated by Sollac in France [30] (see Figure 5-20).

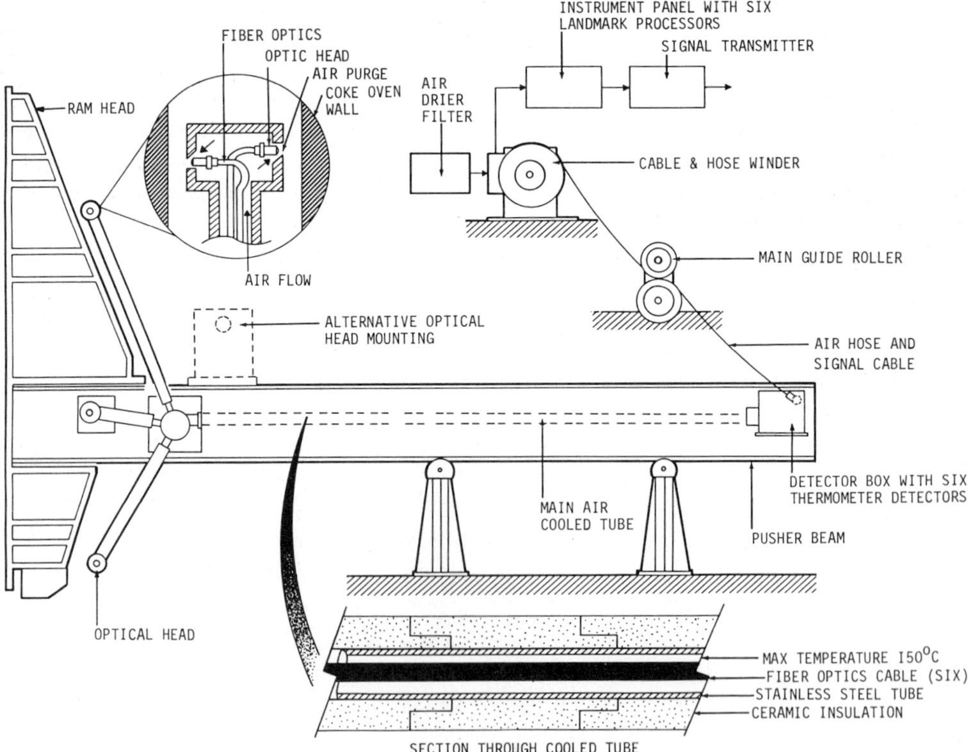

Figure 5-20. Sollac's solution for recording temperature profiles in coke ovens. Six fiber-optic heads are mounted in the pusher arm so that the temperature gradient down both walls of each chamber can be measured at three heights every time the pusher is used.

5.8.1.3 Billet or Slab Temperatures in the Spray Chamber and After Leaving the Casting Machine

Measurement Objectives and Problems

Considerable studies, mainly in Japan, have shown that product quality, surface cracking, and inclusion levels can be controlled if the temperature profile is correctly controlled during the spray cooling process. Problems are due to the presence of water droplets near the target, the high temperature and dirty environment, the damage caused if a break-out of liquid steel occurs, and the presence of scale on certain alloys.

Solutions

Several manufacturers have supplied special equipment to minimize the above problems. Two representative systems are described.

1. A short-wavelength thermometer is housed in a robust water-cooled jacket and purge unit which can be fitted onto the roll mounting structure and placed near the metal. A dual air-purge system minimizes the spray interference and keeps the lens clean.
2. Figure 5-21 shows schematically a small optical head housed in a simple purged tube and a robust multiple-fiber light guide which enables the detector and electronics to be housed outside the chamber. A quick-release system on the optic head facilitates removal during roll changes [31].

Both systems require signal processing, with peak picking or peak sampling, to minimize errors due to the presence of scale. Even peak picking is not sufficient for some carbon steels if measurements are needed between the top rolls. For these difficult alloys and positions a scraper method has been introduced to remove the scale at the measurement point. However, it is generally possible to avoid the heavily scaled area.

Both ratio and single-waveband (short-wavelength) thermometers have been tested. In general, the simpler single-waveband model is satisfactory. Method 2 has the major advantage of minimizing the cost of repair if damage occurs near the strand. Both methods require careful control of the air purge position in order to avoid overcooling at the measurement point.

Measurement outside the spray chamber is considerably easier and can be treated as a steel strip rolling application.

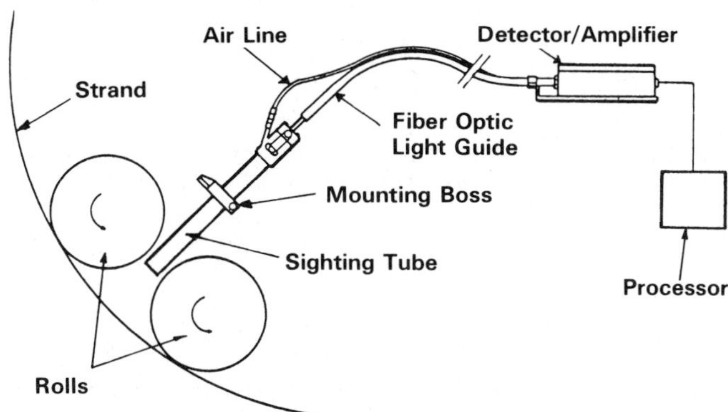

Figure 5-21. Billet or slab temperature measurement in the spray chamber with a fiber-optic thermometer.

5.8.1.4 Billet or Slab Reheating Furnaces

Measurement Objectives and Problems

A knowledge of the steel surface temperature at all stages of the heating process enables the burner controls to be optimized for energy conservation, and excessive scaling to be avoided. Furnaces can be controlled by computer model, but actual measurements are highly desirable if unplanned stoppages occur.

In the final soaking zone the measurement is straightforward, as the conditions are nearly black body. Almost any model of thermometer would be suitable. However, in the preheat and heating zones there are two major problems. The most significant problem results from the radiation reflected from the steel. This radiation originates from the much hotter walls and flames near the target and, even with the high emissivity value of the oxidized steel, can result in falsely high readings of several hundred degrees. The second problem is to avoid errors due to emission from the combustion gases in the line of sight.

Solutions

In applications where the product being heated is stationary, it is not possible to shield the measurement point as a cold spot would result. The most satisfactory method is to measure and compensate for the reflected component.

Trials have shown [32] that this can be satisfactorily achieved provided that the measurement is made with a carefully selected long-wavelength thermometer. It can be shown from Wien's Equation (5-12) that, for any pair of product and furnace temperatures, the departure of the measured from the true temperature is greatly reduced if a long wavelength is chosen. An example is shown in Figure 5-22.

The choice of the wavelength is very restricted as it is also necessary to avoid any waveband where the hot combustion gases will absorb and emit energy. Thermometers with a narrow band at 3.9 μm are optimum compromises for this type of application.

It has been demonstrated that the incident energy can either be computed from a knowledge of the general furnace temperature measured by a thermocouple in the roof near the measurement point or directly measured by a second radiation thermometer viewing the roof. The true temperature is then calculated from the output of the two thermometers (ie, radiation-radiation or thermocouple-radiation) and a knowledge of the emissivity of the metal. The latter can normally be assumed to be 0.82 for most alloys.

The thermometers must be positioned so as to minimize the interference by burners. In gas-fired furnaces this is not a problem, however; in oil-fired furnaces it is often necessary to instal a long sighting tube to within 0.5– 1 m of the steel.

This technique has been very successful in solving a very difficult application problem. It can be applied to some other applications in which reflected radiation is the major problem.

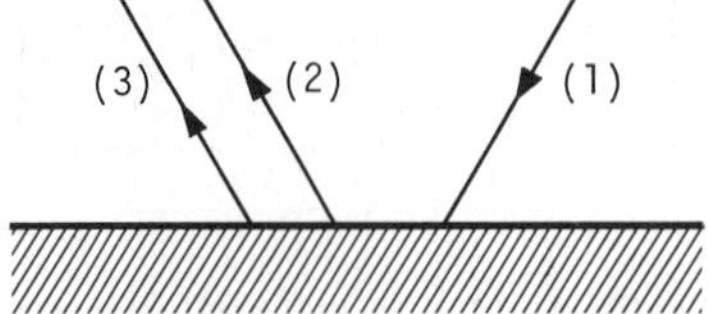

Figure 5-22.
Reflection effect in reheating furnaces. It is shown that the measurement error decreases as the wavelength increases.

5.8.1.5 Rolling Mills

Measurement Objectives and Problems

This is a major application area for radiation thermometers. It is imperative to measure the product temperature at several positions both to ensure the correct metal properties and to protect the plant. Cold steel can seriously damage expensive rolls. Problems stem from the presence of water droplets and steam on the metal in the line of sight and the presence of scale patches.

Solutions

There are many types of rolling mill and the conditions vary greatly depending on the mill design and on the position on a particular mill. However, in 95% of applications a sighting position can be chosen in which the interfering factors can be reduced to an extent that enables a single-waveband short-wavelength thermometer with suitable peak picking and other signal processing to give satisfactory readings (±5–10 °C).

Several techniques are worthy of note:

1. If an acceptable measurement position cannot be found which is adequately free of water droplets, these can be swept clear, prior to the measurement point, by means of a cross-jet of high-pressure water.

2. In severe cases it is possible to design a mounting system which can be mounted to view upwards. A ruggedized version of the probe developed for the spray chamber has been found to give satisfactory freedom from dirt on the lens (see Figure 5-23).

3. If the position of the metal on the table cannot be controlled to ensure that the field of view of a thermometer is always filled, then several alternative techniques are available:

3a. View the metal at its edge with a specially designed thermometer that gives a parallel sight path for the required distance. Choose a sighting position along a roll so that the vertical position of the metal can be assured (see Figure 5-24).

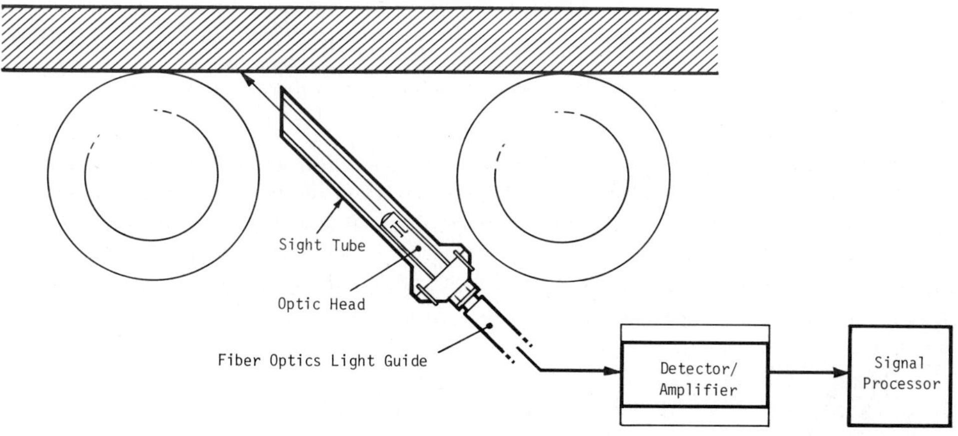

Figure 5-23. Thermometer mount for measurements in rolling mills under severe conditions.

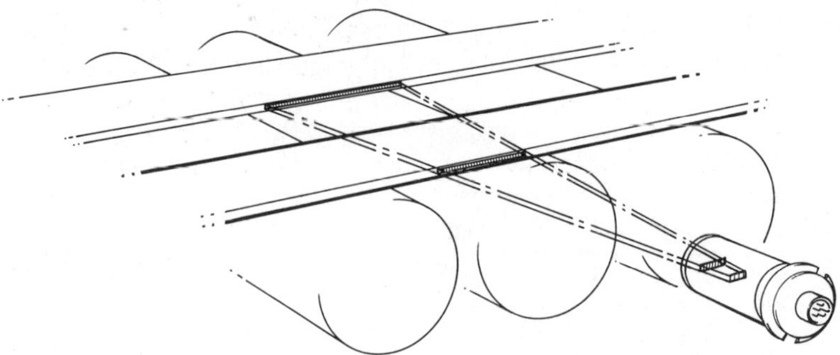

Figure 5-24. Thermometer installation to account for changes in the position of the metal in rolling mills.

3b. *Rod Mill (Method I):* If the diameter of the rod is known, it is possible to design a thermometer with a rectangular field of view at its focusing distance. The dimensions are chosen such that the rod is always within the long dimension of the rectangle (see Figure 5-25). The processor must be informed of the rod diameter and will calculate the rod temperature. This method is simple, but care must be taken to ensure that the rod is always at the focal plane of the thermometer, as this is the only position at which the sensitivity is constant over the full field of view.

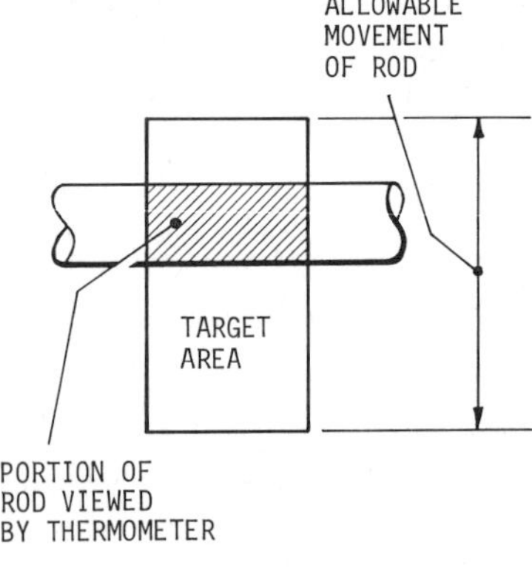

Figure 5-25. Use of thermometer with rectangular field of view in rod mills.

3c. *Rod Mill (Method 2) Cooling Table:* If the rod can be restrained to ensure that it is always in the field of view of a ratio thermometer, the reading will be independent of its position (see Section 5.4.2.2).

3d. *Rod Mill (Method 3) Cooling Table:* If the rod is coiled on a moving table for controlled cooling (eg, Morgan mill), it is possible to sight a narrow-angle thermometer at a position where the several layers of rod provide an adequate target [33].

3e. *Strip Mill-Temperature Gradients:* Special thermometers and methods have been developed to measure the lateral temperature gradients in rolled strip:

1. For the higher temperature range (>600 °C), thermometers incorporating an array (typically) of silicon detectors are available.

2. An alternative approach, that is applicable over a wider range of temperature, is to incorporate a high-speed thermometer in an optical scanning unit which provides a series of angled scans across the strip.

Both these methods must incorporate suitable data-logging and display computers to supply the data required by the user.

5.8.1.6 Bright Strip (Annealing, Galvanizing, Coating, etc.)

Measurement Objectives and Problems

Temperature is a critical parameter in nearly all bright-strip processes. The demand for accurate measurement is continuing to grow and the many basic difficulties have led to the proposal and use of a wide variety of methods. No ideal universal solution has yet been produced. Problems occur because the emissivity of the strip is low and occasionally variable (this is the major problem), the low temperatures often preclude the use of short-wavelength thermometers (eg, silicon and germanium), and the surroundings are frequently hotter than the strip.

Solutions

There is no single solution and, in fact, several applications have still not been satisfactorily resolved. The solutions listed here include some methods that are well tried in certain applications and others that have been used in a limited number of cases, but which illustrate principles that can be considered.

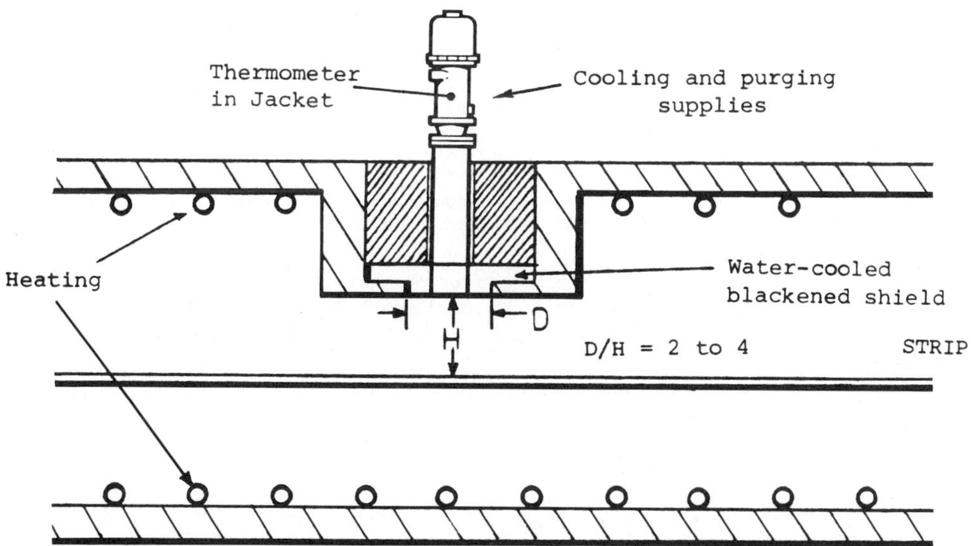

Figure 5-26. Use of cooled shield to eliminate reflection errors in the temperature measurement of bright strips.

1. Continuous Annealing – Controlled Atmospheric Furnaces (Catenary and CAL Furnaces):

1 a. In several processes the steel is partially oxidized (even in the controlled atmosphere) and at short wavelengths the emissivity is above 0.5 and is reasonably stable. The major problem is the elimination of reflection errors. In these instances, very satisfactory performance can be achieved by the use of a cooled shield either built into the furnace (see Figure 5-26) or added in the form of a water-cooled probe [11]. The thermometer chosen should be the shortest possible wavelength single-waveband type that can meet the minimum temperature requirement.

Because the strip is moving rapidly, the cooling effect of the cooled plate does not significantly affect the temperature of the strip, but it does enable the reflected radiation to be eliminated.

The minimum size and spacing requirements for the shield must be carefully calculated. They depend on the degree to which the strip is a diffuse or specular reflector. A typical ratio of diameter to spacing of 2–4 will cover most products. If the size is not acceptable, then the product reflective properties should be studied to enable a calculation to be made of the minimum size. In multi-stage processes this technique is often applied in the connecting sections.

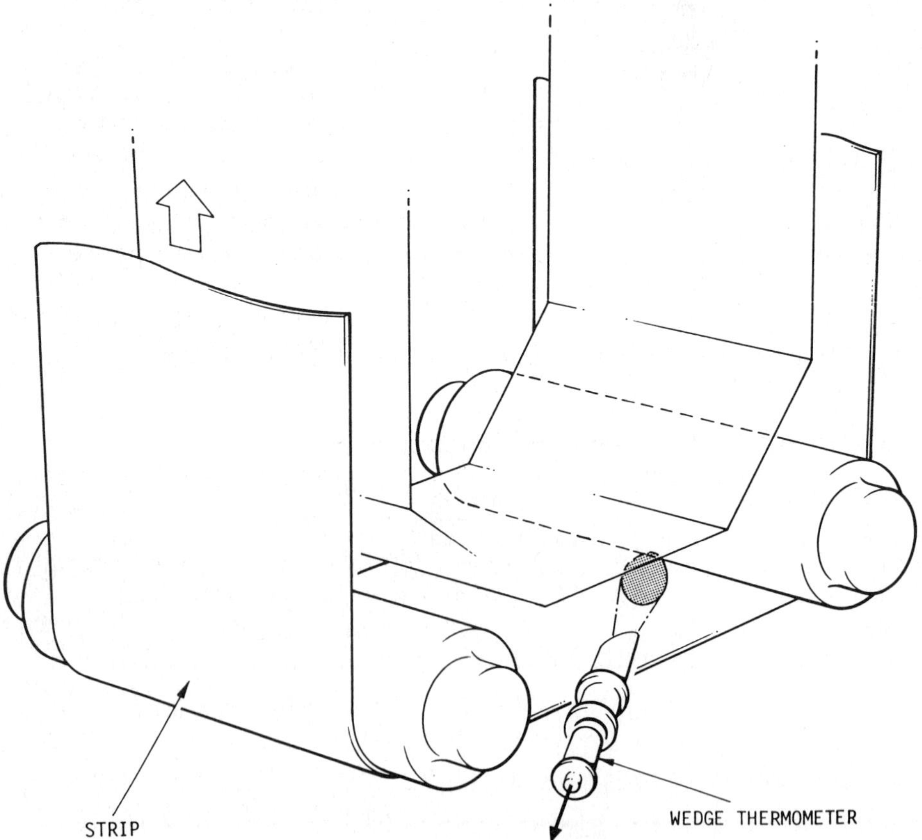

Figure 5-27. Alternative solution to cooled shield for the temperature measurement of bright strips.

1b. A second technique can offer a mechanically simpler solution in certain furnaces. Again, the shortest wavelength single-waveband thermometer is used (although it has been shown that the wavelength is not critical). It is sighted on the strip just after having passed over the roll (see Figure 5-27). In this position multiple reflections between the strip and the roll create an effectively high apparent emissivity (nearly black-body condition). The emissivity control should be set to 1.00.

Tests at Swedish Steel [34] have shown that the measurement at this point gives a reading typically 20 °C higher at 720 °C than that measured a few meters after the roll. This is due to the surface layer of the material rapidly reaching equilibrium with the hotter roll and equally rapidly reaching equilibrium with the bulk temperature after leaving the roll. However, this difference is very reproducible and once the value has been established the "wedge" temperature provides a very stable control point.

2. Special Techniques for Bright Strip: Many techniques have been proposed and tested to minimize the problems of low and variable emissivity, including the polaradiometer and auxiliary-reflector methods described in Section 5.4.3.

It is very revealing that although all the above methods have been proved to give adequate results, none has so far received broad industrial acceptance. This is probably due to a combination of cost and complexity. The user who has an urgent or cost-effective need to solve a particular problem and the technical competence to instal and support the equipment can find a successful method, even for difficult applications. However, other users are reluctant to follow this lead and prefer simpler although perhaps less accurate solutions.

5.8.1.7 Induction Heating

Measurement Objectives and Problems

The objectives are to achieve a desired temperature profile history and a desired product temperature for future processing. Problems stems from the significant levels of electromagnetic fields that can exist near the coil, and limited access.

Solutions

Both of the above problems can be overcome by the use of a short-wavelength single-waveband fiber-optic thermometer. The optical head, and hence the lens diameter, are much smaller than those of the typical standard thermometer. The thermometer can be sighted through a very small gap created between adjacent coils. The detector and electronics can be positioned sufficiently far from the coil to eliminate electrical pick-up. Correct earthing procedures should be followed.

5.8.2 Glass Processes

Measurement Objectives and Problems

The glass industry is the second major user of radiation thermometers in many processes. It is not possible to itemize all applications, so here we shall concentrate on the major problems and solutions. Problems stem from the transparency of glass, which obviously precludes the use of any type of short-wavelength thermometer if surface temperatures are required. Other problems relate to the specific application and some are covered in the solutions.

Solutions

The absorbing and reflective properties of glasses vary greatly in the infrared region (see Figure 5-28). It is therefore possible to obtain several different measurements of temperature by radiation methods:

- at *long* wavelengths it is possible to interpret the reading as being a surface temperature;
- at *short* wavelengths it is possible to look through a thin window of "white" glass and measure a hot background temperature, or to look into the bulk of a bath of molten glass;
- at *intermediate* wavelengths it is possible to measure subsurface temperatures;
- using *two* wavebands it is possible to estimate the temperature gradient in the glass.

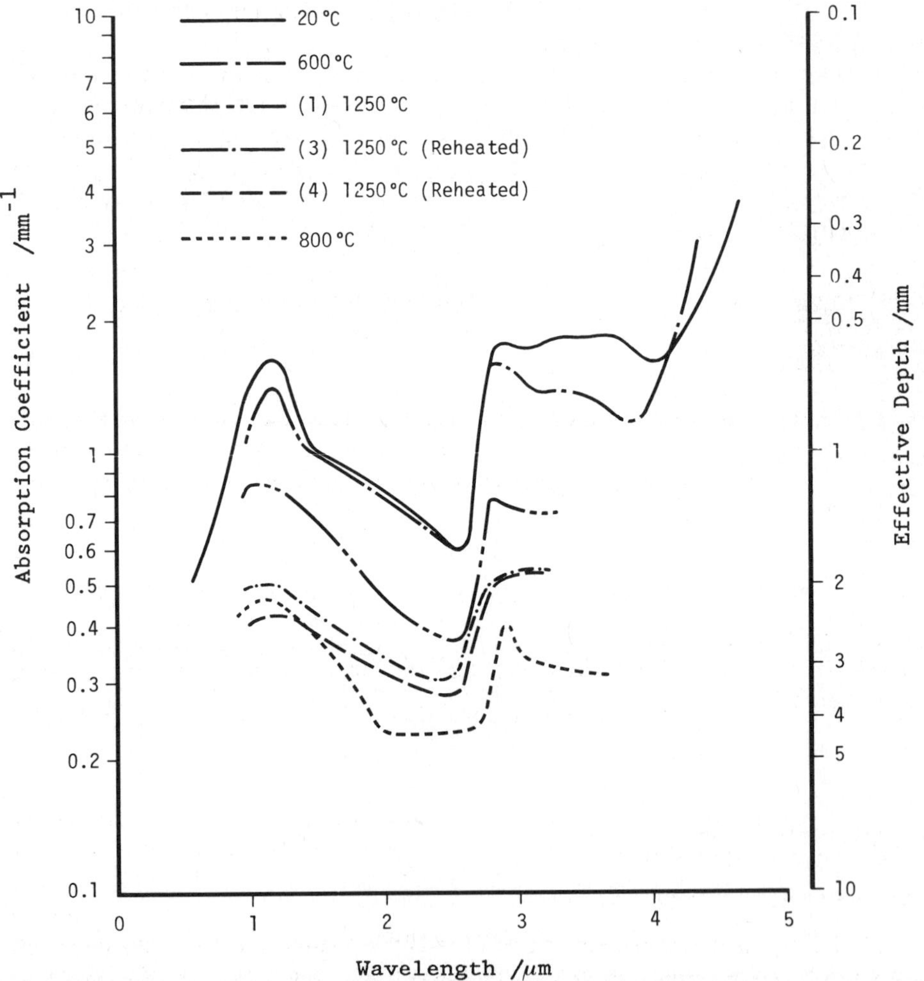

Figure 5-28. Effect of temperature on the absorbing properties of green glass.

5.8.2.1 Concept of Depth of Measurement

The concept of depth of measurement is useful for interpreting the measurements of glass temperatures.

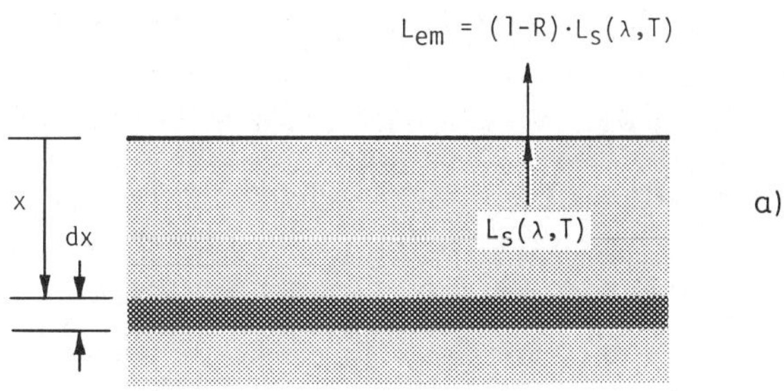

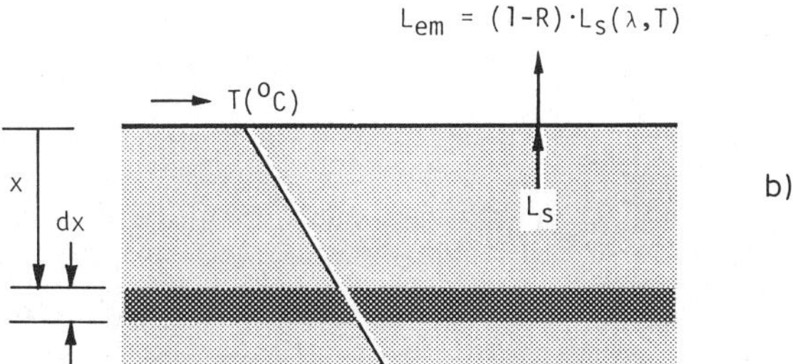

Figure 5-29.
Models of semi-infinite glass bath. (a) Isothermal; (b) nonisothermal.

Consider the model of a semi-infinite isothermal bath of glass in which the absorption coefficient (absorptance per unit thickness) is $a_\lambda\,(T)$ at wavelength λ and temperature T (see Figure 5-29 a). The absorptance $\alpha_\lambda\,(T)$ of the elemental slab of thickness $\mathrm{d}x$ can be determined from Beer's law:

$$\alpha = 1 - \exp(-a\,\mathrm{d}x) \tag{5-34}$$

where the subscripts have been omitted.

Expanding the exponential term into its series and ignoring high-order terms, Equation (5-34) becomes

$$\alpha = 1 - (1 - a\,\mathrm{d}x) = a\,\mathrm{d}x\,. \tag{5-35}$$

According to Kirchhoff's law, the emissivity of the slab will be the same as its absorptance at the same temperature and wavelength, ie,

$$\alpha = \varepsilon = a\,\mathrm{d}x\,. \tag{5-36}$$

The normal radiance at the surface $L_{s,\lambda}\,(T)$ due to the emission from the elemental slab can now be calculated by allowing for the absorption in the glass (thickness x) between the slab and the surface. The exiting radiance $L_{em,\lambda}(T)$ will be reduced by reflection at the surface, ie,

$$(L_{em})_{dx} = (1 - \rho_s) L_s = (1 - \rho_s)\, a \exp(-a x)\, L_b\, dx \tag{5-37}$$

where L_b is the black-body radiance at the same temperature and wavelength and ρ_s is the Fresnel reflectance.

The radiance from all elements in the bath can be found by integration:

$$L_{em} = \int_0^\infty (1 - \rho_s)\, a \exp(-a x)\, L_b\, dx = (1 - \rho_s) L_b\,. \tag{5-38}$$

The fraction F of the radiance that has originated from all elements between 0 and x can be calculated as follows:

$$F = \frac{\int_0^x (L_{em})_{dx}\, dx}{\int_0^\infty (L_{em})_{dx}\, dx} = 1 - \exp(-ax)\,. \tag{5-39}$$

This simple equation can be used to calculate the origin of the radiation being measured by a thermometer. It can also be used to calculate the effective emissivity of an infinite slab of glass, but the concept of emissivity of a semitransparent material can lead to misunderstandings. It is nearly always preferable to consider the absorption and surface reflection effects separately.

A very useful and simple concept is to consider that a thermometer measures the temperature of any glass at a distance beneath the surface which can be termed the *effective depth* (D_{eff}).

Consider the same model as above but specify that there exists a small temperature gradient (see Figure 5-29b). Let the gradient be such that the ratio of the radiance of an elemental slab at a depth x to a similar slab at the surface (depth zero) is given by

$$\frac{L_b(x)}{L_b(0)} = 1 + kx \tag{5-40}$$

where k is a constant.

Using Equation (5-37), the exiting radiance resulting from the slab at depth x is

$$(L_{em})_{dx} = (1 - \rho_s)\, a \exp(-ax)\, (1 + kx)\, L_b(0)\, dx\,. \tag{5-41}$$

By integration, the radiance L_{em} from all slabs is

$$L_{em} = \int_0^\infty (L_{em})_{dx}\, dx = (1 - \rho_s)(1 + k/a)\, L_b(0)\,. \tag{5-42}$$

This can be seen by comparison with Equation (5-38) to be the radiance that would exist if the bath was isothermal with a temperature such that

$$L_b = (1 + k/a)\, L_b\,(0)\,. \tag{5-43}$$

Therefore, we know that a correctly calibrated thermometer (in which due correction has been made for the reflection loss) would, when viewing the bath with a gradient, record the temperature equal to that a distance $1/a$ under the surface. We can therefore define the effective depth from any combination of thermometer and glass such that:

$$\mathrm{D}_{\mathrm{eff},\lambda}\,(T) = 1/a_\lambda\,(T) \tag{5-44}$$

where a is the optical absorption coefficient. Figure 5-28 gives data for typical glasses.

It is also possible to determine the percentage of the energy that originates from any particular depth, in relation to the effective depth (see Table 5-8).

Table 5-8 can be used to determine the thinnest glass on which accurate temperature can be determined with any thermometer. For example, with a typical clear bottle glass a thermometer operating at 5 μm will collect energy from a 0.2 mm depth. For very thin glass it is preferable to use a thermometer operating at 7.9 μm, which collects all the energy from less than 0.05 mm. A 7.9 μm model also has an advantage that the reflectance of the glass is close to zero and hence the emissivity is close to unity.

Table 5-8. Percentage of total radiance originating from various depths in glass

B*	0.11	0.22	0.36	0.51	0.69	0.92	1.2	1.6	2.3	4.0	4.6
%	10	20	30	40	50	60	70	80	90	98	99

* B is depth as fraction of effective depth

5.8.2.2 General-Purpose Models

a. Glass Surface Temperature

Most measurements are made using a waveband close to 5 μm, typically between 4.8 and 5.2 μm. However, as stated above, if very thin glass is being measured there is an advantage in using 7.9 μm thermometers.

Typical applications for the 5 μm surface-temperature models include measurement in the tin bath and in the lehr of a float-glass line, in drawing towers and temperature toughening furnaces, and in blowing frames.

At low temperatures below the range of a 5-μm model (eg, below 200 °C) it is necessary to use longer wavelength models.

An 8–14-μm thermometer is used in measuring stacking temperatures for sheet and plate glass. It must be noted that the emissivity is 0.83 for this waveband, and that the 17% reflection from adjacent objects cannot be neglected.

b. Temperature-Gradient Measurement or Subsurface Measurement

It is desirable to know the temperature gradient in addition to the surface temperature when glass sheets are being toughened by the air-cooling method. This can be achieved by viewing the same area of the sheet with two thermometers. Wavebands near 5 and 3.9 μm can give the temperature at the surface and at an effective depth of ca. 0.5–1 mm below the surface. Details of the method were given by Beatty and Coen [35].

c. Bulk Temperature of Molten Glass

Measurement and control of the temperature of liquid glass in the foreheart of container-manufacturing machines can (and have) been made by a wide variety of thermometers as the technology has improved. It is now common practice to use a short-wavelength (silicon detector) thermometer which gives good control due to the high rate of change of output with temperature. Additional advantages are that there is minimum interference from burner gases, and the measurement is made at a sensible depth under the surface. The most recent development is to use a short-wavelength fiber-optic thermometer. It is possible by this method to avoid the use of cooling water and to reduce the size of the sighting tube that has to be used.

5.8.2.3 Special Applications

a. Gob-Temperature Measurement

Special Problems: The intermittent nature of the target, the presence of oil spray and the partial transparency of small gobs are particular problems.

Solutions: Spot measurements were made using a DFP (Disappearing Filament Pyrometer), but more accurate measurements are now made using a portable short-wavelength thermometer with peak-holding circuitry. The gob must be viewed at a sufficiently steep angle to ensure that the transmitted energy originates from the nozzle. This can be checked visually as being the area of greatest radiance (brightest). Permanent installations can be fitted using the same waveband, signal processing, and installation precautions.

b. Mold Temperatures

Measurement Objectives and Problems: The temperature profile on the inside faces of the molds is critical for product quality. The correct profile is achieved by the design of integral air-cooling ducts in the molds.

Special problems are high-speed operation and intermittent access, variable emissivity of the mold and the occurrence of high temperatures of the glass, creating reflection errors.

Solutions: This application required the development of a special-purpose hand-held probe and remote signal processor [36].

The equipment shown in Figure 5-30 consists of a fiber-optic energy collector, a robust push-button switch, and a processor/indicator unit. The hand-held probe is designed to allow rapid access to the molds as they open and provides a reflection shield to protect the measurement point from incident radiation. The switch is actuated at the instant the probe is correctly seated on the measurement point and the probe is then rapidly widthdrawn. A sample-and-hold circuit enables the temperature at the instance of pressing the switch to be displayed.

Figure 5-30. Special equipment for measuring mold temperature.

Once again the use of short-wavelength enables acceptable readings to be obtained. One benefit of short wavelength is shown clearly in Figure 5-31, which gives measurements of the emissivity of a typical mold during its lifetime [37]. It can be seen that the oxide has formed during pretreatment sufficiently to absorb (and hence emit) at short wavelengths but not at long wavelengths. Consequently, the emissivity is high and stable at 1 μm throughout the life of the mold, but highly variable at 10 μm.

c. *Bath-Temperature Measurement*

Special probes have been designed to enable a thermometer to measure the temperature at known positions in the deep bath of glass in the melting furnace.

The probe comprises twin sheaths of molybdenum and alumina mounted through the base of the furnace to protrude into the glass. A miniature short-wavelength thermometer (with or

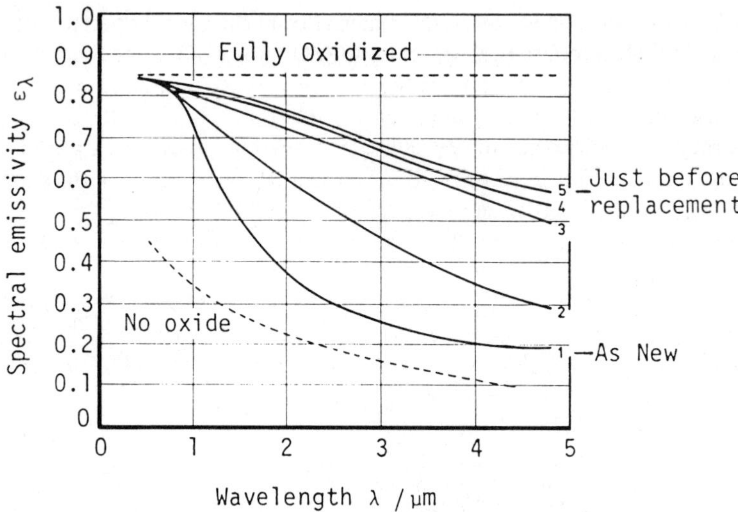

Figure 5-31. Spectral emissivity of mold. It is shown that emissivity increases during its lifetime.

without fiber-optic transmission) sights up the inside sheath to measure the temperature at the closed end. Molybdenum is chosen for the outer sheath owing to its long life in molten glass. The inner alumina tube plus a nonoxidizing gas purge protects the molybdenum from oxidation at its inner surface.

d. Communication fibers

The relatively new processes for producing silica fibers for optical communications have demanded new techniques of temperature measurement. The high temperatures and the process mechanics have precluded any method other than radiation thermometry.

d1. Preform Production

Three different methods have been developed for the production of the "rod" of the base material. All involve the deposition of the silica-based solids from controlled vapor phases at very high temperatures (1800–2200 °C).

The inside vapor deposition method (IVDP) demands the control of the temperature at the moving hottest zone of a pure silica tube, inside which the solids are progressively deposited. The two measurement problems relate to the proximity of the high-intensity flame and the transmission of the silica.

Experiments have shown that the oxy-hydrogen flame gives very little absorption/emission problems so the way is clear to choose the optimum wavelength. Thermometers using wavebands at 3.9, 5, and 8–14 µm have been tested, and both 3.9 and 5 µm are in use. Theoretically, 3.9 µm should be preferable as it measures the temperature in the bulk of the tube rather than at the surface. In addition, it should be less affected by any water-vapor emission from the flame. The thermometers are sighted at the hottest part of the rotating tube and are traversed together with the burner.

The two other methods involve the deposition of the material from the vapors injected into the flame, onto the outside of a silica rod. In these processes (OVDP and MCVD), the thermometers view the deposited material as it is progressively laid on the rod. For this application it is essential to use a waveband at which silica is opaque.

This application illustrates the importance of the choice of waveband when viewing semitransparent particulate matter. An effect identical with that seen with snow results in the surface having a very high reflectance in any waveband where the material is transparent. The basic materials (water and silica) have very low reflectance (4–5%), but the multiple reflection effect of a mass of crystals results in the very high reflectance seen with snow at visible wavelengths.

The best solution for these processes is to use the long-wavelength 8–14 μm band despite the very low percentage per degree Celsius change at high temperatures (note: at 2000 °C, a 1% output change corresponds to a 20 °C temperature change).

d2. Drawing Towers

In this production process the preforms are heated to softening temperatures (typically 2200 °C) and the material is drawn to the required fiber size. Induction or resistive heating methods are employed. In the former a susceptor of zirconia is used, in the latter a carbon heating element.

In both applications a narrow-angle short-wavelength thermometer gives the best control signal. In the resistive heating furnace, where a vacuum is used, it is necessary to sight through a sealed silica window at the graphite heating element. An emissivity setting of 0.9 will allow for window reflection loss and the emissivity of the graphite which is partly enhanced by reflection from the hot sighting tube.

In the zirconia furnace an emissivity of 0.7 is usually employed, although if absolute accuracy rather than reproducibility is required it will be necessary to check the effective emissivity of the particular construction.

5.9 Conclusions

The use of radiation thermometers has been growing in recent years, partly as a result of the development of new infrared thermometers that can now meet the measurement needs in the many industries that operate in the middle-low temperature range.

However, the main reason for the success of radiation thermometers is probably the wide selection of thermometer types that is available. It is therefore now possible to select the optimum type for each application.

Although not excluding the need for further developments in the instrumentation, the problem of greatest concern seems to be that of the education of the user. It has been demonstrated in the preceding sections that the measurement results can be greatly influenced by the degree of knowledge of the conditions of measurement and of the available solutions.

Up to now application assistance has been almost entirely provided by the instrument manufacturers. It is hoped that in the future a more consistent contribution will come from the users themselves. With this aim, treatments of this kind should certainly be useful. Further contributions can come from the wide distribution of application notes prepared by the manufacturers and from the development of detailed and clear standards, as was recently done, for example, by the Japanese Standards Association [38].

5.10 References

[1] Touloukian, Y. S., DeWitt, D. P., in: *Thermophysical Properties of Matter,* Vols. 7 and 8, Touloukian, Y. S., Ho, C. Y. (eds.); New York: Plenum Press, 1972.
[2] DeWitt, D. P., Incropera, F. P., in; *Theory and Practice of Radiation Thermometry,* DeWitt, D. P., Nutter, G. D. (eds.); New York: Wiley, 1988, pp. 21-89.
[3] Quinn, T. J., Martin, J. E., *Philos. Trans. R. Soc. London* **316** (1985) 85-189.
[4] Kostkowski, H. J., Lee, R. D., in: *Temperature, Its Measurement and Control in Science and Industry,* Vol. 3, Brickwedde, F. G. (ed.); New York: Reinhold, 1962, pp. 449-481.
[5] Gardner, J. L., *Appl. Opt.* **19** (1980) 3088-3091.
[6] Yates, H. W., Taylor, J. H., *NRL Report 5453,* 1960, US Naval Research Laboratory, Washington, DC.
[7] Pyatt, E. C., *Br. J. Appl. Phys.* **5** (1954) 200-204.
[8] Hornbeck, G. A., in: *Temperature, Its Measurement and Control in Science and Industry,* Vol. 3, Brickwedde, F. G. (ed.); New York: Reinhold, 1962, pp. 425-428.
[9] Coates, P. B., *High Temp. High Press.* **20** (1988) 443-448.
[10] Drury, M. D., Perry, K. P., Land, T., *J. Iron Steel Inst.* **169** (1951) 245-250.
[11] Iuchi, T., in: *Applications of Radiation Thermometry,* Richmond, J. C., DeWitt, D. P. (eds.); Philadelphia: ASTM, 1985, STP 895, pp. 121-150.
[12] Kelsall, D., *J. Sci. Instrum.* **40** (1963) 1-4.
[13] Toyota, H., Yamada, T., Nariai, Y., in: *Temperature, Its Measurement and Control in Science and Industry,* Vol. 4, Plumb, H. H. (ed.); Pittsburgh: Instrument Society of America, 1972, pp. 611-618.
[14] Murray, T. P., *Rev. Sci. Instrum.* **38** (1967) 791-798.
[15] Svet, D. Ya., *Optical Methods of Measuring True Temperature;* Moscow: Nauka, 1982 (in Russian).
[16] Dils, R. R., *J. Appl. Phys.* **54** (1983) 1198-1201.
[17] Brown, M. E., *Proc. Symp. on Major Problems on Present-day Radiation Thermometry,* IMEKO TC12; Moscow, 1986, pp. 51-61.
[18] Ruffino, G., *Appl. Opt.* **10** (1971) 1241-1245.
[19] Rosso, A., Righini, F., *Measurement* **3** (1985) 131-136.
[20] Jones, T. P., Tapping, J., *Metrologia* **18** (1982) 23-31.
[21] Sakuma, F., Hattori, S., in: *Temperature, Its Measurement and Control in Science and Industry,* Vol. 5, Schooley, J. F. (ed.); New York: American Institute of Physics, 1982, pp. 421-427.
[22] Battuello, M., Lanza, F., Ricolfi, T., *Metrologia* **27,** (1990) 75-82.
[23] Ricolfi, T., Lanza, F., *High Temp. High Press* **15** (1983) 13-20.
[24] Battuello, M., Ricolfi, T., *Measurement* **5** (1987) 189-191.
[25] Quinn, T. J., Lee, R. D., in: *Temperature, Its Measurement and Control in Science and Industry,* Vol. 4, Plumb, H. H. (ed.); Pittsburgh: Instrument Society of America, 1972, pp. 395-411.
[26] Jones, T. P., Tapping, J., *Metrologia* **15** (1979) 135-141.
[27] Coates, P. B., in: *Theory and Practice of Radiation Thermometry,* DeWitt, D. P., Nutter, G. D. (eds.); New York: Wiley, 1988, pp. 773-819.
[28] Barber, R., Brown, M. E., in: *Applications of Radiation Thermometry,* Richmond, J. C., DeWitt, D. P. (eds.); Philadelphia: ASTM, 1985, STP 895, pp. 39-60.
[29] Shevis, A., Melvin, H., Coke Oven Managers Association, Northern Section, 1981, unpublished.
[30] *Sollac company, Dunkerque, France.*

[31] Amory, D. C., Beynon, T. G. R., in: *Thermal and Temperature Measurement in Science and Industry,* London: Institute of Measurement and Control, 1987, pp. 275–293.
[32] Beynon, T. G. R., Ridley, I., *High Temperature Technology* **2** (1984) 75–87.
[33] *"Stelmor" Rod Temperature Control,* Morgan, Worcester, UK, unpublished.
[34] Ridley, I., Beynon, T. G. R., in: *Thermal and Temperature Measurement in Science and Industry;* London: Institute of Measurement and Control, 1987, pp. 295–310.
[35] Beattie, J. R., Coen, E., *Brt. J. Appl. Phys.* **11** (1960), 151–157.
[36] Barber, R., in: *Proc. International Congress on Glass XII;* New York: North-Holland, 1980, pp. 903–908.
[37] Rucklidge, J. M., in: *Proc. IEEE Industry and General Applications Group, 6th Annual Meeting;* Cleveland, O, 1971, pp. 717–719.
[38] *Japanese Industrial Standard, JIS C 1612,* 1988, Japanese Standards Association, Tokyo.

6 Noise Thermometers

Heinz Brixy, Forschungszentrum Jülich GmbH, Jülich, FRG

Contents

6.1 Introduction

In measuring temperatures with contact thermometers, the problem may generally arise that the *sensor is changed by influences of the environment,* ie, the ambient effects change the measured quantity used in the sensor in an unpredictable and uncontrollable manner. This is the case, for example, in high neutron fields, in highly aggressive atmospheres, and in any event at high temperatures (>1000 °C). In practice, this means that the temperature characteristics of the sensor are changed and that both the operating range and time are limited by this drift. This is especially true of thermocouples and resistance thermometers [1–7].

Particularly at high temperatures, *any sensor undergoes changes* with time owing to the high temperature itself and therefore *a measurement method is desirable which tolerates such changes,* ie, which is not affected by such changes. *A method capable of meeting these demands is noise thermometry.* Here, the random, statistical thermal agitation of the electrons in the conduction band of, for example, metal conductors is used for measurement. This electron movement, perceptible as a voltage fluctuation across a resistor, is a function of the absolute temperature and it is thus in principle possible to determine the temperature with the aid of suitable mean values (mean values because these are statistical processes). This phenomenon is known as the thermal noise of electrical resistors. Figure 6-1 shows the oscillogram of the amplified thermal noise of a simple metal film resistor (voltage fluctuations around zero with a linear average value of zero).

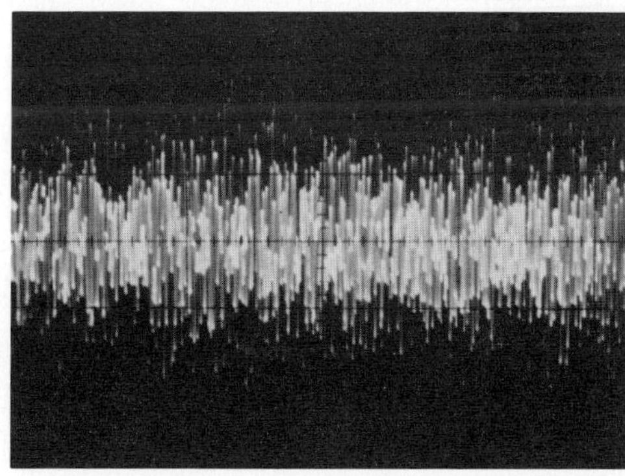

Figure 6-1.
Oscillogram of (band-limited) thermal noise.

6.2 Fundamentals of Noise Thermometry

Special requirements are imposed on thermal noise sensors from measurement principles, signal transmission, etc. Therefore, the fundamentals of noise thermometry will be briefly reviewed.

6.2.1 Nyquist's Theorem

Noise thermometry is quantitatively based on a relationship established by Nyquist [8] in 1928 from general thermodynamic considerations without any reference to the atomistic structure of matter and which was experimentally confirmed in the same year by Johnson [9]. According to this theorem, a voltage fluctuating statistically around zero occurs at the ends of a passive electrical network (or sensor). For the mean square of this voltage, $\mathrm{d}\overline{V^2}$, in a frequency interval $\mathrm{d}f$, the following equation holds:

$$\mathrm{d}\overline{V^2} = 4k_\mathrm{B}T\,\mathrm{Re}\,(Z(f))\,\mathrm{d}f \tag{6-1}$$

or, as is also frequently written:

$$S = 4k_\mathrm{B}T\,\mathrm{Re}\,(Z(f)) \tag{6-1 a}$$

where k_B is the Boltzmann constant, T the absolute temperature, $\mathrm{Re}\,(Z)$ the real part of the complex impedance Z, f the frequency, and S the spectral power density or power spectrum of thermal noise

Equation (6-1) describes the noise of any passive unloaded sensor, irrespective of any special conduction mechanism. The thermal noise, often called Johnson noise, is independent of the internal structure of the sensing resistor (eg, physical state or chemical composition) and the nature of the charge carriers. Hence Equation (6-1) is valid for solid or liquid metals, carbon films, semiconductors, and also electrolytes. This provides great freedom in the choice of materials for the construction of sensing resistors.

In ranges where quantum effects are of significance (ie, $hf \approx k_\mathrm{B}T$, h being Planck's constant), Equation (6-1) must be multiplied by a correction factor:

$$p(f) = \frac{hf}{k_\mathrm{B}T}\left(\exp\frac{hf}{k_\mathrm{B}T} - 1\right)^{-1}. \tag{6-2}$$

However, in all cases of practical interest this factor is equal to 1, ie, at temperatures above 1 mK and frequencies below 1 MHz.

The correction factor leads to a decrease in the power spectrum at high frequencies so that the total mean square noise remains finite. Another reason for the finite power of thermal noise is the unavoidable capacitances and inductances of the sensor which also cause a drop at high frequencies.

In practice, it is useful to design the sensor in such a way that the real part of the impedance is independent of frequency and only consists of an ohmic resistance R. The following is thus obtained by integration between a lower (f_l) and an upper (f_u) frequency limit:

$$\overline{V^2} = 4k_\mathrm{B}TR\,\Delta f \qquad \Delta f = f_\mathrm{u} - f_l\,. \tag{6-3}$$

If one measures the mean square noise voltage, $\overline{V^2}$, of a resistor in a frequency interval Δf and the resistance R is also measured, then the absolute temperature of the resistor is determined.

Equation (6-3) is the Nyquist equation in its simple form, which nevertheless holds for all practical applications of noise thermometry.

Equation (6-3) shows that all environmental influences (eg, from the atmosphere or nuclear radiation) and all influences from mechanical or thermal pretreatment can only change the resistance R. Also, all the material properties and their changes only enter into R. This resistance can, however, be determined at any time by a simple measurement and thus all environmental influences, etc., are precisely determined. *The noise thermometer therefore shows no drift* (as occurs, for example, with thermocouples or resistance thermometers), *because all changes of the sensor are measured,* ie, they are known and hence do not cause any measurement error.

These features distinguish noise thermometry from other temperature measuring methods, thus ensuring accuracy and long-term stability. Moreover, in contrast to most thermometers, which have to be calibrated, *the noise thermometer is an absolute thermometer* which directly indicates the thermodynamic temperature.

With such interesting features, the question arises of why noise thermometry is not widely used for metrological and industrial applications. The reason lies in the extremely small amplitudes of thermal noise voltages and the difficulty of electronically processing them. At the same time it must be ensured that external electromagnetic interferences are not superimposed on the useful signal noise. Both difficulties could only be overcome with the development of efficient electronic components in recent years.

Equation (6-3) gives an impression of the magnitude of the noise voltage to be expected if 300 K is chosen for the temperature, 100 Ω for the resistance and 100 kHz for the frequency band. With these values we obtain for the root of the mean square noise voltage

$$\sqrt{\overline{V^2}} \approx 4 \cdot 10^{-7}\ \mathrm{V}\,. \tag{6-4}$$

6.2.2 Methods of Noise Thermometry

Figure 6-2 shows a basic diagram that can be used in principle to measure the noise voltage. The following equation holds for the mean square noise voltage at the output of the integrator:

$$\overline{V^2} = 4k_B TR \int_0^{\infty} \left| A(f) \right|^2 \mathrm{d}f + \overline{V_e^2}\,, \tag{6-5}$$

where $A(f)$ is the frequency-dependent gain and V_e is the internal noise of the amplifier, corresponding to an equivalent noise resistance R_e.

According to Figure 6-2 and Equation (6-5), two measurements are necessary for the determination of the noise temperature (if one does not consider the amplifier gain $A(f)$ and the internal noise $\overline{V_e^2}$, which can be determined separately): the measurement of the noise volt-

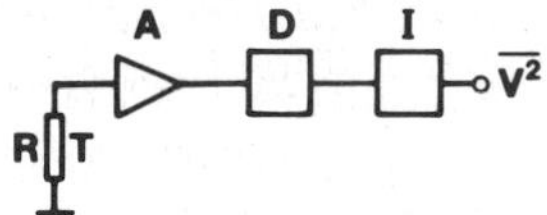

Figure 6-2.
Basic diagram for measurement of the noise voltage (noise resistor R at temperature T; A, amplifier; D, square-law detector; I, integrator; $\overline{V^2}$, mean square noise voltage).

age $\overline{V^2}$ with a high-input impedance voltage-sensitive preamplifier and the measurement of the noise resistance R with a d.c. or a.c. measuring instrument.

A second method of determining the noise temperature results from a modified version of the usual Nyquist equation for the noise current of a resistor R:

$$\overline{I^2} = 4k_B T\Delta f/R \,. \tag{6-6}$$

From the measurement of the noise current, $\overline{I^2}$, with a low-input impedance current-sensitive preamplifier and the measurement of R, one also obtains the noise temperature.

A third method of determining the noise temperature results from the determination of the noise power of a resistor. Multiplication of Equations (6-3) and (6-6) gives the thermal noise power, P:

$$P = 4k_B T\Delta f \,. \tag{6-7}$$

Here the noise voltage, $\overline{V^2}$, and the noise current, $\overline{I^2}$, have to be measured for the determination of the noise temperature. This method does not depend on the noise resistance R, which does not have to be known.

These three methods have often been realized with different signal-processing techniques and, mostly, highly sophisticated electronic circuits. A historical review up to 1982 for both metrological and industrial applications was given by Blalock and Shepard [10].

A fourth basic method of noise thermometry, which can be used only at very low temperatures and employs quantum devices such as Josephson junctions, is briefly described in Section 6.8.3.

In the following, a measurement process is described which is most suitable for both industrial and metrological applications. This measurement process is based on the comparison of the noise voltages of two resistors and according to the first method, noise voltages and resistances or rather the ratios of the two are measured. This measurement technique was first reported by Garrison and Lawson [11] and used later in a great variety of arrangements [12–31]. For the elimination of the unwanted amplifier noise a correlation technique is often applied, feeding the noise of the sensing resistor into two amplifiers connected in parallel [32–39]. The combination of the comparison with the correlation technique [6, 18, 19, 21, 22, 30] leads to a measurement process with good conditions for the whole field of application of noise thermometry.

As shown in Figure 6-3, the sensing resistor R_S at the unknown temperature T_S and the reference resistor R_R at the known reference temperature T_R are connected in turn to the imputs of the amplifiers. For the temperature to be measured one obtains from Equation (6-5)

$$T_S = \frac{\overline{V_S^2}}{\overline{V_R^2}} \cdot \frac{R_R}{R_S} \cdot T_R \tag{6-8}$$

where $\overline{V_S^2}$ and $\overline{V_R^2}$ are the square noise voltages of the sensing and reference resistors. $\overline{V_e^2}$ is eliminated by cross correlation (see below). If R_R is adjusted until

$$\overline{V_S^2} = \overline{V_R^2} \tag{6-9}$$

then the unkown temperature becomes

$$T_S = \frac{R_R}{R_S} \cdot T_R \,. \tag{6-10}$$

The temperature to be measured, T_S, is therefore equal to the ratio of two resistances times a reference temperature, T_R, specifying the scale. A fixed point can be chosen as the reference temperature, eg, the triple point of water (273.16 K), but also any other known temperature, eg, room temperature, which particularly simplifies process temperature measurements.

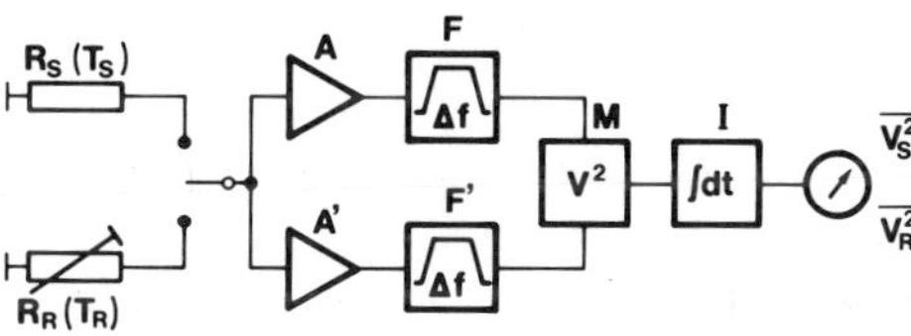

Figure 6-3. Block diagram of the comparison and correlation noise thermometer (R_S, sensing resistor; T_S, temperature of the sensing resistor, temperature to be measured; R_R, reference resistor; T_R, temperature of the reference resistor, reference temperature; A, A′, amplifiers; F, F′, filters; M, multiplier; I, integrator, $\overline{V_S^2}$, mean square noise voltage of the sensing resistor; $\overline{V_R^2}$, mean square noise voltage of the reference resistor).

Equation (6-10) shows that the noise thermometer is an absolute thermometer since only the defining point on the temperature scale (T_R) and a dimensionless scale factor (R_R/R_S) are used and no reference to any other physical quantity is needed.

The arrangement of two parallel amplifiers with subsequent multiplication and integration results in the cross correlation of uncorrelated signals for the internal noise of the amplifiers; amplifier noise is thus eliminated. Consequently, only the averaged products resulting from the noise voltages of the sensing resistor and the reference resistor, for which the circuit according to Figure 6-3 means autocorrelation, appear at the output of the integrator. (Conclusions from finite measurement time τ will be dealt with in Section 6.2.4.) It is not necessary to know the gain or bandwidth of the amplifiers and the linearity or stability of gain does not exhibit problems either if the comparison technique is used.

Similarly to the described method of measuring the noise voltage and resistance, the method of measuring the noise power has also found wide application [40–48]. The power method has advantages if the real part of the impedance of the sensor is frequency dependent [48], but this method also has some disadvantages. Different approaches to noise thermometry signal processing have been reported [49, 50].

6.2.3 Measurement Process for Determining Noise Temperatures

A number of aspects become important in practice when measuring noise temperatures under both industrial and laboratory conditions which are directly related to the sensor construction. These are the signal wires in the sensor and the electrical potential of the sensor.

A noise temperature measurement process which takes this into account is shown in Figure 6-4 [6, 19, 30].

The coupling of the noise resistors can be seen in detail. The lead arrangement has two channels and reaches directly to the sensing and reference resistors, and moreover for both the outgoing and also the return leads. It is thus possible to eliminate the noise voltages of all parasitic resistors, ie, the noise voltages of the switches and connection leads, in the same way as the internal noise of the amplifiers by cross correlation.

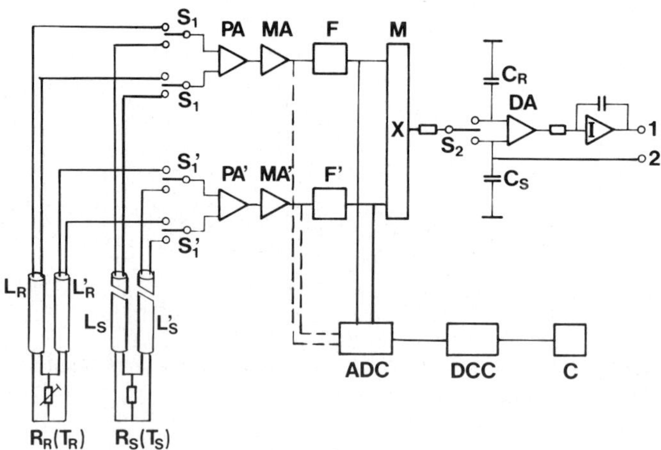

Figure 6-4. Detailed block diagram of a practical noise thermometer system (R_S (T_S), sensing resistor at temperature T_S; R_R (T_R), reference resistor at temperature T_R; L_S, L_S', leads (cables) to the sensing resistor R_S; L_R, L_R', leads (cables) to the reference resistor R_R; S_1, S_1', S_2, switches; PA, PA′, preamplifiers; MA, MA′, main amplifiers; F, F′, filters; M, multiplier; C_S, C_R, storage capacitors for the noise signals of R_S and R_R, respectively; DA, differential amplifier; I, integrator; ADC, analog to digital converter; DCC, digital cross correlator; C, computer and control unit; AD conversion, behind F, F′ (or MA, MA′ with digital filtering)).

The two-channel arrangement of the connection cables between R_S or R_R and the electronic device means that the preamplifiers must be designed as differential amplifiers. Differential amplifiers make a high common mode rejection of external interference signals possible.

In a large number of applications the sheath of the sensor is inevitably connected to ground at a remote point; this ground potential is in many cases different from the ground potential of the device. The preamplifiers are therefore supplied from a separate power pack with a shielded transformer. The signal cable's sheath (sensor sheath) can now be connected to the independent potential of the preamplifiers. The signal connection from the preamplifiers to the successive stages takes place by means of either transformer or differential amplifiers. This type of signal and ground connection in the input section has proved effective in practice. Behind the filters, F, F′ (or main amplifiers MA, MA′) the analog signals are digitized and then digitally processed. For historical reasons, analog signal processing behind the filters is also shown.

6.2.4 Measurement Time and Statistical Accuracy

An important point for noise thermometry, which among other things has significance for selecting the value of the noise resistance, the design of the sensor and measurement set-up, is the function of the measurement time with respect to the desired accuracy. Owing to the stochastic nature of thermal noise, the precision of noise measurements is largely restricted by the statistics. The measured values of noise voltages, and thus the associated temperature values, therefore involve statistical errors. The following expression holds for the relative error of a measurement [51]:

$$\frac{\Delta T}{T} \sim \frac{1}{\sqrt{\Delta f \cdot \tau}} . \tag{6-11}$$

where Δf is the width of the frequency band used and τ the measurement time.

It follows from this equation that at a constant bandwidth the measurement time must, for example, be extended by a factor of 100 if the precision is to be increased by a factor of 10. Since the power spectrum of the noise in the frequency band used should be white (see Section 6.2.5) and the bandwidth therefore cannot be arbitrarily increased, the accuracy of noise temperature measurements is also determined by the measurement time.

Equation (6-12) gives the precise equation for the relative statistical error for the comparison noise thermometer system in Figure 6-4 in a symmetrical form [19]:

$$\frac{\Delta T}{T} = \sqrt{\frac{2}{\Delta f \cdot \tau}\left(2 + \frac{B}{A} + \frac{C}{A} + \frac{BC}{A^2}\right)} . \tag{6-12}$$

In spite of the elimination of amplifier noise by cross correlation, the power spectra B and C of the amplifier chains' internal noise appear once again; they increase the relative error. The power spectrum of the noise resistance itself (R_S or R_R, which are equal at balance) is designated by A. The internal noise of the preamplifiers therefore plays a decisive role with respect to measurement time and one must endeavor to keep it as small as possible. This is particularly true if small sensing resistances have to be used (eg, when measuring high temperatures, see Section 6.4.2).

An equivalent noise resistance, R_e, of 24 Ω for a differential preamplifier has been reported [34]. It should be possible to obtain values of R_e as low as ca. 10 Ω by connecting several FETs (field-effect transistors) in parallel and cooling them to an optimum temperature.

If the resistances of the connection cables between the sensing resistor and electronic device become relevant in comparison with the equivalent noise resistance of the amplifiers then they must be taken into account accordingly in B and C with respect to the averaging time for V_S^2.

With a direct measurement of the noise voltage with stationary switches, ie, not using the comparison method, a measurement time is needed for the same precision which is shorter by a factor of 4:

$$\frac{\Delta T}{T} = \sqrt{\frac{1}{2 \cdot \Delta f \cdot \tau}\left(2 + \frac{B}{A} + \frac{C}{A} + \frac{BC}{A^2}\right)} . \tag{6-13}$$

In this case, however, the electronic device has to be calibrated, which can be carried out with the aid of the reference resistor. The time saved in the direct measurement is needed for the previous calibration. Equations (6-12) und (6-13) hold in all cases, although shorter times are often reported in the literature.

6.2.5 Signal Transmission

In the case of conventional temperature-measuring sensors, such as thermocouples and resistance thermometers, the measured signals are DC voltages, in connection with which the transmission behavior of the lines (cables) is principally governed by the ratio of series resistance to insulation resistance. The noise thermometer's measured signal in practical applications consists of frequencies between about 5 and 300 kHz. The frequency-dependent transmission characteristics of the signal cables must therefore be taken into consideration in noise thermometry.

In the comparison method, the white noise spectrum of the reference resistor is transmitted to the preamplifier input via such short lines that it appears undistorted (in the above-mentioned frequency range). Since this undistorted reference spectrum serves as a measure of the sensor noise spectrum, the white noise spectrum of the sensor must also be transmitted to the preamplifier without distortion in the frequency band used (this also holds for methods other than the comparison method).

Figure 6-5 shows a typical cable arrangement. The front line section with the sensing resistor consists at high temperatures of, eg, a four-wire mineral-insulated cable. The line section at a lower temperature is designed with a two-wire symmetrical plastic-insulated cable. The sensing resistor is represented in the equivalent circuit diagram as the series circuit of a noiseless resistance R_S with a noise voltage source V_S.

In laboratory measurements with usually short cable lengths, the effects on transmission are less significant, but in industrial measurement, where the sensor position and the electronic measuring device are, in general, at some distance from each other, the noise spectrum of the sensing resistor can be considerably affected.

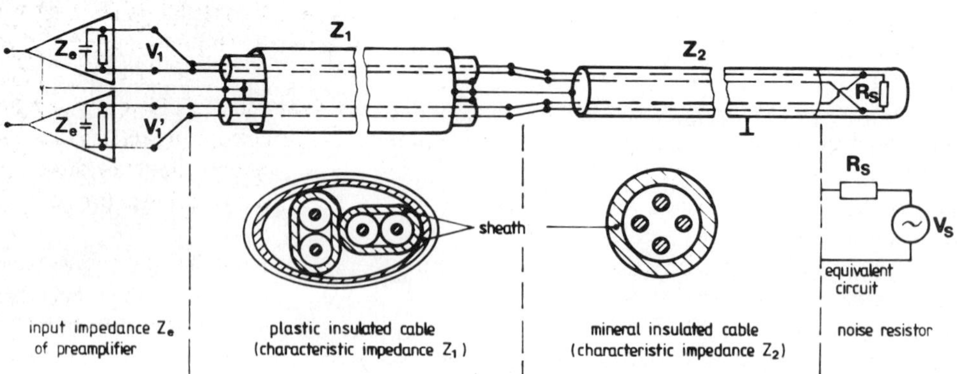

Figure 6-5. Typical transmission cable arrangement in noise thermometry.

By a special matching of the noise resistance (R_S) and the transmission cables, the transmission error in the frequency band used can be kept sufficiently small ($<10^{-3}$), if necessary, by reducing the upper band limit [30, 52]. Thus, for example, measurements with accuracies of $1 \cdot 10^{-3} - 2 \cdot 10^{-3}$ have been carried out with a cable length of 50 m (upper band limit 50 kHz; averaging time ca. 10 min) [53].

Another possibility for avoiding transmission errors is to instal the preamplifier with the corresponding components as close to the sensor as ambient conditions permit (remote preamplifier). The signals can then be transmitted over long distances, usually to the control room. A further possibility is to calculate or measure the overall system frequency response and to compensate for the distortions [45, 48]. In general, the cables should be as short as possible.

6.2.6 Rejection of Interferences

The most severe problem in noise thermometry which requires particular attention with measurements in both industrial plants and the laboratory is the detection and the avoidance or rejection of external interferences. These interferences can be of electromagnetic or mechanical (eg, vibrational) origin. Vibrations can cause changes in the distances of the lead wires and internal components of the sensor (microphonic noise) and they can cause Barkhausen noise in ferromagnetic lead wires and resistors (see Section 6.4.3), but electromagnetic interferences (EMI) are usually much more important.

Noise thermometer sensors generate extremely small signals which, moreover, have a large bandwidth. Therefore, EMI induced in the sensor and/or the transmission line due to electric, magnetic, or electomagnetic fields, and electric currents flowing in the cable shields, can cause significant errors. The interference signals are superimposed on the sensor's signal in both channels and are correlated and usually in-phase so that they cannot be eliminated by the correlation technique. A relative measurement error then results for the noise temperature, which is given by the ratio of interference power to useful noise power in the frequency band used.

Careful attention must be paid to the choice of cables (dual, shielded, twisted-pair arrangement of the four lead wires, see Figure 6-5) and their installation (remote from power or control lines and units), and particular shielding techniques must be applied (preferably double shielding, laying of the cables in Cu and/or Fe or stainless-steel tubes). It has been shown in practice that the shields should be of as low impedance as possible. Ground loops must, of course, also be avoided or eliminated.

In many cases interference signals in the transmission cables cannot be avoided by shielding techniques alone. There are then in principle two possibilities of eliminating interference signals: blanking in the time domain and filtering in the frequency domain. They are complementary to each other and they are sufficient. Which of the two possibilities (or both) is to be applied depends on the type of interference signals [30, 54].

A high common-mode rejection of the differential preamplifiers also has a very favorable effect with respect to parasitic interferences because it eliminates part of the interferences that appear as common-mode signals on the transmission lines [30, 38]. Another possibility of removing the EMI noise power is to measure it quantitatively and subtract it from the useful noise power [46, 48]. However, as the EMI effect is in general not readily amenable to correction, the best strategy is to eliminate it by the measures described above.

The problem of interference rejection is different for each measuring task and must therefore be solved individually each time.

6.2.7 Requirements for the Sensor

Owing to the extremely small amplitude and the wide band width of the noise signals, the transmission and processing are more complex and elaborate here than in other temperature-measuring methods. For the same reasons, the sensor has strong connections to transmission and processing and vice versa. Therefore, requirements are made on the sensor from the characteristics described in the previous sections.

The cross correlation in signal processing to eliminate the parasitic noise of the transmission lines (R_L) requires four lead wires to the sensing resistor. The correlation technique is particularly important at high temperatures because, owing to decreasing insulation resistances of the sensors, small values for the sensing resistances (R_S) have to be chosen and, therefore, the ratio of the useful signal noise to the parasitic cable noise becomes disadvantageous. In addition, the lead resistances are at undefinable temperatures so that their noise cannot easily be eliminated by other measures. The low R_S values at high temperatures, further, make low equivalent noise resistances of the preamplifiers and low R_L values desirable for realizing short measurement times or, conversely, the R_S values selected must not be too small because of measurement time (see Equation (6-12)). The necessity for an undistorted signal transmission (Section 6.2.5) requires special matching of the R_S value and cable [52] or, if this is not possible, a reduction of the upper band limit. However, such a band limitation increases the measurement time.

If low temperatures are to be measured, high R_S values are necessary. Then the cables must be short, ie, the preamplifier has to be installed near to the sensor.

To realize an optimum EMI rejection it is advantageous to employ sensors with the noise resistor R_S insulated from ground. The efficiency of the shielding here is usually better the higher the insulation resistance can be kept.

At very high temperatures the sensor resistance commonly becomes unstable even for short times. Then it is favorable to employ the direct measurement method (with previous calibration of the electronic device) and not the comparison method. This reduces the needed measurement time by a factor of 4. Additionally, the band width should be kept as wide as possible.

Overall, the selection and matching of a sensor for a certain measuring task are an optimization problem to be solved when the measuring set-up is being designed and installed.

6.3 Combined Thermometry

Apart from its undisputed advantages, described in Section 6.2.1, noise thermometry also displays certain disadvantages. For example, one disadvantage is that the electronic devices for processing the very small noise signals are sophisticated and expensive. A further disadvantage is that the measured signal is composed of stochastic signals and the temperature is proportional to the averaged value. A certain time is necessary for averaging, namely the more precise the measurements required, the longer it takes (Section 6.2.4).

6.3.1 The Combined Thermocouple-Noise Thermometer Sensor (TC-NT Sensors)

Significant advantages are therefore provided by a sensor which combines a noise resistor with two thermocouples, resulting in an extremely adaptable thermometer [55]. It consists, as shown in Figure 6-6, of a sheathed double thermocouple with two hot junctions. The noise resistor is positioned between the two hot junctions and insulated from the cladding.

This thermometer can be used both as a thermocouple (TC) and as a noise thermometer (NT). *It therefore combines the advantages of both instrument types.* It functions as follows. During most of the operating time, the emf of the TC is used for determining and recording the temperature due to the simple and rapid way in which measurements can be carried out. There is now, however, the possibility of associating the correct absolute temperature with the emf, which can change its value at constant temperature for a number of reasons, as often as required by means of the thermal noise of the resistor. It is therefore possible to calibrate or recalibrate the thermocouple in the sensor at any time after installation, ie, in situ. The advantage of this in situ calibration is that all the parasitic thermovoltages resulting from temperature gradients across inhomogeneous thermocouple wires are included in the calibration and thus measurement errors are avoided. With this combined sensor it is now possible to detect and eliminate quantitatively in practice the inhomogeneity errors of thermocouples, because this can only be done by in situ calibration. In many respects, therefore, the TC-NT has unique properties.

The four TC wires, which serve at the same time as signal wires for noise measurement, offer the opportunity of eliminating the resistance of the wires by cross correlation when measuring the noise temperature. This is of particular importance here because of the sometimes considerable resistance of the usual TC wires (eg, NiCr/Ni wires). Moreover, when using combined sensors only one electronic device for processing the noise voltages is necessary. With this device it is possible to recalibrate from time to time any number of sensors.

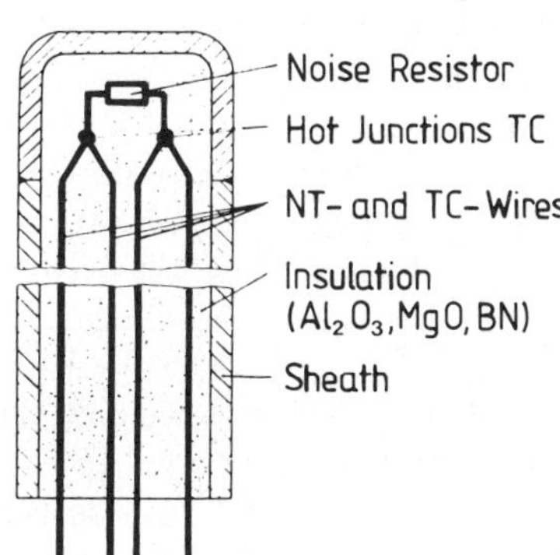

Figure 6-6.
Combined thermocouple-noise thermometer sensor.

It is in principle also possible to carry out a simultaneous measurement with a combined thermocouple-noise temperature sensor, since the thermocouple provides a direct voltage whereas the noise resistor provides a fluctuating voltage which can be fed out by means of a capacitor.

6.3.2 Combined Resistance-Noise Thermometry

According to the Nyquist equation (Equation (6-1)), any electrical resistor can be used as a noise thermometer. This also holds, of course, for the resistor of resistance thermometers (RT), which can therefore serve as noise thermometers. In the same manner as described above for the combined TC-NT sensor, noise thermometry can be employed for in situ calibration and recalibration of RTs. However a problem may arise with the usual RT sensors, because the d.c. resistance is not generally the same as the noise resistance (ie, the real part of the sensor impedance $\mathrm{Re}\,(Z\,(f))$ within the pass band of the measurement). The difference between the two depends on the resistor construction. A reduction of the upper band limit can be a solution in some cases. The application of the noise power method for processing the noise signals generally solves the problem, because in the power method the resistance does not need to be known [40].

In practice, restrictions are imposed on the application of RTs as noise thermometers owing to the difficulty of achieving a satisfactory match with the transmission cable.

If the four lead wires of a resistance thermometer are made of TC material, all three methods (TC, RT and NT) can be employed with one sensor.

6.4 Sensors for Noise Thermometry and Combined Thermocouple-Noise Thermometry

6.4.1 General Aspects

In the same way as a thermocouple and also a resistance thermometer, the noise thermometer sensor is a contact thermometer with all the resulting requirements such as good thermal coupling of the sensor to the temperature to be measured, etc.

In principle, a sensor consists of any electrical resistor whose ends are connected to two or four wires insulated against each other. Neither temperature coefficients nor any other variations in the value of the resistance are, in principle, of any significance. Pure metals and alloys are particularly suitable as resistor materials.

In contrast to resistance thermometers and thermocouples, the materials can therefore be largely freely selected. The particular choice can thus be adapted to the special measurement problem so that ambient conditions, material compatibilities, thermal expansion, etc., are less important.

6.4.2 Value of Noise Resistance and Temperature Range

The value of the noise resistance can be selected between approximately 1 Ω and several kilohms, the actual value being dependent on the experimental conditions, namely the higher the temperatures the smaller the resistances have to be. The reason for this is the decreasing insulation resistance of the sensor with increasing temperature and the necessity for avoiding insulation errors, similarly to the case with thermocouples and resistance thermometers (see also Section 6.4.4).

At very low temperatures (ca. 4 K), metal film resistors are preferably used with values up to 20 kΩ [34, 35]. A high sensor resistance ensures that the noise power (which is proportional to RT) retains an appropriate value. Short connecting cables have to be used here for an undistorted signal transmission which can be performed easily in laboratory measurements.

In the range from room temperature up to about 1200 °C, mostly resistance values between 100 and 10 Ω are chosen. Such values are adequate for a good match with the transmission cable. At very high temperature (above 1500 °C), only a few ohms are feasible for the sensor owing to insulation problems, as mentioned above. It may be of historical interest that Johnson in 1928 employed resistance values up to a few megaohms [9].

6.4.3 Materials for NT and TC-NT Sensors

Great constancy of the sensor resistance is not of importance in the first instance when using NT sensors, since the resistance can be measured at any time (however, constancy during averaging is necessary). In temperature measurements where the sensor position and electronic device are at a considerable distance from each other, as is usually the case in large-scale industrial plants, the sensor resistance and the connection cable must be matched with each other in order to achieve good transmission behavior (see Section 6.2.5). It is then appropriate to use materials with a low temperature coefficient of resistance in order to make a sufficiently good match possible over a wide temperature range.

An alloy of 80% Ni and 20% Cr has proved to be particularly suitable. This alloy does not change its specific resistance by more than 3% in the wide temperature range from 4 K to ca. 1600 K. Moreover, it can be employed in air up to the upper temperature limit owing to its very stable surface oxide layer. Most of the metal film resistors used at low temperatures also have NiCr resistance layers.

Metals or alloys with a high melting point are employed for temperatures above 1500 K (W, Re, Mo, Rh, Pt, etc.). Owing to the relatively high temperature coefficient of resistance of these materials, the cable must be matched with the resistance value at the temperature to be measured.

Compatible materials are in general used for combined sensors, eg, an NiCr alloy is chosen as resistor material together with NiCr TCs, Pt, Rh or a PtRh alloy with PtRh TCs, and W, Re or a WRe alloy with WRe TCs. For insulation the same materials are employed as for TCs and RTs at high temperatures, eg, Al_2O_3, MgO, BeO, HfO_2, BN, etc.

In a similar way the materials for sheaths and protection tubes are chosen to be compatible with the respective environment and, if possible, with the TC and NT materials.

Compatibility with environmental conditions means for noise sensors not only compatibility with the existing atmosphere and the surrounding materials, but also electromagnetic compatibility. Electromagnetic compatibility in this context means that the materials chosen for the sensor resistor and the lead wires are insensitive to external alternating magnetic fields. In ferromagnetic materials, such alternating magnetic fields, if they are high enough, can generate Barkhausen noise; the power spectrum of this noise does decrease with rising frequency, but may be relevant in the whole frequency range of noise thermometry if it is strong enough (pink noise).

High alternating magnetic fields arise especially in plants which are operated with high alternating currents from mains (eg, large high-temperature furnace plants). If the sensor resistor is made of a ferromagnetic material, then magnetic interferences with 100-Hz repetition frequency occur in the sensor noise at temperature below the Curie point (Figure 6-7). The same is true of ferromagnetic wire materials in cables because the external alternating magnetic field causes to a large extent correlated interference noise, so that the correlation technique has no effect. As the cables are also positioned in cold regions, ie, below the Curie point, it is not possible in such cases to employ, for example, combined NiCr TC-NT sensors. Even materials which have such a low content of Fe or Ni as to be usually regarded as non-magnetic may show Barkhausen noise in strong alternating magnetic fields.

Ferromagnetic materials are often applied for effective shielding of external alternating magnetic fields; in extremely strong fields, however, the Barkhausen noise generated in the shield can couple into the lead wires; in such cases, non-magnetic shieldings have to be applied.

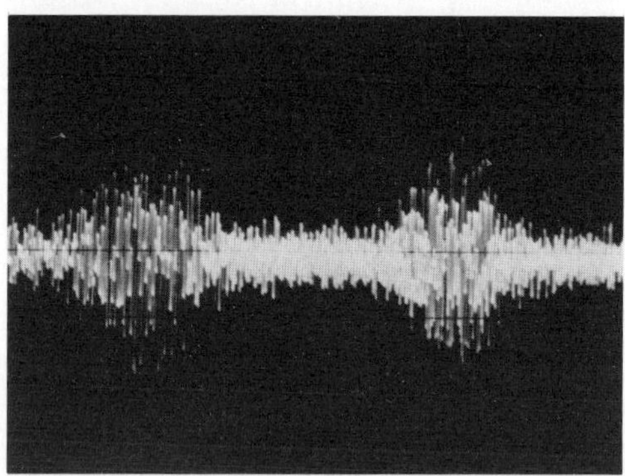

Figure 6-7.
Oscillogram of the noise of a combined sensor with NiCr/Ni thermocouples, superimposed by Barkhausen noise from the Ni wires (magnetic antinodes induced by alternating magnetic fields of a. c. current from 50 Hz mains; time window, 20 ms).

Apart from alternating magnetic fields, Barkhausen noise can also be generated by mechanical shocks or vibrations. If the sensor or the cable is subject to such stresses, ferromagnetic materials must also be avoided here.

6.4.4. Sensors at High Temperatures

As with thermocouples, mineral-insulated sheathed cables are also used for noise thermometers in the hot zone when measuring high temperatures. If the insulation resistances of mineral-insulated cables are no longer sufficient, then hard-sintered, rigid ceramics (eg, Al_2O_3, BN, HfO_2, BeO) are used for the insulation. As in the case of all contact thermometers, the relatively low electrical resistance of the insulating materials also causes difficulties (electrical shunting error) for the NT at temperatures above about 1500 °C (depending on the length of the hot zone). In the case of the NT, the insulation resistance should be higher than the sensing resistance by a factor of at least 6 [30]. This means that in selecting the sensing resistance value this latter may only amount to a few ohms.

Other problems, such as material compatibility and thermal expansion, are less important for simple noise sensors than for thermocouples since there is greater freedom in selecting the materials and it is in principle possible to construct the sensor (apart from the insulation) from one material, ie, the possibility exists of fabricating a practically single-material sensor (eg, of graphite). Further, the sensor material can be adapted to the actual ambient conditions.

As mentioned in Section 6.2.7, the resistance value often varies at high temperatures within relatively short times owing both to the high temperature itself (eg, recrystallization, evaporation) and to influences from the environment (eg, diffusion, chemical reaction). Then the direct measurement method (Section 6.2.4, Equation (6-13)) must be employed owing to the shorter averaging times and, possibly, a higher statistical error must be tolerated. The application of both measures then allows an on-line, quasi-continuous readout of the temperature to be made.

6.5 Description of Actual NT and TC-NT Sensors

The general possibility of using any electrical resistor for noise thermometry has been restricted in some respects in the previous sections, for various reasons.

From a multitude of feasible forms some construction types have been developed with favorable characteristics and proved successful in practice. Thus, noise resistors with two or four lead wires are employed, but four lead wires together with the correlation technique have advantages. The noise resistor can be grounded or insulated from ground; owing to the better possibilities of interference rejection, however, mainly insulated resistors are used. Further, the resistors should be designed for small inductances and capacitances to show white noise in the frequency band of measurement. It is often desirable to keep the dimensions of the sensor small. With noise sensors geometrical dimensions can be realized in such cases that are comparable to those of the corresponding thermocouples or resistance thermometer. However, minimum diameters of thermocouples below ca. 1.5 mm are difficult to realize with noise sensors.

6.5.1 Construction Types of Noise Resistors

For reasons of historical documentation, in Figure 6-8 a noise resistor and sensor is shown which had already been used in 1949 for noise temperature measurements up to 1050 °C. The sensing resistor is made from 2.5-μm platinum Wollaston wire, welded to heavy platinum leads 1.3 mm in diameter [11].

A more durable construction is exhibited by the noise resistor in Figure 6-9. A metal wire is threaded in a meandering way through a multi-hole ceramic body, thus forming a stable sensing element with both low inductance and low capacitance.

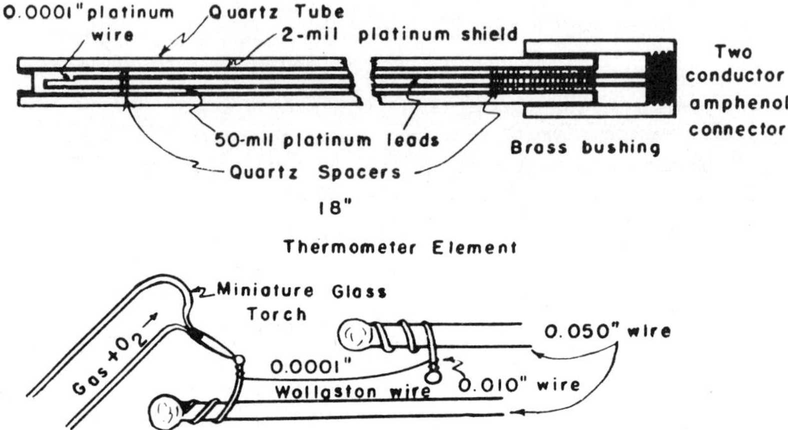

Figure 6-8. "Historical" noise resistor and sensor for high temperatures (from [11]).

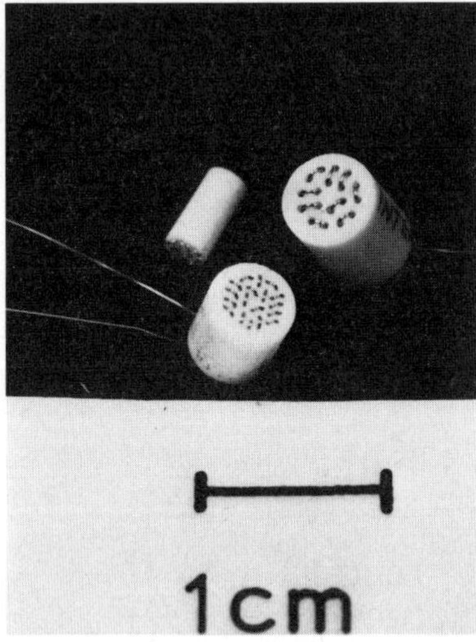

Figure 6-9.
Threaded noise resistors. Resistor, eg, metal wire (NiCr8020, diameter 0.15 mm) in a 36-hole ceramic (Al_2O_3, diameter 4.6 mm), resistance value 17 Ω.

Another construction of a noise sensor for high temperatures is shown in Figure 6-10 [44]. It is a combined sensor in a two-wire configuration with an Re coil as a noise resistor and one WRe thermocouple. To eliminate microphonic noise caused by vibration of the internal components, fine insulation powder (HfO_2) is compacted in all voids around the sensing coil and lead wires. Another method for reducing microphonic noise is to swage the sensor down and in this way compact the insulators, which are crushable.

In general, all the different construction designs of resistance thermometers and metal film resistors can also be used for noise resistors, but care must be taken to keep the inductances and capacitances low to ensure white noise in the frequency pass band.

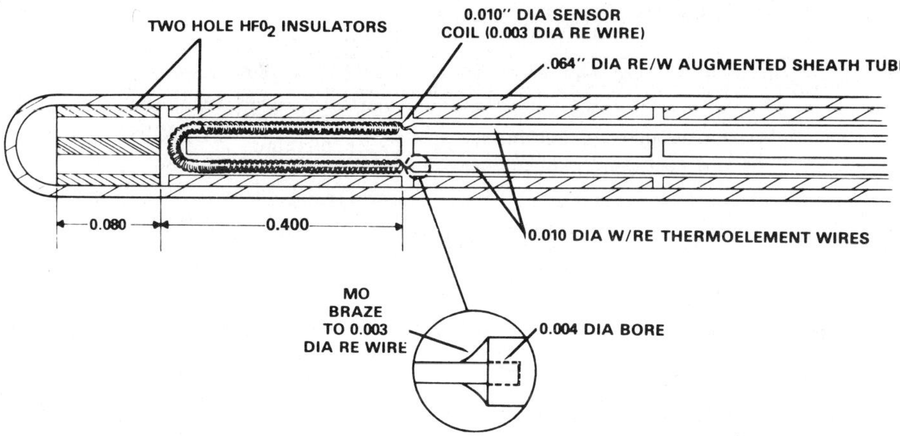

Figure 6-10. Illustration of 2200 °C JNPT (Johnson noise power thermometer) design using 0.010-in o. d. sensor coil (from [44]).

6.5.2 Sensors for Different Temperature Ranges

Commercial film resistors can be employed from very low temperatures (a few kelvin) up to several hundred kelvin. Resistors with a non-inductive design, due to a special serpentine pattern, are available from different manufacturers.

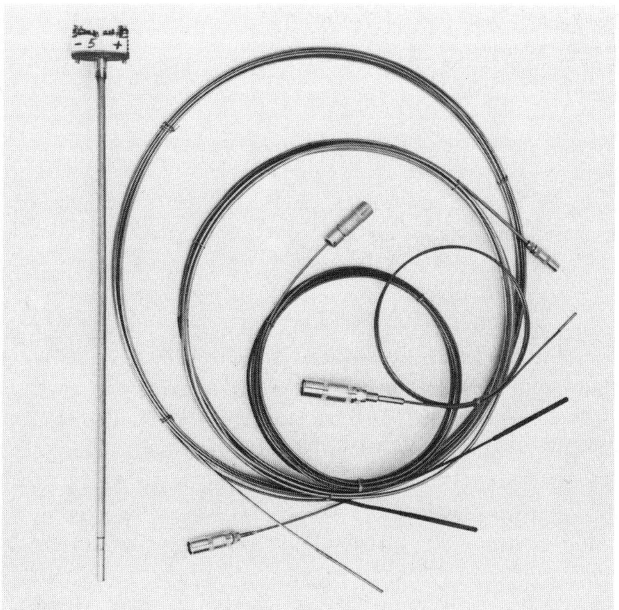

Figure 6-11. Different types of combined TC-NT sensors with mineral-insulated sheathed cables.

Resistors of the construction type in Figures 6-9 and 6-10 are suitable for the whole temperature range if the materials are selected appropriately. For high temperatures ceramic insulators (Section 6.4.3) are used and high-melting metals and their alloys serve as resistor wires, in addition to carbon fibers and filaments. The electrical shield necessary in most cases can be made from a refractory metal or graphite. Most important in this connection is the compatibility of the materials. The dimensions of the insulators of the construction type shown in Figure 6-9 reach down to about 1 mm diameter (eg, six-hole tube of Al_2O_3). Concerning the wire diameter, filaments down to ca. 0.01 mm are used in practice. However, for high temperatures the wire diameters should not be chosen to be less than 0.1 mm for long-term durability reasons.

For temperatures up to about 1200°C mineral-insulated sheathed cables (MI cables) are employed, similar to those used for thermocouples. Figure 6-11 shows combined sensors of different construction and dimensions. The sensors consist of four-wire MI cables with the noise resistor between the two hot junctions of the TCs in the tip (see also Figure 6-6). Such sensors are as robust as ordinary TCs; they cannot be distinguished externally from TCs.

6.6 Sensors for Metrological Applications

The Nyquist law is based on very fundamental considerations and gives a simple relationship for the thermodynamic temperature; it therefore provides in principle a firm basis for establishing a temperature scale. Noise thermometry has been used for temperature scale metrology up to now in the temperature range 1–1360 K [23–25, 27–29, 31, 34–39, 56]. At low temperatures metal film resistors with, for example, a serpentine pattern are employed as sensors, whereas at higher temperatures Pt resistance thermometers with non-inductive performance and sensor constructions as in Figure 6–9 with wires of Pt, PtRh and NiCr alloys are employed.

Important aspects for sensor construction are a negligible skin effect, small capacitances and inductances, or a balance of these two, to ensure white noise in the frequency band of measurement [19, 56]. Special attention has to be paid to electromagnetic interferences which must be reduced well below 10^{-4}; for this purpose an appropriate sensor construction (effective shielding, dual twisted pair arrangement of the lead wires) can be helpful. At high temperatures leakage my become a problem [30]. The ratio of insulation to sensor resistance must be sufficiently high, up to the upper temperature limit. Therefore, small sensor resistances (R_S) have to be chosen. With the comparison method, linearity errors of the electronic device can be avoided by inserting into the lines of the reference resistor (R_R) resistance values that generate the same parasitic noise as is generated by the sensor lines, so that at balance the useful and parasitic noise of the R_R circuit has the same magnitude as that of the R_S circuit. As the correlation technique works effectively with short cables, correlation errors are negligible [52]. Using sensors with the described characteristics, uncertainties down to 10^{-4} have been realized at high temperatures (>500°C) and to some 10^{-5} at low temperatures.

For the future, the noise thermometer appears very promising even beyond 1000°C because it has the exceptional advantage over other methods that any changes in the sensor are of no significance.

6.7 Development Trends for Noise Sensors

As described in Section 6.5.2, metals and alloys with high melting points (Pt, Rh, W, Re, etc.) have been applied in the production of noise resistors for temperatures above 1200 °C. However, the often low durability of the noise resistors manufactured from thin wires or films is disadvantageous in practice. Moreover, a protective atmosphere or vacuum is often needed for their application.

6.7.1 Ceramics as Materials for Noise Resistors

The Nyquist equation in its general form (Equation (6-1)) is valid for any material, ie, including ceramics. In contrast to metallic resistors, ceramic noise resistors can be manufactured much more robustly, and they can also be employed in chemically harsh (eg, oxidizing) atmospheres.

When the comparison method is used to measure the noise temperature, however, some requirements have to be fulfilled by ceramic resistors. The noise power spectrum must be white in the frequency range used (up to ca. 200 kHz), and the resistance should be independent of an applied small voltage (eg, measuring voltage, thermovoltage). A low temperature coefficient of resistance in the temperature range used (>1000 °C) is also desirable. Other requirements are a high melting point, high thermal and chemical stability, a low vapor pressure in the temperature range applied, and a small resistivity so that specimens can be manufactured that are easy to handle [57–60].

Most ceramics are insulators, but many can be manipulated by doping to make them show electronic conductivity, the order of magnitude of which can be adjusted with the nature and amount of the dopant. In this way, usable specimens with resistance values between about 20 and 1 Ω in the temperature range 1000– 2000 °C can be manufactured.

6.7.2 Compact Construction of Noise Resistors

As mentioned above, a robust and compact type of construction is possible for ceramic resistors. Disk-shaped or cylindrical specimens can be fabricated. The manufacturing process is simple and the samples can be produced cheaply.

In a similar manner, compact, cylindrical resistors can be fabricated from, eg, NiCr powders which are mixed with an insulation powder raising the resistivity. Robust noise resistors can also be fabricated by sintering, eg, coiled wires (as in Figure 6-10) in suitable insulation powders. Moreover, it seems possible to enclose wire noise resistors with insulation powders using a plasma spray process.

6.8 Applications of NT and TC-NT Sensors

Noise thermometry, as a complex method, is employed especially in areas where its exceptional characteristics (no decalibration or drift and high absolute accuracy) are required. The high costs of the electronic device may be seen in relative terms, however, if combined sensors are used; only one such device is then necessary for recalibrating any number of sensors from time to time.

6.8.1 Applications in the Laboratory and Industrial Plants

It is well known that in nuclear reactors thermocouples and resistance thermometers undergo decalibrations and exactly in such plants accurate and reliable measurements are necessary over long periods. Therefore, NT and combined TC-NT in addition to RT-NT sensors, in which the TCs and RTs are recalibrated in situ, have been used here (see also Table 6-1) [18, 19, 21, 22, 30, 41, 43–48, 61–63]. With the application of remote preamplifiers the measurement signals could be transmitted over distances of 150 m [61] and 600 m [62] to the control room.

A wide field of application for noise sensors is to be found in the high-temperature range ($>1000\,°C$), since at these temperatures any sensor exhibits drifts. Combined TC-NT sensors have been used here for basic investigations of other temperature sensors, eg, drift tests of WRe and B_4C/C TCs as a function of the insulation material and pressure [7]. Noise sensors are also useful when material data or process parameters are to be determined in the high-temperature range, since here exact information about the temperature is essential. If they are applied in plants operated with high a.c. currents, eg, hot isostatic pressure (HIP) plants [7], ferromagnetic materials have to be avoided in the signal lines (see Section 6.4.3). Other fields of application are metal and glass melts and chemical and petrochemical plants (see Table 6-1), if economic profit can be expected (eg, by higher yields from a process) by precise temperature measurement.

Table 6-1 gives a survey of sensor data and conditions of employment. It gives an indication of the numerous possible variations for applying NT and combined TC-NT sensors. In all cases accurate, undisturbed transmission of the measured signal was checked by selecting various frequency bands and controls in the time and frequency domain. The measurement accuracy achieved was better than 0.5% in all cases. These measurements, carried out under industrial conditions and in part in the presence of high electromagnetic interference fields, indicate the applicability of noise thermometry under plant conditions.

6.8.2 Applications in Temperature Scale Metrology

Since noise thermometry is a thermodynamic method it has been used for metrological purposes, in particular for determining fixed point temperatures.

With sensing resistances of about 10 kΩ, measurements of ^4He at various vapor pressures (near 4 K) were performed with a total uncertainty of only 0.3 mK [34, 35]. In the range

Table 6-1. Examples of sensor data and conditions of employment of NT and TC-NT sensors (RPA, Remote PreAmplifier; KWO and GKN, pressurized water reactors with 350 and 850 MW_{el}; AVR, high-temperature experimental reactor, 15 MW_{el}).

Plant	Sensor		Mineral-insulated cable		Transmission cable		
	NT-TC (materials)	R_S (cold) (Ω)	∅ (mm)	Length (m)	Length (m)	Impedance (Ω)	T (°C)
KWO Obrigheim (core region)	NiCr20	25; 46	2	9	15 (+20 test)	50; 120	310
GKN Neckarwestheim (primary circuit)	NiCr30-NiCr/Ni	50	6	0.4	10–50; RPA: 200	125	320
AVR Jülich (top reflector)	NiCr20-NiCr/Ni	25	3; 6	3	20–30; RPA: 600	50	700–1150
Methane reforming plant (He heater)	NiCr20-NiCr/Ni	17	4; 6	3	10	35	1000
Liquid air plant	NiCr20	120; 300			6	150	−170
Molten glass	PtRh30-PtRh30/PtRh6	7; 10	1 m Al_2O_3 tube, 15 mm ∅		5 (+20 test)	35; 50	1200–1400
HIP plant	WRe26-WRe5/WRe26	2	0.5–1 m in W tube of 7 mm ∅		3	20	1000–2000

100–500 K uncertainties between 10^{-4} and 10^{-5} were reported using metal film resistors of several kΩ [23, 27, 28, 38]. Between the Sb and the Ag fixed point (630–962 °C), platinum resistance thermometers with resistance values near 600 Ω at 960 °C were employed with an uncertainty of $2 \cdot 10^{-4} - 3 \cdot 10^{-4}$ in the measured temperatures [24, 25, 29].

With NiCr and Pt resistors of the type of construction in Figure 6-9, uncertainties of $2 \cdot 10^{-4} - 3 \cdot 10^{-4}$ for a single measurement and $5 \cdot 10^{-5}$ for the mean values (of more than ten measurements) at the Zn (419 °C) and the Ag (962 °C) fixed points could be achieved (sensing resistance 58 Ω at 419 °C and 38 Ω at 962 °C) [56]; with an improvement on the configuration given in [56] an uncertainty at the Zn point of $5 \cdot 10^{-5}$ could be attained for the single measurement.

6.8.3 Noise Thermometry at Very Low Temperatures

At very low temperatures, below a few K, the noise power of the amplifier becomes so high in comparison with the noise power of the sensing resistor that reasonable measurements are no longer possible with conventional noise thermometry. The low-noise amplification properties of the Josephson junction can be used to eliminate this problem [64]. Two different Josephson junction devices, a resistive SQUID (superconducting quantum interference device) [64, 65] and a SQUID magnetometer [66, 67], with amplifier temperatures of several millikelvin and less than 0.1 mK, have been employed to measure temperatures from a few millikelvin to 4 K (up to 20 K is possible) with inaccuracies of 1–0.5% (0.1% is possible). The resistors are made preferably of silicon-bronze (eg, $Cu_{99}Si_1$) with resistance values of typically 10^{-5} Ω.

6.9 Comparison of Noise Thermometers with Other Sensors

The noise thermometer covers the entire technical temperature range from about 1 K (or 1 mK with SQUID application) to 2500 K (in future probably also higher), ie, a wider range than any other sensor.

Table 6-2. Comparison of noise thermometer (NT) and thermocouple (TC).

Type	Advantages	Disadvantages
NT	Absolute thermometer	Very low signal
	High absolute accuracy (applicable as standard)	Complex electronic processing
		Sensitive to EMI
	No decalibration, no drift (not impaired by changes of the sensor)	For high accuracy: long measurement time
	Material independent	Expensive
TC	Simple, easy to handle	Secondary thermometer
	Relatively insensitive to EMI	Drift of the emf, decalibration (impaired by changes of the sensor)
	Cheap	No high absolute accuracy

In contrast to all other sensors, the noise sensor may in principle undergo any change during application, without a reduction in accuracy. Noise sensors are therefore universally applicable for both industrial and metrological purposes. They probably have the highest development potential.

In Table 6-2 the noise thermometer and thermocouple are compared.

With the application of combined TC-NT sensors, ie, if the TCs are recalibrated in situ from time to time, the advantages of both are gained *while largely avoiding the disadvantages.* In many respects the TC-NT sensor thus has the characteristics of an ideal thermometer.

6.10 References

[1] *Report WASH-1067,* U.S. Atomic Energy Commission, Washington, DC, 1966.

[2] *Temperature, Its Measurement and Control in Science and Industry,* Herzfeld, C.M. (ed.); New York: Reinhold, 1962.

[3] *Temperature, Its Measurement and Control in Science and Industry,* Plumb, H. H. (ed.); Pittsburgh: Instrument Society of America, 1972.

[4] *Temperature, Its Measurement and Control in Science and Industry,* Schooley, J. F. (ed.); New York: American Institute of Physics, 1982.

[5] *International Colloquium on High-Temperature In-Pile Thermometry, Petten, The Netherlands, December 1974* Hardt, P. von der, Zeisser, P., Mason, F. (eds.) Commission of the European Communities, Luxembourg, 1975.

[6] Brixy, H., *Report Jül-2051 (transl.),* Kernforschungsanlage Jülich, Jülich, April 1986.

[7] Brixy, H., Oehmen, J., Hecker, R., Müller, G., paper presented at the *International Conference on Hot Isostatic Pressing of Materials, Antwerp, April 1988.*

[8] Nyquist, H., *Phys. Rev.* **32** (1928) 110.

[9] Johnson, J. B., *Phys. Rev.* **32** (1928) 97.

[10] Blalock, T. V., Shepard, R. L., in: *Temperature, Its Measurement and Control in Science and Industry* Vol. 5, Schooley, J. F. (ed.); New York: American Institute of Physics, 1982, p. 1219.

[11] Garrison, J. B., Lawson, A. W., *Rev. Sci. Instrum.* **20** (1949) 785.

[12] Hogue, E. W., *Report NBS 3471,* July 1954 National Bureau of Standards, Washington, DC.

[13] Pursey, H., Pyatt, E. C., *J. Sci. Instrum.* **36** (1959) 260.

[14] Patronis, E. T., et al., *Rev. Sci. Instrum.* **30** (1959) 578.

[15] Fink, H. J., *Can. J. Phys.* **37** (1959) 1397.

[16] Tillinger, M. H., *Report AD 467990,* June 1965. Defense Documentation Center for Scientific and Technical Information, Cameron Station Alexandria, Virginia.

[17] Crovini, L., *Ric. Sci.* **37** (1967) 1238.

[18] Brixy, H., *Nucl. Instrum. Methods* **97** (1971) 75.

[19] Brixy, H., *Report Jül-885-RG,* Kernforschungsanlage Jülich, Jülich, 1972.

[20] Actis, A., Cibrario, A., Crovini, L. in: *Temperature, Its Measurement and Control in Science and Industry;* Plumb, H. H. (ed.); Pittsburgh: Instrument Society of America, 1972, 355.

[21] Brixy, H., Hecker, R., Overhoff, T., in: *Symposium on Nuclear Power Plants Control and Instrumentation,* Report IAEA/SM-168/F-1, IAEA Vienna, 1973.

[22] Brixy, H., Rittinghaus, K. F., Gärtner, K. J., Hecker, R., paper presented at the *International Colloquium on High-Temperature In-Pile Thermometry, Petten, The Netherlands, December 1974.* Hard, P. von der, Zeisser, P., Mason, F., (eds.), Commission of the European Communities, Luxembourg, 1975, p. 775.

[23] Pickup, C. P., *Metrologia* **11** (1975) 151.

[24] Crovini, L., Actis, A. in: *Temperature Measurement, Institute of Physics Conference Series* Vol. 26, Billings, B. F., Quinn, T. J. (eds.); The Institute of Physics, London and Bristol, 1975, p. 398.

[25] Crovini, L., Actis, A., *Metrologia* **14** (1978) 69.
[26] Pepper, M. G., Brown, J. B., *J. Phys. E.: Sci. Instrum.* **12** (1979) 31.
[27] Pickup, C. P. in: *Proceedings of the 6th International Conference on Noise in Physical Systems,* NBS Publication 614, April 1981, National Bureau of Standards Washington, DC.
[28] Pickup, C. P.,in: *Temperature, Its Measurement and Control in Science and Industry* Vol. 5, Schooley, J. F. (ed.); New York: American Institute of Physics, 1982, p. 129.
[29] Crovini, L., Actis, A., in: *Temperature, Its Measurement and Control in Science and Industry* Vol. 5, Schooley, J. F. (ed.); New York: American Institute of Physics, 1982, p. 133.
[30] Brixy, H., Hecker, R., Rittinghaus, K. F., Höwener, H., in: *Temperature, Its Measurement and Control in Science and Industry* Vol. 5, Schooley, J. F. (ed.); New York: American Institute of Physics, 1982, p. 1225.
[31] Crovini, L., Actis, A., Galleano, R., paper presented at the *International Colloquium on Temperature Measrements in Industry and Science, Beijing, April 1986.*
[32] Shore, F. J., Williamson, R. S., *Rev. Sci. Instrum.* **37** (1966) 787.
[33] Storm, L., *Z. Angew. Phys.* **28** (1970) 331.
[34] Klein, G., Klempt, G., Storm, L., *Metrologia* **15** (1979) 143.
[35] Klempt, G., Storm, L. in: *Proceedings of the 6th International Conference on Noise in Physical Systems,* NBS Publication 614, April 1981, National Bureau of Standards, Washington, DC.
[36] Klempt, G., in: *Temperature, Its Measurement and Control in Science and Industry* Vol. 5, Schooley, J. F. (ed.); New York: American Institute of Physics, 1982, p. 125.
[37] Pickup, C. P. in: *Noise in Physical Systems and 1/f Noise,* Savelli, M., et al. (eds.); Amsterdam: Elsevier, 1983, p. 409.
[38] White, D. R., *Metrologia* **20** (1984) 1.
[39] White, D. R., Pickup, C. P., *IEEE Trans. Instrum. and Meas.* **IM-36** No. 1 (1987) 47.
[40] Borkowski, C. J., Blalock, T. V., *Rev. Sci. Instrum.* **45** (1974) 151.
[41] Shepard, R. L., et al., paper presented at the *International Colloqium on High-Temperature In-Pile Thermometry, Petten, The Netherlands, December 1974.* Hardt, P. von der, Zeisser, P., Mason F. (eds.), Commission of the European Communities, Luxembourg, 1975, p. 737.
[42] Blalock, T. V., Borkowski, C. J., *Rev. Sci. Instrum.* **49** (1978) 1046.
[43] Decreton, M. et al., *High Temp. High Press.* **12** (1980) 395.
[44] Cannon, C. P., *IEEE Trans. Nucl. Sci.* **NS-28** No. 1 (1981) 763.
[45] Blalock, T. V., Horton, J. L., Shepard, R. L., in: *Temperature, Its Measurement and Control in Science and Industry* Vol. 5, Schooley, J. F. (ed.); New York: American Institute of Physics, 1982, p. 1249.
[46] Decreton, M. C., in: *Temperature, Its Measurement and Control in Science and Industry* Vol. 5, Schooley, J. F. (ed.); New York: American Institute of Physics, 1982, p. 1239.
[47] Billeter, T. R., Cannon, C. P., in: *Temperature, Its Measurement and Control in Science and Industry* Vol. 5, Schooley, J. F. (ed.); New York: American Institute of Physics, 1982, p. 1245.
[48] Shepard, R. L., Blalock, T. V., Roberts, M. J., *Report EPRI RP 2254-1,* February 1988. Electric Power Research Institute, Palo Alto, California.
[49] Brodskii, A. D., Savateev, A. V., *Meas. Tech. USSR* **5** (1960) 397 (translated from *Izmer. Tekh.* **5** (1960) 21).
[50] Imamura, M., Ohte, A., in: *Temperature, Its Measurement and Control in Science and Industry* Vol. 5, Schooley, J. F. (ed.); New York: American Institute of Physics, 1982, 139.
[51] Rice, S. O., *Bell Syst. Tech. J.* **23** (1944) 282.
[52] Höwener, H., *Report Jül-2107,* December 1986 Kernforschungsanlage Jülich, Jülich.
[53] Brixy, H., Hecker, R., Lehmann, M., Weingarten, J., paper presented at the *NEA Specialists' Meeting on In-Core Instrumentation and the Assessment of Reactor Nuclear and Thermal/Hydraulic Performence, Fredrikstad, Norway, October 1983.*
[54] Müller, G., Ph. D. Thesis, RWTH Aachen, 1989.
[55] Brixy, H., *Ger. Patent 2263469,* 1975; *US Patent 3956936,* 1976.
[56] Fechner, H., *Report Jül-2254,* 1988; Kernforschungsanlage Jülich, Jülich.
[57] Brixy, H., Mallinckrodt, D. von, Lynen, A., *High Temp. High Press.* **15** (1983) 139.
[58] Mallinckrodt, D. von, Häusser, H., Brixy, H., *High Temp. High Press.* **17** (1985) 639.
[59] Häusser, H., Mallinckrodt, D. von, Brixy, H., paper presented at the *8th German-Yugoslav Meeting on Materials Science and Development, "Ceramics and Metals" Kranju, Yugoslavia, 1987.*

[60] Häusser, H., Ph. D. Thesis, RWTH Aachen, 1989.
[61] Weingarten, J., Brixy, H., paper presented at the *OECD-NEA Specialists' Meeting on In-Core Instrumentation and Reactor Core Assessment, Cadarache, France, June 1988.*
[62] Brixy, H., Hecker, R., Oehmen, J., Müller, G., Rittinghaus, K. F., Wegener, H. P., Zimmermann, E., paper presented at the *Annual Meeting on Nuclear Technology, German Nuclear Society/German Atomic Forum, Travemünde,* FRG, May 1988.
[63] Höwener, H., Brixy, H., Hecker, R., Wakayama, N., Tsuyuzaki, N., paper presented at the *OECD-NEA Specialists' Meeting on In-Core Instrumentation and Reactor Core Assessment, Cadarache, France, June 1988.*
[64] Kamper, R. A., Zimmerman, J. E., *J. Appl. Phys.* **42** (1971) 132.
[65] Soulen, R. J., Von Vechten, D., in: *Temperature, Its Measurement and Control in Science and Industry* Vol. 5, Schooley, J. F. (ed.); New York: American Institute of Physics, 1982, p. 115.
[66] Webb, R. A., Gifford, R. P., Wheatley, J. C., *J. Low Temp. Phys.* **13** (1973) 383.
[67] Hudson, R. P., Marshak, H., Soulen, R. J., Utton, D. B., *J. Low Temp. Phys.* **20** (1975) 1–39.

7 Time / Frequency Thermometers

H. Ziegler, H. J. Aulfes, H. Quint, Universität GH Paderborn, FRG

Contents

7.1 Overview of Frequency/Time Sensors for Temperature

7.1.1 Sound Velocity and Temperature Dependence

7.1.1.1 Gases

The velocity of sound propagated through ideal gases is dependent on temperature (Figure 7-1) according to the equation

$$c = \sqrt{\frac{\gamma RT}{M}} \tag{7-1}$$

where

c = velocity of sound
γ = specific heat ratio (constant pressure/constant volume) (C_p/C_v)
R = universal gas constant
M = molecular weight of the gas
T = temperature (Kelvin)

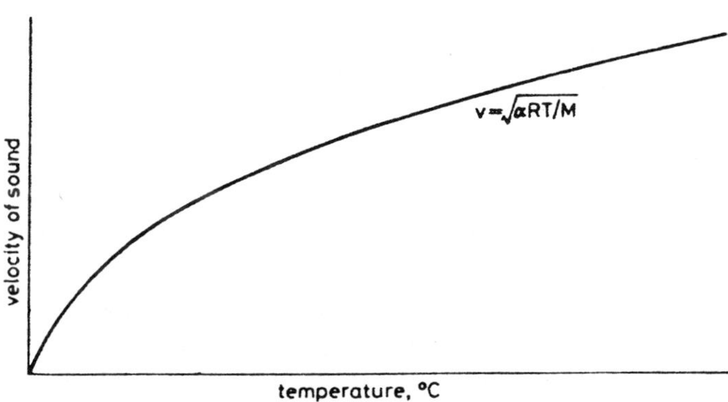

Figure 7-1. Variation of velocity of sound with temperature [1].

This simple relationship is often used for the determination of both specific heat and temperature by measuring the sound velocity of a gas. Note that the sound velocity of an ideal gas does not depend on the pressure. At very high frequencies the sound velocity will depend on the frequency (sound dispersion) due to molecular relaxation and hence Equation (7-1) is no longer valid [2, 3]. An example of one type of dispersion is the behavior of CO_2 around 25 kHz [3]. For Van der Waals gases and other non-ideal gases the situation is generally more complicated.

Equation (7-1) is derived from the following relationship:

$$c = \sqrt{\frac{\partial p}{\partial \rho}} \tag{7-2}$$

where

p = gas pressure
ρ = gas density

This is the general expression for the velocity of small-amplitude, plane waves in a fluid or gas with neither viscosity nor heat conduction [2, 3]. Equation (7-1) follows from Equation (7-2) for the case of adiabatic sound propagation in an ideal gas, where the Poisson equations are valid.

7.1.1.2 Liquids

For liquids Equation (7-2) for the speed of sound is often written as

$$c = \sqrt{\frac{K_s}{\rho}} \tag{7-3}$$

where

$$K_s = \rho \frac{\partial}{\partial_\rho} p(\rho, S) \tag{7-4}$$

is the adiabatic bulk modulus (for an ideal gas, $K_s = \gamma p$) [4]. Since the density depends on both temperature and pressure, the velocity of sound in liquids consequently also depends on temperature and pressure (Figure 7-2).

For a fixed temperature T, c increases nearly linearly with pressure, but for a fixed pressure p it increases to a maximum and subsequently decreases with increasing temperature. There is no *a priori* knowledge of the variation of the bulk modulus and density with temperature, so $c(T)$ in a liquid must be calibrated against other standards [5]. For the range of water temperatures between 0 and 20°C and pressures between 1 and 100 atm (10^5 and 10^7 Pa), the following empirical equation may suffice [4]:

$$c = 1447 + 4.0\,\mathrm{d}T + (1.6 \cdot 10^{-6})\,p \tag{7-5}$$

where

$\mathrm{d}T$ = T − 283.15 (temperature relative to 10°C)
p = absolute pressure in Pa
c = velocity of sound in m/s

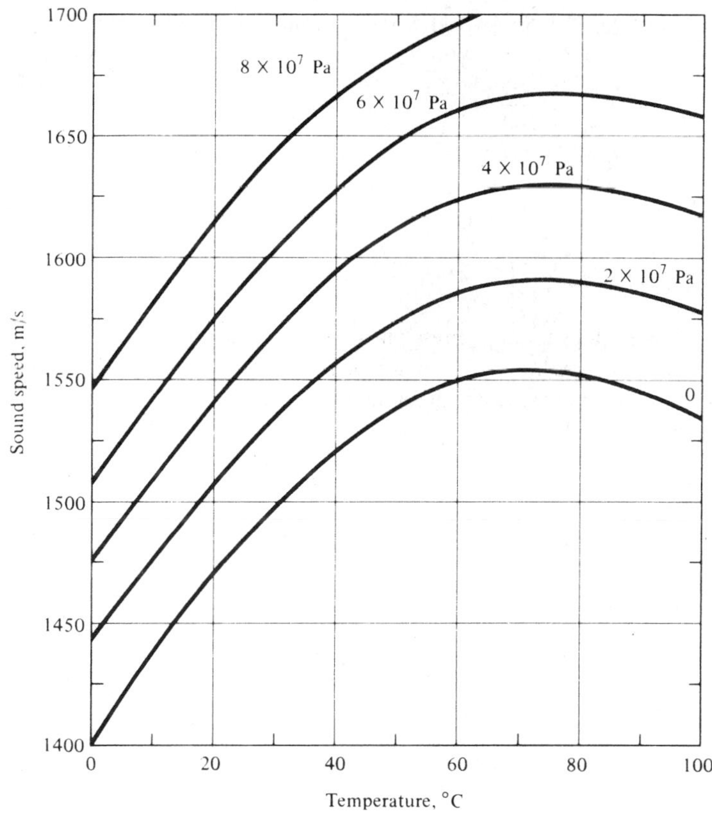

Figure 7-2. Temperature and pressure dependence of the velocity of sound in distilled water [4].

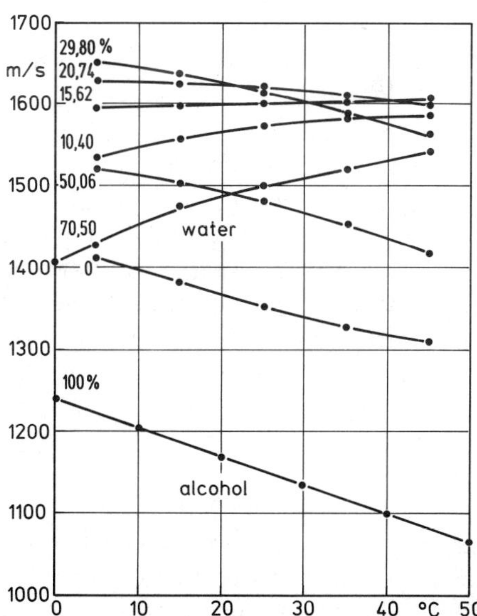

Figure 7-3.
Temperature dependence of the velocity of sound in a water-alcohol mixture [6].

For a mixture of water and other liquids, tuning of the temperature coefficient is possible by varying the proportions of the components. Figure 7-3 shows the temperature dependence of the velocity of sound in a water-alcohol mixture.

The content of alcohol is expressed in weight percent. Figure 7-3 shows that the velocities of the water-alcohol mixtures are generally not between those of the pure components. Close to 17.35 wt.-% of alcohol the temperature coefficient is zero [7]. In aqueous solution the velocity of sound increases with increasing concentration.

7.1.1.3 Solids

In solids the velocity of sound is determined by both the elastic stiffness and the density:

$$c = \sqrt{\frac{E}{\rho}} \tag{7-6}$$

where

E = elastic stiffness of solid (Young's modulus)
ρ = density of solid

This equation determines the velocity of sound of longitudinal waves in bars. The velocity in solids of infinite extension normally differs from Equation (7-6) and must be calculated by means of the Poisson number:

$$\frac{C_{\text{inf}}}{C_{\text{Bar}}} = \sqrt{\frac{(1-\mu)}{(1-\mu-2\,\mu^2)}} \tag{7-7}$$

where

μ = Poisson number (for metals $0.2 < \mu < 0.45$)
C_{inf} = velocity of sound in infinite solids
C_{Bar} = velocity of sound in bars

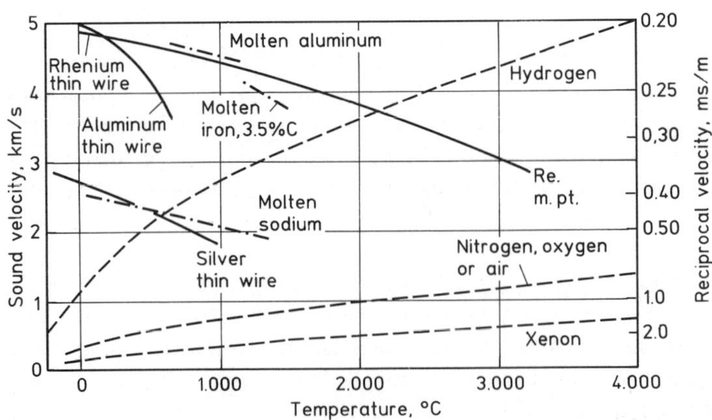

Figure 7-4. Temperature dependence of the velocity of sound in different metals [5].

The values for the velocity of sound in solids vary by an order of magnitude between 1000 and 10000 m/s. In metals the velocity decreases with increasing temperature (Figure 7-4). The slope is of the order of -0.5 m/s per K (rhenium) [8].

7.1.1.4 Single Crystals

The velocity of sound in crystals is also obtained by Equation (7-6). The important difference is the dependence of the elastic stiffness on the direction of sound propagation. For SiO_2 (quartz) the elastic stiffness is of the order of 29–130 N/m^2. These values correspond to velocities of sound of 3300–7000 m/s, depending on the quartz cut angle.

The temperature coefficient is also dependent on the quartz cut angle. Further details are given in Section 7.3.

7.1.2 Measurement of Sound Velocity

The following sections describe the three most important groups of techniques for the measurement of the velocity of sound.

7.1.2.1 Pulse Measurement

The measurement of the velocity of sound is possible by observing the time difference in the arrival of a sound signal at two different points. If $t_2 - t_1$ is the sound travel time and $x_2 - x_1$ the travel distance between the two points, then c is given by

$$c = \frac{(x_2 - x_1)}{(t_2 - t_1)} \,. \qquad (7\text{-}8)$$

To obtain more precise measurement results it is necessary to have a sufficiently large measuring distance and a high-resolution time measurement. In liquids the best results are achieved with short, high-frequency ultrasonic pulses ($\geq$15 MHz) which are produced and detected by a quartz sensor. In this case the measuring distance can be reduced to only a few centimeters. This method allows an accuracy of better than 1% [6]. A special application of the pulse method is a system called sing-around. The block diagram in Figure 7-5 illustrates the principle of operation.

The pulse generator energizes the transducer T to transmit an acoustic pulse through the liquid contained in a vessel. A transducer on the opposite side of the vessel is used as a receiver. A gated amplifier on the receiving side amplifies the received pulse, which is then used to trigger the pulse generator, initiating a new pulse. This loop runs continuously. The frequency of the pulse generator is then measured using a frequency counter. This frequency, called the sing-around frequency (f_{SAF}), is highly stable [10] and temperature sensitive [11].

The reciprocal of f_{SAF} gives the travel time through the liquid plus t_e, where t_e is the sum of the electrical decay and other time errors. The total time delay is

$$\frac{1}{f_{SAF}} = t_e + \frac{l}{c} \tag{7-9}$$

where

f_{SAF} = measured sing-around frequency
t_e = time error
l = distance between transducers
c = velocity of sound

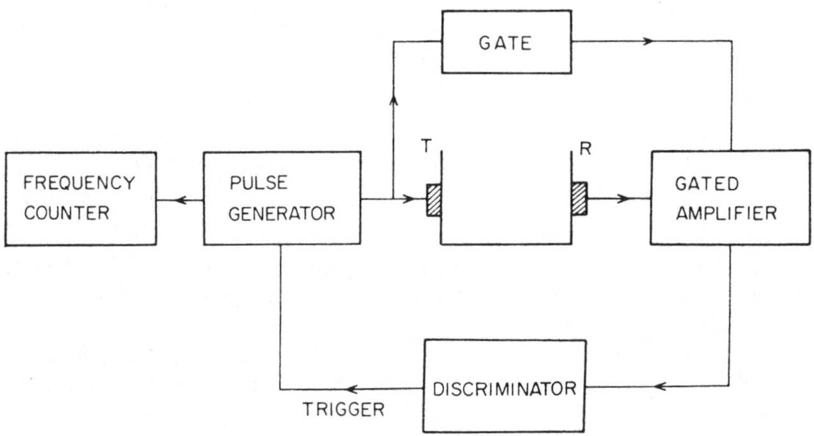

Figure 7-5. Block diagram of the sing-around velocimeter [9].

t_e must be obtained from calibration measurements with well known liquids. If both the distance between the transducer and the pulse generator and t_e are known, the velocity in the liquid is obtained with Equation (7-3) [9–11].

The limit of the pulse measurement methods is given by the dispersion of the medium ($\partial c/\partial f \neq 0$). Owing to dispersion, the pulse shape changes with time and therefore it is impossible to find a reproducible trigger point in the pulse burst.

7.1.2.2 Phase Measurement

This system uses a fixed-frequency signal propagated between two transducers spaced a known distance apart. The block diagram in Figure 7-6 illustrates the principle of operation. The oscillator supplies the driving power to the transmitting transducer, which launches an ultrasonic wave into the medium. The wave is received by the receiving transducer, amplified, amplitude limited, differentiated, and applied to a phasemeter.

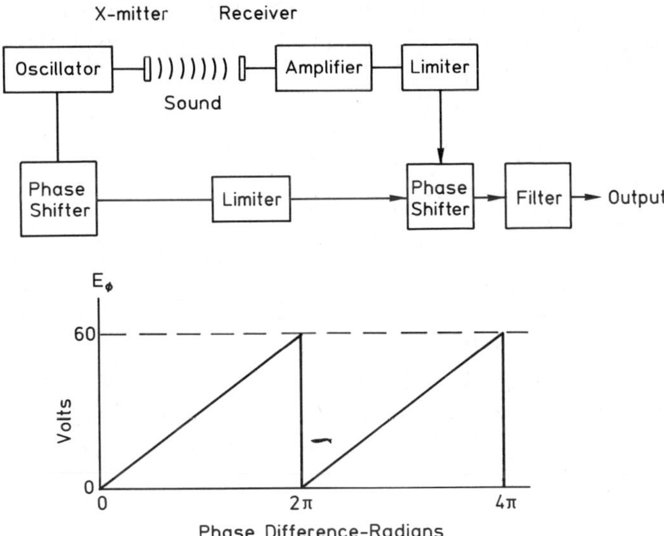

Figure 7-6. Block diagram of the phase velocimeter [3].

By filtering out the high-frequency components, a sawtooth output voltage is obtained as a function of phase difference. The output voltage repeats for every 360° phase shift, as shown in Figure 7-6. The radian phase shift between transducers is

$$\Phi = \frac{2\pi l f}{c} \tag{7-10}$$

where

l = distance between transducers
c = velocity of sound
f = frequency of sound

When the distance between the transducers and f are known, the velocity of sound is easily obtained with Equation (7-10).

7.1.2.3 Feedback/Oscillator

The third group of techniques for sound velocity measurement are the oscillator or feedback methods.

An acoustic resonator (eg, quartz resonator [12–18] or gas-filled spherical resonator [19–21] is exited to continuous vibrations with the aid of an oscillator. The vibration frequency is a function of both geometrical dimensions (eg, thickness of the quartz resonator or radius of the gas-filled spherical resonator) and the constants of the resonator material. The vibration quality factors of acoustic resonators are very high. Gas-filled spherical resonators have Q

values of the order of 2000–10000 [20]. Hence they can be used with very simple instrumentation to measure the speed of sound in a gas with an accuracy of 0.02% [21]. Quartz resonators have Q values up to 10^6.

7.2 Time Interval Sensors

Two principles of measurement can be distinguished in ultrasonic thermometry. A known sensor can be introduced into the thermal enclosure and its thermal velocity monitored. This type of sensor is called the "foreign sensor in the medium". In the second method ultrasonic waves travel through the medium itself, and this is the "medium itself sensor". In principle there are four basic elements in an ultrasonic thermometer: (a) sensor; (b) transducer; (c) buffer; and (d) electronics (Figure 7-7).

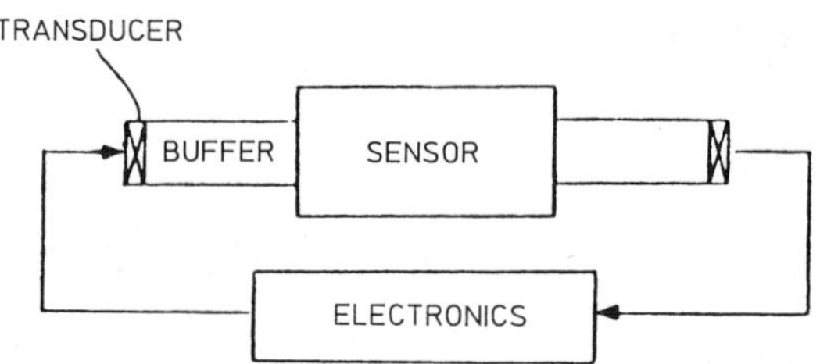

Figure 7-7.
Basic ultrasonic thermometer instrumentation system [5].

The sensor is the element in which the velocity of sound is measured as a function of temperature. The transducer is an electroacoustic converter which launches acoustic pulses and then after receiving the same pulses converts them back into electrical signals. Transducers and the electronics are thermally insulated from the hot sensor area by acoustic buffers. The electronic equipment consists typically of a pulse generator, receiver, signal processor and a display.

7.2.1 Solid-State Acoustic Sensors

7.2.1.1 Nonresonant Sensors

A pulse-echo system [5, 8, 22–29] exploits the temperature dependence of the transit time of an ultrasonic signal through a heated material of definite path length. The travel time is measured and varies with temperature owing to the velocity of sound. This sensor type is called a nonresonant sensor.

The principle of this type of sensor is simple: a single pulse is transmitted through the sensor and partly reflected by the discontinuities in the sensor and at the end of the sensor. The reflected pulses are detected by a transmitting coil (see Figure 7-8). This pair of pulses have travelled different distances, the pulse reflected from the end of the sensor having travelled a longer distance. The transit time between these two pulses is a function of the temperature of the sensor area. The temperature sensitivity is essentially determined by the following parameters:

- accuracy of the time interval counter,
- sensor material,
- sensor path length (sensor design),
- temperature range,
- number of echo reverberation selected, and
- vibration mode.

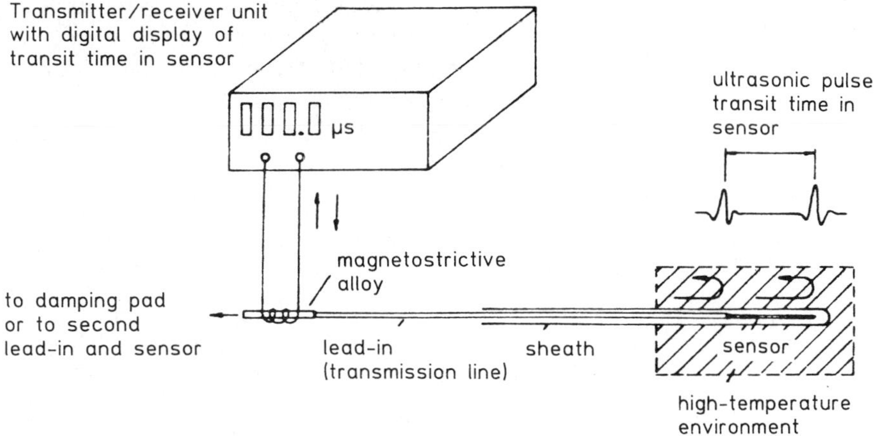

Figure 7-8. The principle of a nonresonant sensor [28].

For example, Figure 7-9 shows the calibration graph for a thoriated tungsten (2% ThO_2) sensor with a 50 mm long path, measured with a commercial pulse generator and time interval counter (Panatherm 5010; ± 0.1 μs resolution, 3 μs pulse width). The reproducibility and temperature resolution above 1800 °C are about ± 20 K [28].

Such acoustic sensors have the fundamental advantage over other thermometers of a wider choice of sensor materials depending mainly on the temperature range. Commonly used refractory materials are [5]:

temperature range	material
up to 800 K	aluminium
up to 1350 K	stainless steel
up to 1900 K	sapphire
up to 2400 K	ruthenium, molybdenum
up to 3000 K	thoriated tungsten and tungsten/rhenium

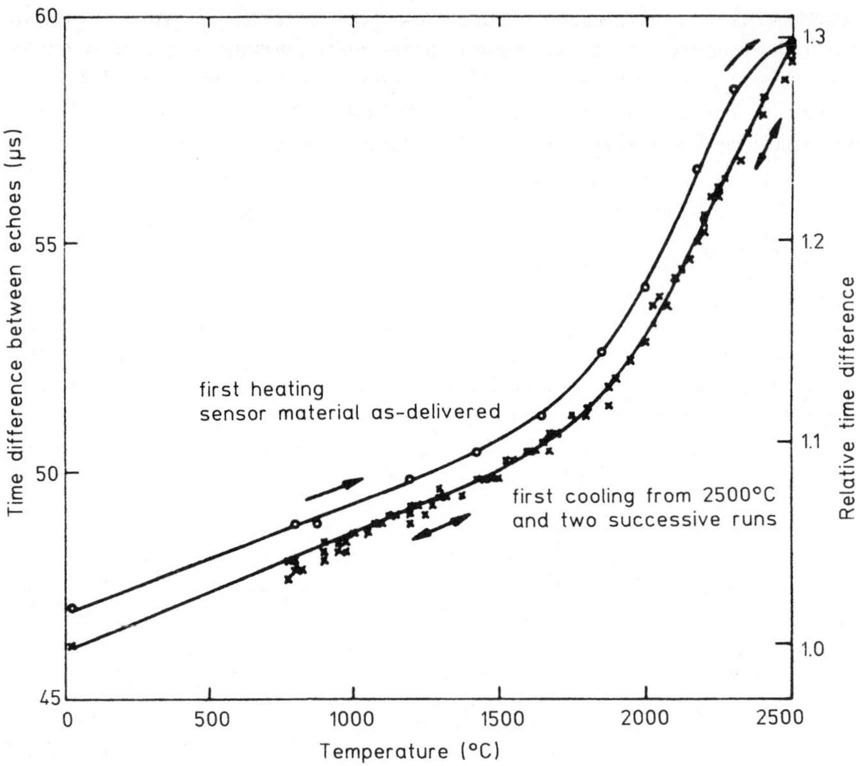

Figure 7-9. Calibration graph for a W-2% ThO_2 sensor, 50 mm long [28].

The sensor design is mostly a thin-wire sensor. It is very important that the sensor body is a monolithic, one-piece, weld-free construction of a single material. The typical dimensions depend, of course, on the temperature range. They range from 0.03 to 3 mm in diameter and from 1.3 cm to 3 m in length [5]. A special application of the nonresonant pulse-echo sensor is in the measurement of temperature profiles [5, 22, 30]. To illustrate the principle, consider a sensor such as that shown in Figure 7-10.

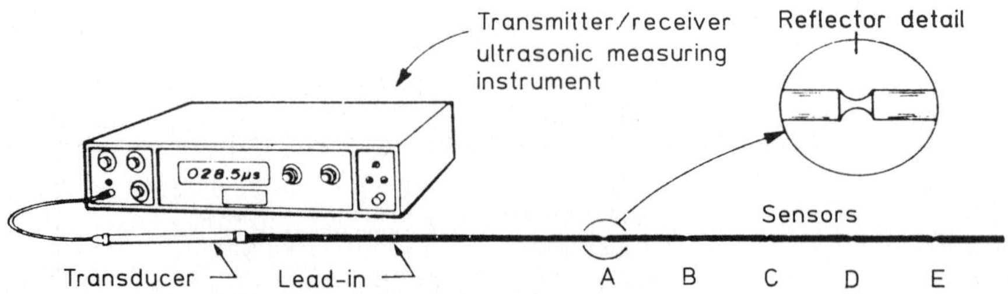

Figure 7-10. Schematic diagram of a multizone thermometer [30].

This is a single sensor line with many (17) pulse reflectors (notches). The echoes from the notches are gated sequentially so that the time in each sensors element is read sequentially. The transit time between selected pair of echoes corresponds to the temperature differences between reflection points. The evaluation yields a temperature profile over the sensor length. The echo pair from the first and last notches yield the average temperature.

7.2.1.2 Resonant Sensors

Resonance thermometers depend on the resonant frequency, the mechanical dimensions and the velocity of sound in the body. The resonant frequency of such a body will change with temperature. An example of this type is the well known quartz thermometer [12–18], described in detail in Section 7.3.

This section describes another resonant solid acoustic thermometer. Figure 7-11 shows a general diagram of the instrument.

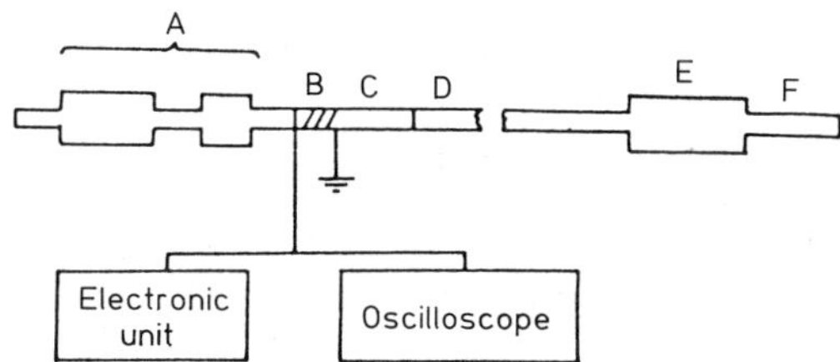

Figure 7-11.
General diagram of an acoustic resonant thermometer [27].

Frequency bursts are generated in the probe by the magnetostrictive transducer (C), and travel through the lead-in line (D) to both the coupler (E) ($\lambda/4$ long) and the resonator (F) ($\lambda/4$ long). The resonator takes up energy from the pulse, stores it, and re-radiates it in proportion to the amount stored. Hence, the echo consists of the signal from the coupling mismatch and the re-radiated signal. Note that the two signals are in antiphase, and therefore the echo is the difference between the two signals. Figure 7-12 shows the echo for the resonant condition.

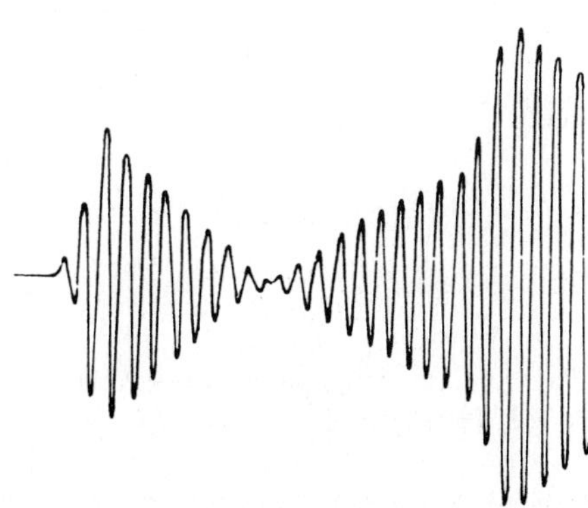

Figure 7-12.
The echo for the resonant condition [31].

If the resonator is in resonance there is a well defined crossover in the echo return. This crossover is the best criterion for finding the resonant frequency automatically. The frequency-temperature characteristics are found by previous calibrations and hence the temperature is obtained.

7.2.2 Gas Sensors

The acoustic gas thermometer is an inherently simple instrument. It was principally developed in the USA in the late 1950s and early 1960s.

For ideal gases the two important relationships between temperature, pressure, and gas velocity are:

- c is proportional to $\sqrt{T}$,
- c is independent of pressure P.

Considering a nonideal gas, ie, assuming that the volume of the gas molecules and intermolecular forces are not zero, the equation of state is given by the Van der Waals equation:

$$(V - b)(P + a_v) = MRT \tag{7-11}$$

where

$$a_v = \frac{a}{V^2}$$

V = gas volume
P = gas pressure
R = universal gas constant
M = molecular weigth of the gas
a_v, b = functions of molecular constants

When b/V and a_v/V are small (eg, ca. 0.1) the following equation for the acoustic sound velocity can be derived using the Van der Waals equation:

$$c = \sqrt{\frac{\gamma RT}{M}} \cdot \left[1 + \frac{KP}{RT}\right] \tag{7-12}$$

where

K = function of a_v, b, and T.

For this case the following important relationships are valid:

- c is no longer simply proportional to $\sqrt{T}$ although this is still a good approximation,
- c is proportional to pressure.

Calculated values for a_v/p and b/V are given in Table 7-1 [24].

Table 7-1. Calculated values for α_v and b for gases at various temperature and pressures [24].

Pressure (atm)		1		100	
Temperature (K)		273	1000	273	1000
Helium	b/V	1.1×10^{-3}	2.9×10^{-4}	1.1×10^{-1}	2.9×10^{-2}
	a_v/P	7.3×10^{-6}	4.5×10^{-8}	7.3×10^{-4}	4.5×10^{-6}
Hydrogen	b/V	1.2×10^{-3}	3.2×10^{-4}	1.2×10^{-1}	3.2×10^{-2}
	a_v/P	4.1×10^{-5}	2.5×10^{-7}	4.1×10^{-3}	2.5×10^{-5}
Nitrogen	b/V	1.7×10^{-3}	4.8×10^{-4}	1.7×10^{-1}	4.8×10^{-2}
	a_v/P	2.0×10^{-4}	1.2×10^{-6}	2.0×10^{-2}	1.2×10^{-4}

7.2.2.1 *Acoustic Gas Interferometers*

A practical instrument for the determination of the absolute temperature from sound velocity measurements is the acoustic helium interferometer. Plumb et al. [2, 26, 32] developed a sensor for the temperature range 2–20 K. In this instrument a quartz crystal, operating at its resonant frequency, is both an emitter and transmitter of ultrasonic waves. The emitted radiation passes through the gas, is reflected from the surface of a piston, and returned to the

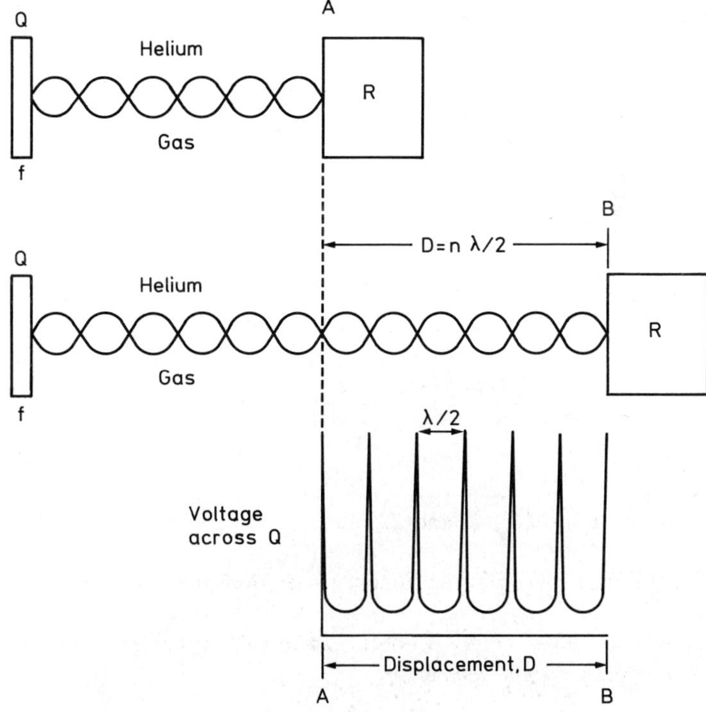

Figure 7-13. Schematic diagram of the basic method for measuring the velocity of sound in a gas [26].

quartz crystal. As the reflecting piston surface is moved toward or away from the quartz crystal, each time the sound travel path length is a multiple number of half-wavelengths, the helium gas column is set into resonance. Because the quartz resonator must give off maximum resonance energy, the condition of resonance can be detected by measurement of the voltage of the quartz oscillator. The voltage varies with $\lambda/2$ and is maximum if the travel path length is a multiple of $\lambda/2$. Figure 7-13 shows the basic principle of an acoustic interferometer.

This measuring approach sets very high requirements on mechanical manufacture. For example, the deplacement of the piston must be controlled with an accuracy of better than 1 μm [2].

7.2.2.2 *Characteristics of Acoustic Gas Thermometers*

The measurement of the acoustic velocity is effected by measuring the travel time of a sound pulse through a known distance in the gas. The measured temperature is, of course, related to the temperature distribution in the gas volume. The best results are obtained when the temperature profile is reasonably uniform [15]. The response time of this sensor is very small compared with the "foreign sensor in the medium". This is simply given by the time required for the pulse to traverse the acoustic path:

$$t = \frac{d}{c} \tag{7-13}$$

where

t = response time of the sensor
d = acoustic path length
c = velocity of sound

Values of the response time are given in Table 7-2.

Table 7-2. Approximate thermometer characteristics for an acoustic travel path of 9 cm [24].

Gas	Temperature (K)	Velocity of Sound (m/s)	Measurement Time (μs)	Temperature Sensitivity (μs/K)	Pressure* Sensitivity (μs/MPa)
Helium	300	1020	90	0.15	0.35
	1000	1860	50	0.025	0.06
Hydrogen	300	1320	68	0.11	–
	1000	2410	37	0.021	–
Nitrogen	300	350	255	0.4	1.5
	1000	645	140	0.07	0.25

* Values obtained experimentally

The general arrangement of the thermometer is illustrated in Figure 7-14. An electronic pulse is used to start a timer and to initiate the generation of an acoustic pulse in the transducer. The pulse is transmitted into the gas volume and is received by the receiving transducer. The received pulse is used to generate a stop signal for the timer.

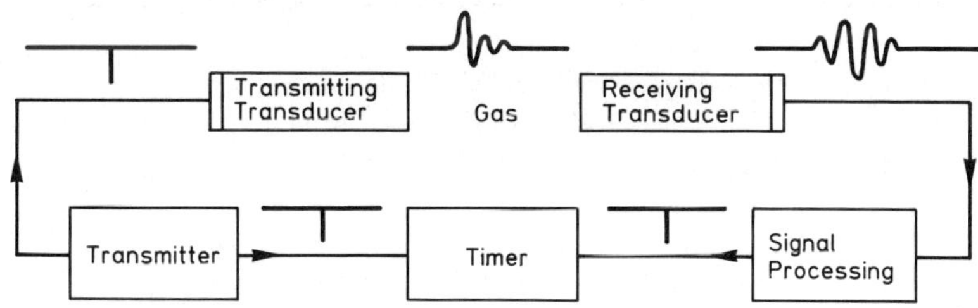

Figure 7-14. The general arrangement of an acoustic gas thermometer [24].

The time interval measured is related to the acoustic velocity by

$$t = t_0 + \frac{d}{c} \tag{7-14}$$

where

t_0 = delay produced by all other components in the measurement system except the travel time.

7.2.3 Applications

Acoustic thermometry is potentially suitable for the measurements of both very high and very low temperatures. The choice of the sensor material is mainly determined by both the temperature range required and the working environment. In nuclear applications, for example, low neutron cross section refractory materials with a high melting point are required for sensor materials [7, 31]. Polycrystalline molybdenum is a useful material for this special application [31].

In non-nuclear applications there are user requirements for thermometers in liquid metal thermometry and burning gas environments. Both the "foreign sensor in the medium" and the "medium itself sensor" concepts are suitable solutions for these problems [22]. If a short response time is needed, then "a medium itself sensor" is a possible system with the additional advantage of very high temperature capability. A gas sensor can be used to measure plasma temperatures up to 8000 K [33, 34].

7.3 Resonant Quartz Sensors

Of the various solid-state acoustic sensors, quartz is easily the most dominant material. Its piezoelectricity allows easy excitation and detection of vibrations. Since it is readily available as single crystals, all its physical properties are extremely well defined and reproducible. The anisotropy of the crystal allows the tuning of most properties simply by changing the angle of cut of the sensor with respect to the crystalline axes.

Hence not only the velocity of sound but also its temperature coefficients depend on the crystalline directions in the vibrating body. Suitable cut quartz sensors can be used to convert temperature into time- or frequency-coded signals.

Among piezoelectric anisotropic single crystals, the quartz excels in its low acoustic absorption and its mechanical, elastic and chemical properties. Its main advantage, however, is 50 years of scientific and technological development in both crystal growth and processing. This is similar to the situation with semiconductors where silicon dominates owing to the great experience gained and continuing developments.

Thus silicon dominates the semiconductor field and silicon dioxide (quartz) the area of micromechanics.

7.3.1 Introduction to Properties of Quartz Crystals

7.3.1.1 General Properties

The single-crystal SiO_2 is called quartz and it is found in numerous modifications. For piezoelectric applications only the low-temperature modification, α-quartz, can be used. Above the transformation temperature $T_{\alpha\beta}$ = 573 °C, α-quartz changes reversibly into the high-temperature modification β-quartz. Some physical properties of α-quartz are listed in Table 7-3. Further relevant information about quartz can be obtained elsewhere [35–38].

Table 7-3. Some physical properties of α-quartz.

Density at room temperature	ρ = 2.649 g/cm^3
Hardness (Moh's scale)	7
Linear temperature expansion coefficient	α_{11} = 13.7 ppm/K
	α_{33} = 7.4 ppm/K
Transformation temperature, α-quartz ⇄ β-quartz	$T_{\alpha\beta}$ = 573 °C
Melting point	T_{MP} = 1710 °C
Specific resistance	10^{15}–10^{16} Ωm
Pressure solidity	2–3 GPa

Every elementary cell consists of three SiO_2 molecules. α-quartz belongs to the symmetry class 32 (Hearmann-Mauguin) or D3 (Schoenflies). It is found in two enantiomorphic forms, shown in Figure 7-15. They are named right and left quartz, according to the rotation of the plane of polarization of light around the optical axis, which is the symmetry axis in the direction of the space diagonal. In cartesian coordinates the z-axis is chosen as the direction of the optical axis and the x-axis as the direction of one of the three electrical axis. The y-axis is frequently called the mechanical axis.

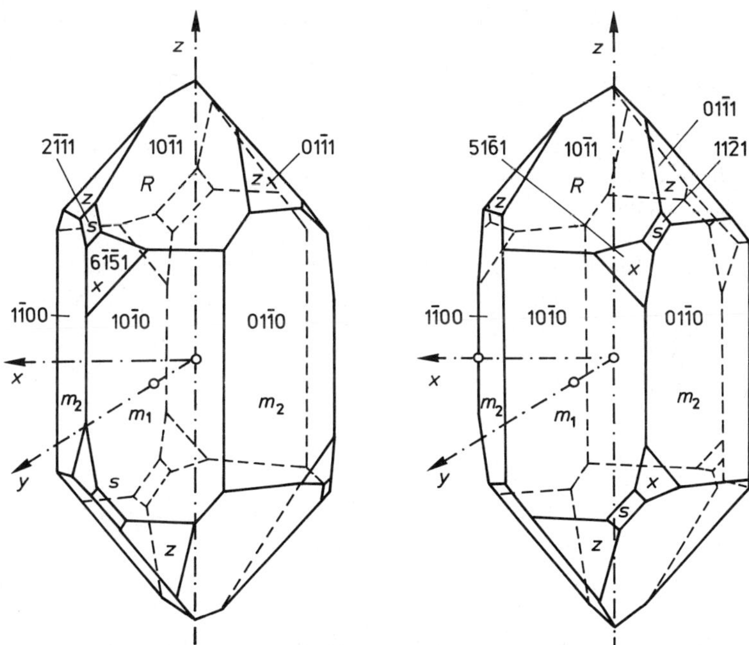

Figure 7-15. The two crystal of α-quartz with the cartesian coordinate system (x-axis = electrical axis, y-axis = mechanical axis, z-axis = optical axis) [39].
(a) Left quartz; (b) right quartz.

7.3.1.2 *Elastic Properties*

A force acting on a solid body generally produces both a rigid body motion and a deformation. The rigid body motion is composed of a translation and a rotation. For a quartz resonator only the deformation is of interest.

The reaction to an external force is described by Hook's law. This law associates the mechanical stress tensor for small deformations and the mechanical strain tensor in a linear way:

$$\sigma_{ij} = c_{ijkl} \cdot L_{kl} \tag{7-15}$$

where

c_{ijkl} = elastic stiffness.

The stress tensor $\boldsymbol{\sigma}$ and strain tensor $\mathbf{L}$ are symmetric tensors of rank two with six fundamental elements. The following notations are used in the literature [40]:

$$\boldsymbol{\sigma} = \begin{bmatrix} X_x & X_y & X_z \\ X_y & Y_x & Y_z \\ X_z & Y_z & Z_z \end{bmatrix} \text{ and } \mathbf{L} = \begin{bmatrix} L_{xx} & L_{xy} & L_{xz} \\ L_{xy} & L_{yy} & L_{yz} \\ L_{xz} & L_{yz} & L_{zz} \end{bmatrix}. \tag{7-16}$$

The components L_{ij} $(i, j = x, y, z)$ are defined by [40]

$$L_{ij} = \frac{1}{2}\left[\frac{\partial \xi_i}{\partial x_j} + \frac{\partial \xi_j}{\partial x_i}\right] \tag{7-17}$$

where

ξ_i = component of particle elongation

The diagonal elements of the strain tensor $\mathbf{L}$ describe a linear expansion along the coordinate axis, and the other tensor elements describe plane shears. In the case of the stress tensor $\boldsymbol{\sigma}$ a capital letter specifies the direction of the force and the normal direction is indicated by

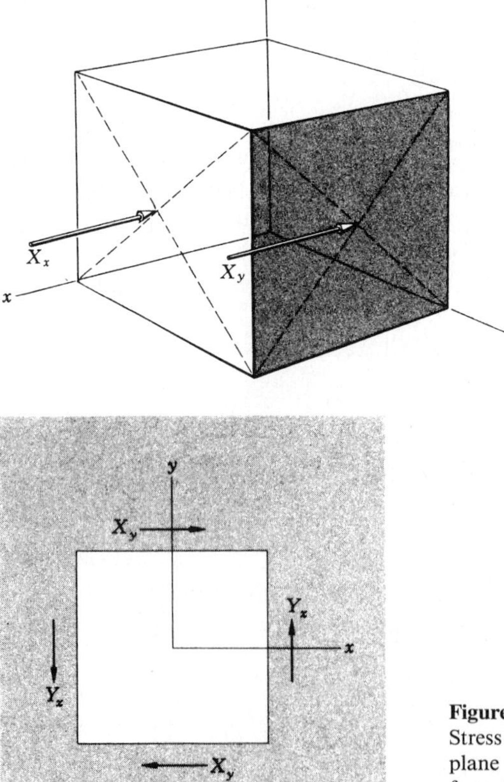

Figure 7-16.
Stress component X_x is a force in the x-direction on a plane with a normal in the x-direction and X_y is a force in the x-direction on a plane with a normal in the y-direction [40].

a small letter. All the tensor components are shown in Table 7-4 and a few stress components are shown in Figure 7-16. The elastic stiffness tensor **C** is of rank four with 81 components. For α-quartz six components are necessary, because of the symmetry properties of its crystal class [43].

Table 7-4. Meaning of stress and strain tensor components.

Component	Meaning of tensor components
X_x	force in x-direction on plane with normal in x-direction
Y_y	force in y-direction on plane with normal in y-direction
Z_z	force in z-direction on plane with normal in z-direction
$X_z = Z_x$	force in x-z plane
$X_y = Y_x$	force in xy plane
$Z_y = Y_z$	force in y-z plane
L_{xx}	linear expansion in x-direction
L_{yy}	linear expansion in y-direction
L_{zz}	linear expansion in z-direction
$L_{xy} = L_{yx}$	linear expansion in x-y plane
$L_{xz} = L_{zx}$	linear expansion in y-z plane
$L_{yz} = L_{zy}$	linear expansion in y-z plane

7.3.1.3 Piezoelectric Properties

In piezoelectric crystals, the elastic and electric properties are related to each other. The elastic variables are the stress and the strain tensor and the electric variables are electric field and electrical displacement. Piezoelectricity is found only in crystals with no center of symmetry for the electrical charge.

The effect is called direct piezoelectricity if a mechanical deformation causes a change in electric polarization. The reverse effect is called reciprocal piezoelectricity.

To explain the piezoelectric effect, a two-dimensional, simplified picture of a SiO_2 structure cell is used (Figure 7-17a).

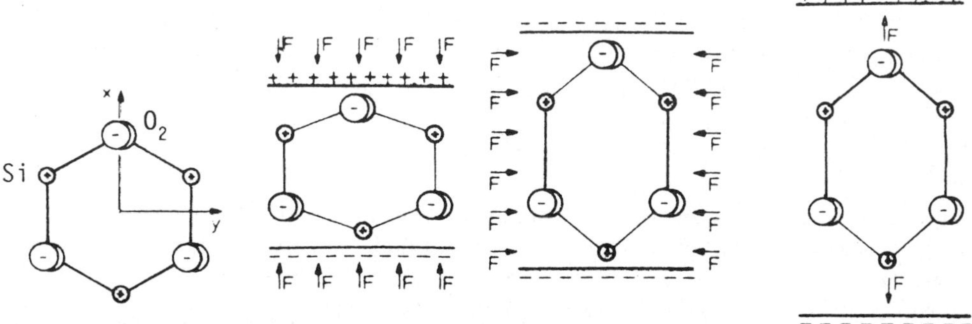

Figure 7-17. Piezoelectric effect [41]. (a) Structure cell of α-quartz (3 positive Si atoms and 6 negative O atoms) shows no symmetry center of charge. (b) Direct effect: external pressure deforms structure cell. (c) Reciprocal effect: external electrical field pushes the electrical charges.

The cell contains three two-counted polar axes which show a symmetry of 120°. Stress in the direction of one of these axes causes a displacement of the positive and negative charges (see Figure 7-17b). If the crystal changes polarization, the surface charge density changes. Figure 7-17c shows that with the reciprocal piezoelectric effect the negative O atom at the top and the positive Si atom at the bottom are attracted, the other atoms are repulsed. This causes an extension along the *x*-axis. If there is a symmetry center of charge, it is impossible to generate a dipole moment caused by mechanical pressure. For quartz resonators both effects are used. The piezoelectric effect is described in detail elsewhere [35, 37–39, 42].

7.3.2 Tensor and Matrix Notation

All tensors of rank two, three and four can be described in a tensor or a matrix notation. The assignment between these two notations is shown in Table 7-5. For example:

- elastic stiffness in tensor notation: c_{ijkl} tensor with $3 \cdot 3 \cdot 3 \cdot 3 = 81$ components;
- elastic stiffness in matrix notation: $c_{\lambda\mu}$ matrix with $6 \cdot 6 = 36$ components.

Table 7-5. Assignment of tensor and matrix indices.

ij or kl	11 xx	22 yy	33 zz	23/32 yz/zy	13/31 xz/zx	12/21 xy/yz
λ or μ	1	2	3	4	5	6

7.3.3 Plate Resonators

A quartz plate resonator is a thin quartz disk cut from a mother crystal under a specified orientation. On both surfaces of the disk metal electrodes are mounted. With a changing electric field a mechanical oscillation can be induced.

In the domain of computer and measuring technology, AT-quartz resonators are most important. The resonance frequency of an AT-quartz is nearly independent of temperature up to 100 °C. Such resonators are used for the generation of high-precision fixed frequencies.

7.3.3.1 Coordinate System and Cuts of Quartz Crystals

In the past there was no standardized coordinate system. Often right- and left-handed coordinate systems were used for right and left quartz crystals. The signs of some material constants depend on the coordinate system, listed in Table 7-6. Some important coordinate systems of the past and the standardized system of today are:

(a) Voight's coordinate system, 1928 [38]:
He used a uniform right-handed coordinate system for both quartz forms.

(b) Coordinate system, IRE Standards, 1945 [43]:
For left quartz a left-handed coordinate system is recommended and for right quartz a right-handed system.

(c) Coordinate system, IRE Standards, 1949 [44]:
The IRE Committee recommends a uniform right-handed coordinate system independent of the form of quartz. In the direction of one of the electric axes the x_1-axis is chosen and the x_3-axis for the direction of the optical axis. The direction of rotation, eg, the AT-cut angle, is not defined.

(d) Bechmann's coordinate system [45]:
This coordinate system is equivalent to the IRE Standards, 1949, system. He defines the AT-cut angle by $\Theta = +35°$.

(e) Cady's coordinate system, 1964 [37]:
Cady uses the coordinate system described under (b).

(f) Coordinate system, IEEE standard on piezoelectricity, 1978 [46]:
This coordinate system is equivalent to the IRE Standards, 1949, system. The direction of rotation is defined and the AT-cut angle $\theta = -35°$ is given.

Table 7-6. Signs of some material constants dependent on the coordinate system.

Described under	a		b, e		c		d		f	
Quartz form	R	L	R	L	R	L	R	L	R	L
Coordinate system	RH	RH	RH	LH	RH	RH	RH	RH	RH	RH
Elastic stiffness, c_{14}	+	+	+	+	−	−	−	−	−	−
Elastic compliance, s_{14}	−	−	−	−	+	+	+	+	+	+
Piezoelectric strain constant: d_{11}	+	−	+	+	−	+	−	+	−	+
d_{14}	−	+	−	−	−	+	−	+	−	+
Piezoelectric stress constant: e_{11}	−	−	+	+	−	+	−	+	−	+
e_{14}	−	−	+	+	+	−	+	−	+	−
AT-cut angle, θ	−35°		−35°		?		+35°		−35°	

R = right quartz RH = right-handed
L = left quartz LH = left-handed

Crystal plate cuts from a single-crystal material can have an orientation relative to the three orthogonal crystal axes. This orientation is described by the starting point, direction of length, rotation axes and rotation angles [46]. For example, an orientation in this nomenclature is described by $(YXwl)\ \Phi/\Theta$. The first letter specifies the starting point, which is a nonrotated plate, called X-, Y- or Z-cut. An X-cut describes a plate in the y-z plane, Y-cut a plate in the x-z plane and Z-cut a plate in the x-y plane. Further letters characterize the rotation axes (l = length, w = width, t = thickness). The following angle defines the rotation around the rotation axes. The angle is positive for clockwise rotation looking in the positive direction of the coordinate axis.

Usually single rotated cuts are described by only the starting plate and rotation angle. For example, $(Y + \Theta)$-cut defines a Y-cut rotated around the x-axis about the angle Θ. Some orientations are shown in Figure 7-18.

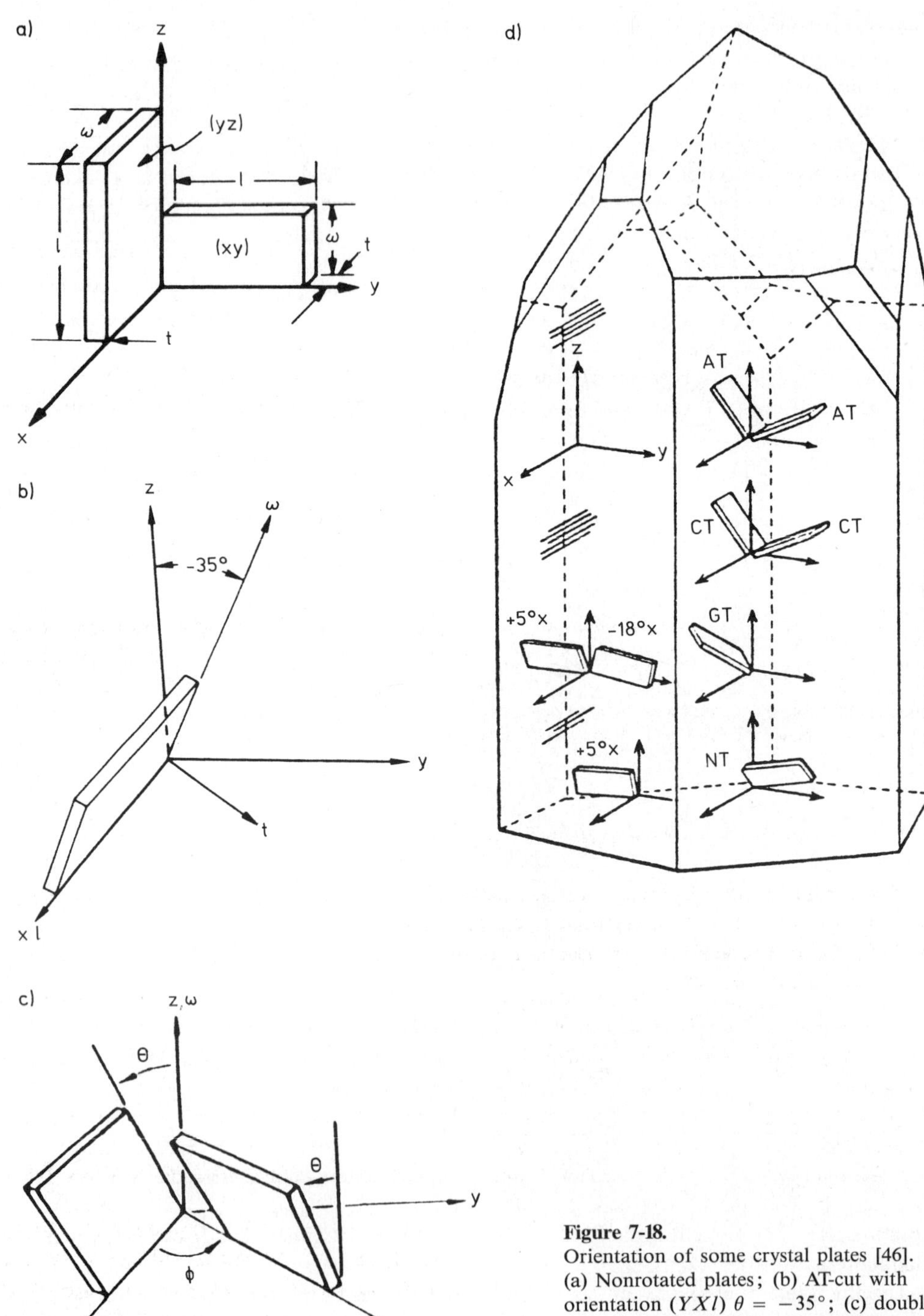

Figure 7-18.
Orientation of some crystal plates [46]. (a) Nonrotated plates; (b) AT-cut with orientation $(YXl)\ \theta = -35°$; (c) double rotated plate with orientation $(YX2l)$ Φ/θ; (d) quartz cuts in mother crystal.

7.3.3.2 Vibration Modes

The vibration modes of a quartz crystal plate resonator are determined by the quartz cut, the geometry of the quartz plate and the electrodes. Possible vibration modes are bending, extension, face shear, and thickness shear modes or a superposition of thesc modes. For temperature sensors with plate resonators, thickness shear resonators are used. Figure 7-19 shows the vibration modes.

Usually the bending mode is called mode A and the shear modes are called modes B and C [45].

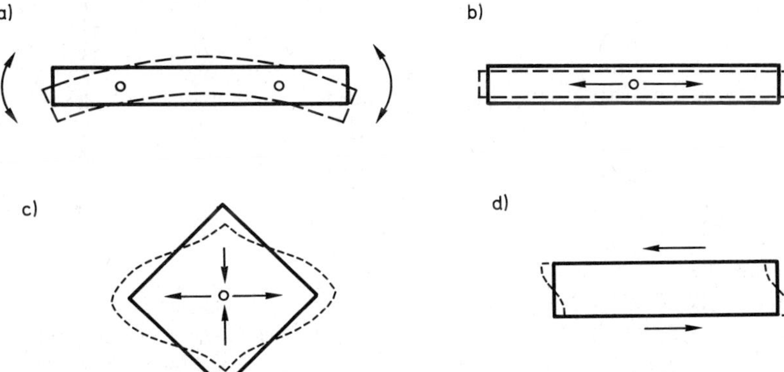

Figure 7-19. Vibration modes of plate resonators [41]. (a) Bending mode; (b) extension mode; (c) face shear mode; (d) thickness shear mode.

7.3.3.3 Overview of Propagation Theory

The general theory of propagation of elastic waves in an anisotropic, elastic solid body of infinite size comes from Christoffel [47, 48]. He showed the occurrence of generally three different propagation velocities of sound. The general differential equations of motion in an anisotropic body are [49]:

$$\begin{aligned}
\rho \frac{\partial^2 \xi_x}{\partial t^2} &= \frac{\partial X_x}{\partial x} + \frac{\partial X_y}{\partial y} + \frac{\partial X_z}{\partial z} \\
\rho \frac{\partial^2 \xi_y}{\partial t^2} &= \frac{\partial X_y}{\partial x} + \frac{\partial Y_y}{\partial y} + \frac{\partial Y_z}{\partial z} \\
\rho \frac{\partial^2 \xi_z}{\partial t^2} &= \frac{\partial X_z}{\partial x} + \frac{\partial Y_z}{\partial y} + \frac{\partial Z_z}{\partial z}
\end{aligned} \tag{7-18}$$

where

ρ = density of quartz
ξ_i = particle elongation

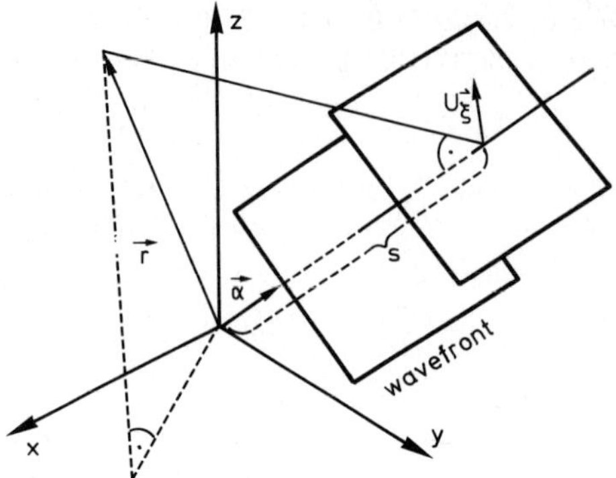

Figure 7-20.
Direction of wave propagation with vector $\vec{\alpha}$. s determines the distance of the wave front from the coordinate origin [49].

The boundary conditions for a quartz plate in the absence of stresses at its surface are [49]

$$\frac{\partial \xi_i}{\partial s} = 0 \quad \text{for} \begin{cases} s = 0 \\ s = h \end{cases} \tag{7-19}$$

Where h is the thickness of the quartz plate. The thickness must be a multiple of the half resonant wavelength λ. The direction of propagation is indicated by a standardized propagation vector $\vec{\alpha}$ (Figure 7-20). From Equations (7-15), (7-17) and (7-19) and the assumption of a standing wave solution of the form

$$\xi_\tau = \xi_0 \cdot \cos\,(n\pi s/h) \cdot \mathrm{e}^{i\omega_\tau t}, \; \tau = \mathrm{A, B, C} \tag{7-20}$$

where

ξ = particle elongation
h = quartz thickness
t = time

Equation (7-18) can be transformed into a homogeneous equation system:

$$\rho v^2 \, \xi_i = \sum_{k=1}^{3} \Gamma_{ik} \, \xi_k \tag{7-21}$$

where

ρ = density of quartz
v = propagation velocity
Γ = Christoffel's elastic stiffness

Christoffel's elastic stiffness is defined by [45, 49]

$$\Gamma_{ik} = \sum_{j=1}^{3} \sum_{l=1}^{3} c_{ijkl} \, \alpha_j \, \alpha_l \tag{7-22}$$

where α_i are cosines of the direction of propagation with respect to the crystalline axes. This equation system is solvable only if Christoffel's system determinant [45, 49] becomes zero, ie,

$$\begin{vmatrix} \Gamma_{11} - \rho v^2 & \Gamma_{12} & \Gamma_{13} \\ \Gamma_{21} & \Gamma_{22} - \rho v^2 & \Gamma_{23} \\ \Gamma_{31} & \Gamma_{32} & \Gamma_{33} - \rho v^2 \end{vmatrix} = 0 \quad (7\text{-}23)$$

ρ being the density of quartz.

The resonance frequency $f_{n\tau}$ is given by

$$f_{n\tau} = \frac{\omega_\tau}{2\pi} = \frac{v_\tau}{\lambda} = \frac{n}{2h}\, v_\tau = \frac{n}{2h} \cdot \sqrt{\frac{c'_\tau}{\rho}} \quad (7\text{-}24)$$

$$\tau = \text{A, B, C, and } n = 1, 3, \ldots$$

where v_τ is the propagation velocity, n the overtone order and τ indicates one of the three solutions of the Christoffel determinant. The elastic stiffness c'_τ depends on the cut of the quartz plate.

7.3.3.4 *Frequency-Temperature Dependence*

The resonant frequency of a quartz plate depends on the elastic stiffness c', the density and the thickness of the plate, and the density and the thickness of the metal electrodes. Since these factors depend on temperature, so does the resonant frequency. Bechmann and Ballato [45] approximated the frequency-temperature behavior with a third-degree polynomial:

$$\frac{f(T) - f_0}{f_0} = \sum_{n=1}^{3} T_f^{(n)} \cdot (T - T_0)^n \quad (7\text{-}25)$$

where

f = resonant frequency
f_0 = resonant frequency at temperature T_0
T = temperature
T_0 = reference temperature
$T_f^{(n)}$ = temperature coefficient of order n

The temperature coefficients are defined by

$$T_f^{(n)} = \frac{1}{n!\, f_0} \cdot \frac{\partial^n f}{\partial T^n}\,. \quad (7\text{-}26)$$

Calculations of temperature coefficients and the coefficients for various quartz cuts have been published [45, 50, 51]. Figure 7-21 shows the temperature coefficients of frequency in first order for double-rotated quartz cuts $(YXwl)\ \Phi/\theta$ and for single-rotated Y-cuts $(YXl)\ \theta$.

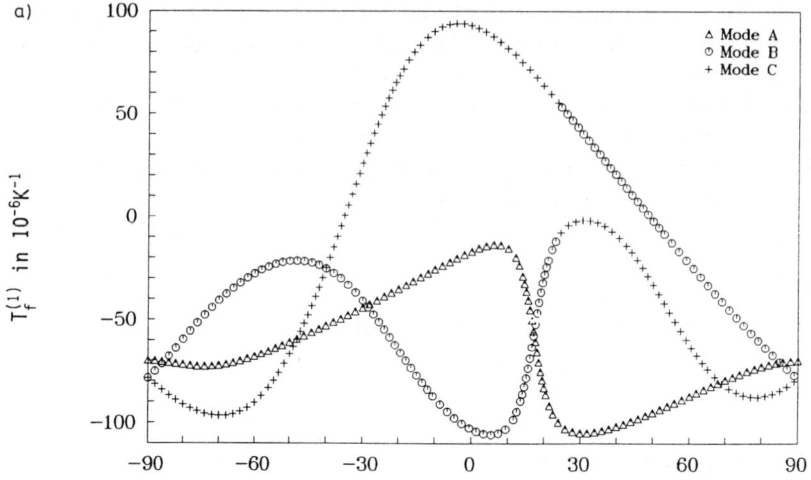

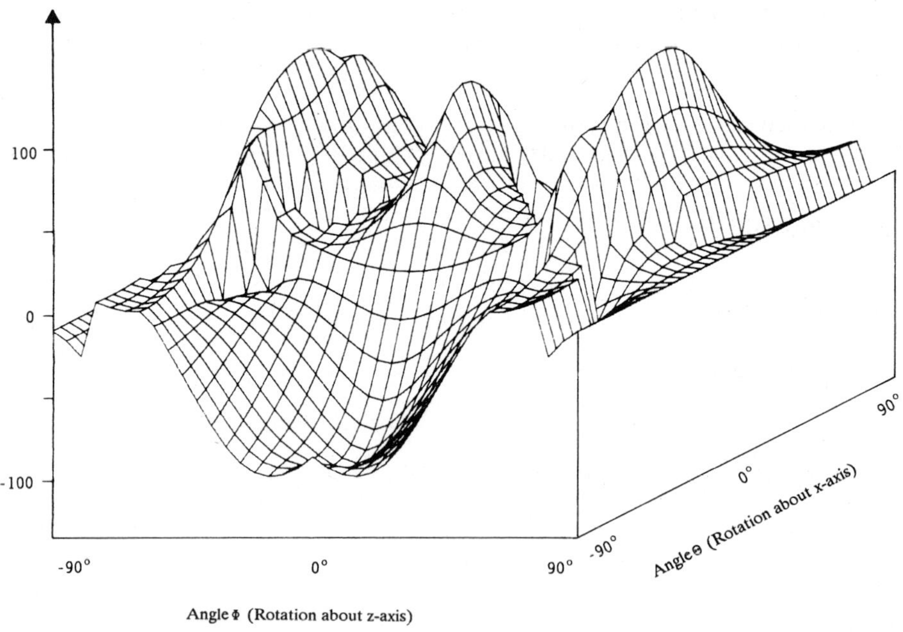

Figure 7-21. First-order temperature coefficient of frequency: (a) for double-rotated quartz cuts ($YXwl$) Φ/θ in mode C [54]; (b) for single-rotated Y-quartz cuts (YXl) θ in modes A, B, C, [45].

A maximum of the first-order temperature coefficient for a vibration in the thickness shear mode C is given for a Y-cut $(YXl)\ \theta$ with a rotation angle of $\theta \approx -4°$. This quartz orientation is called HT-cut. For the HT-cut the temperature coefficients of second and third order differ from zero and produce a small nonlinearity. The temperature coefficients for the HT-cut are [18, 52] $\alpha \approx 90 \cdot 10^{-6}\ \mathrm{K}^{-1}$, $\beta \approx 60 \cdot 10^{-9}\ \mathrm{K}^{-2}$, and $\gamma \approx 30 \cdot 10^{-12}\ \mathrm{K}^{-3}$.

A linear temperature-frequency characteristic curve is given for the double-rotated LC-cut quartz resonator with orientation $(YXwl)\ \Phi = 10.8°/\theta = -9.2°$ (LC means linear coefficient). The temperature coefficients of second and third order are nearly zero and only the first-order temperature coefficient is important [14, 51, 53] $\alpha \approx 40 \cdot 10^{-6}\ \mathrm{K}^{-1}$, $\beta \approx 0 \cdot 10^{-9}\ \mathrm{K}^{-2}$, and $\gamma \approx 0 \cdot 10^{-12}\ \mathrm{K}^{-3}$.

In technical applications, only the HT-cut and the LC-cut are used for temperature sensors with quartz resonators.

7.3.4 Tuning Fork Designs

The quartz resonator is formed as a tuning fork (see Figure 7-22). This quartz design is primarily used for quartz resonators for digital watches. Metal electrodes are mounted on the surface of the forks. Tuning fork quartz resonators vibrate in a torsion mode, where the temperature behavior depends on both the quartz cut and the ratio of width to thickness of the forks.

Designs modified for temperature measurement show a first-order temperature coefficient of frequency which is smaller than that of plate resonators in a thickness shear vibration

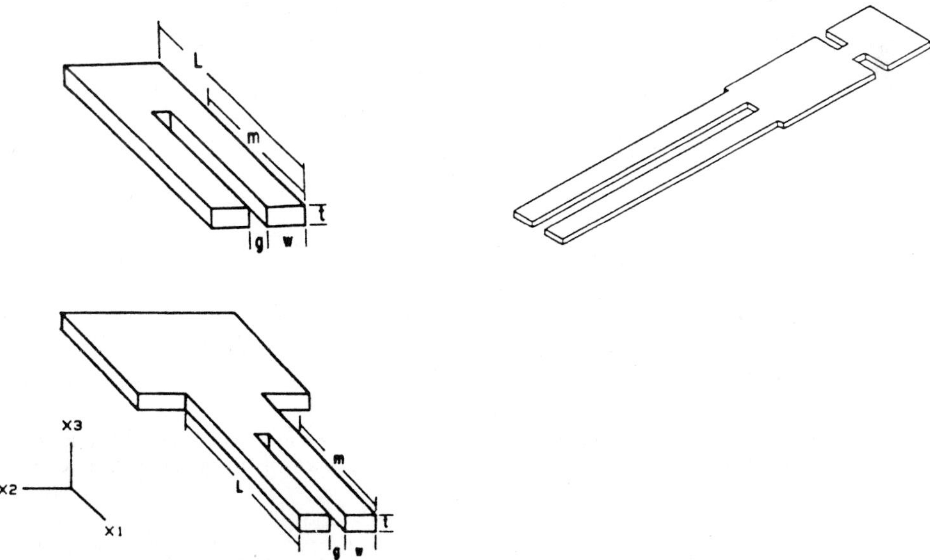

Figure 7-22. Designs of tuning fork quartz resonators [55, 56].

mode. At rotation angles of $\theta \approx 50°$ and $\Phi \approx 0°$ the temperature coefficient has a maximum of $T_f^{(1)} \approx 45$ ppm/K [57]. Through the ratio of width to thickness of the forks the second-order temperature coefficient of frequency can be made small.

Further information about tuning fork quartz resonators can be obtained elsewhere [55–67].

7.3.5 Oscillators

7.3.5.1 *Quartz in the Electric Circuit*

For electric circuits with resonant quartz plates, the quartz resonator can be described by an electrically equivalent circuit. The circuit contains the resistance R_1, the inductance L_1 and the capacitance C_1. C_0 is a static capacitance formed by the electrodes and the support [68, 69]. Figure 7-23 shows the equivalent electrical circuit of a quartz resonator.

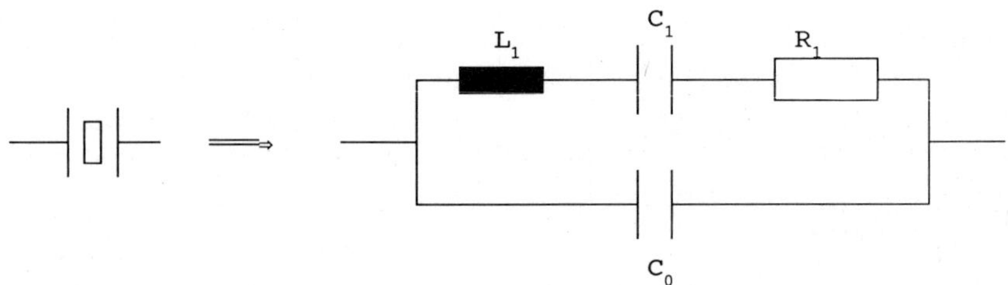

Figure 7-23. Equivalent electrical circuit of a piezoelectric vibrator [41].

For the equivalent circuit there are two resonant frequencies, a frequency f_s at series resonance and a frequency f_p at parallel resonance. These resonance frequencies are related to the electrical parameters according to [41, 49, 68, 69]

$$f_s = \frac{1}{2\pi \cdot \sqrt{L_1 \cdot C_1}} \tag{7-27}$$

$$f_p = \frac{1}{2\pi \cdot \sqrt{L_1 \cdot \dfrac{C_0 \cdot C_1}{C_0 + C_1}}} \tag{7-28}$$

According to Equations (7-27) and (7-28), the difference between the two resonance frequencies is

$$\frac{f_p - f_s}{f_s} = \frac{C_1}{2 \cdot C_0}\,. \tag{7-29}$$

The oscillation frequency of a quartz resonator can be varied between the series and the parallel resonance frequency by a load capacitance C_L (see Figure 7-24).

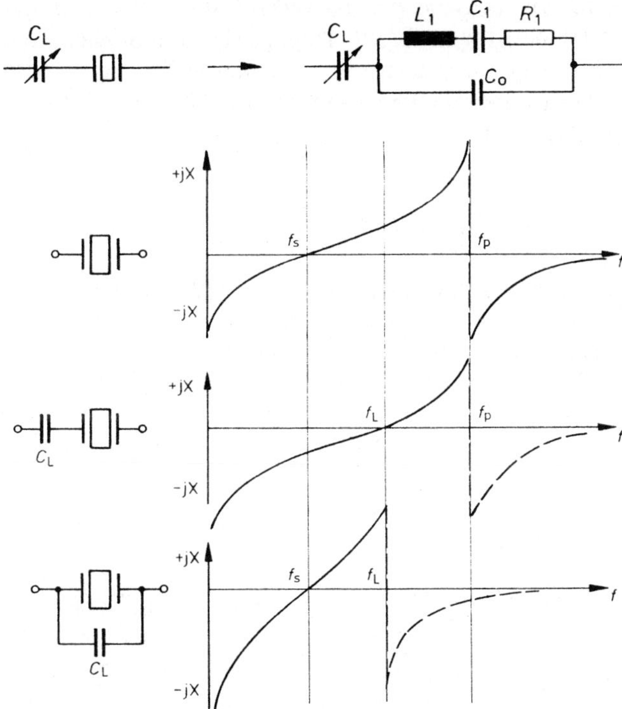

Figure 7-24.
Reactance of a quartz resonator [41]: (a) without load capacitance; (b) with series load capacitance C_L; (c) with parallel load capacitance C_L.

The load resonance frequency as a function of C_L in series or parallel circuit arrangement is [41, 69]

$$f_L = f_s + f_s \cdot \frac{C_1}{2 \cdot (C_0 + C_L)} \,. \tag{7-30}$$

Through the load capacitance, the load resonance frequency is adjustable between f_s and f_p:

$$0 < C_L < \infty \rightarrow f_p > f_L > f_s \,. \tag{7-31}$$

Hence a load resonance frequency offset

$$LO = \frac{f_L - f_s}{f_s} = \frac{C_1}{2 \cdot (C_0 + C_L)} \tag{7-32}$$

results from the load capacitance C_L (see Figure 25a). The differential ratio $\partial LO / \partial C_L$ is called pulling sensitivity, S:

$$S = \frac{\partial LO}{\partial C_L} = -\frac{C_1}{2 \cdot (C_0 + C_L)^2} \,. \tag{7-33}$$

The pulling sensitivity S is given in 10^{-6}/pF and describes the relative frequency change as function of the load capacitance C_L (Figure 7-25b).

The oscillator quality factor Q depends on the frequency, the resonance resistance R_1 and the capacitance C_1:

$$Q = \frac{1}{\omega \cdot C_1 \cdot R_1} = \frac{1}{2\pi \cdot f \cdot C_1 \cdot R_1} . \qquad (7\text{-}34)$$

The quality factor is of the order of 100000 and is much higher than the quality factor of other resonators. With increasing frequency the vibrator quality factor of a quartz resonator decreases, as shown by Equation 7-34.

The resonance resistance R_1 depends on the limited dimension of the quartz resonator plate, the ambient gas, the impurities of the quartz, and the resistance of the electric wires.

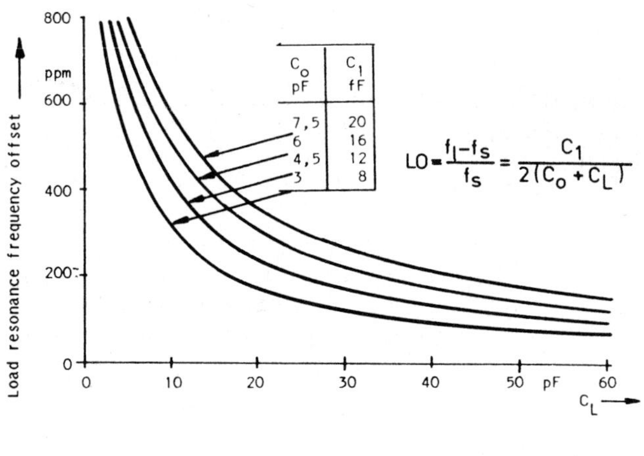

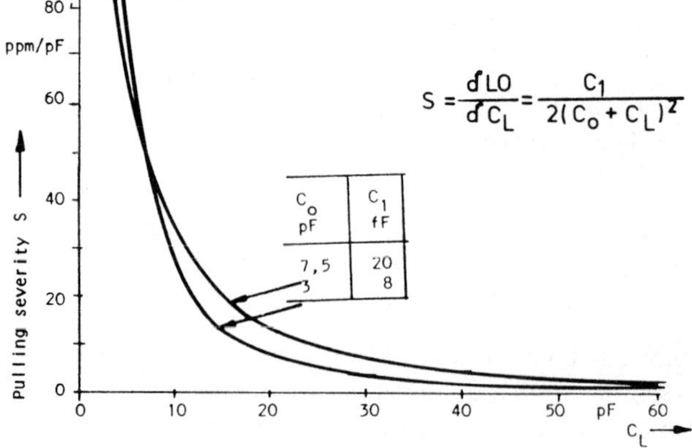

Figure 7-25. (a) Load resonance frequency offset; (b) pulling sensitivity [41].

7.3.5.2 Oscillator Requirements

For a quartz temperature sensor a special oscillator is necessary. This oscillator must satisfy several basic requirements.

In order to reduce self-heating of the sensor, a low-power oscillator is required. A low-power oscillator will also reduce ageing, ie, long-term drift of the resonance frequency.

Since the components of the oscillator do not withstand high temperatures, the distance between the quartz resonator and quartz oscillator must be up to about 1 m. The temperature of the quartz oscillator must not affect the quartz resonator and its resonant frequency. Hence the oscillator and the resonator must be thermally isolated and the high-frequency coaxial cable must be made of a material with a low heat conductivity. Often such materials increase the resonance resistance R_1.

The cable contributes to the load capacitance C_L and the dielectric permittivity ε depends on the temperature, and thus the load capacitance depends on the temperature. Further, the cable produces a phase shift of the oscillator and hence a frequency offset.

7.3.5.3 Oscillator Design

A multitude of quartz oscillator designs have been described in the literature [41, 70–72]. As a general feature, the quartz resonator must be one-side grounded in order to use a coaxial cable to connect the quartz resonator and the quartz oscillator.

7.3.5.4 Oscillator Properties

The oscillator causes a frequency offset of the resonance frequency of the quartz resonator. Therefore, the quartz resonator and quartz oscillator must be calibrated as a single unit. If this offset depends on the quartz frequency, the slope of the frequency-temperature characteristic curve could be affected.

Special care must be taken to prevent oscillation at spurious resonance frequencies and changes in R_1 due to mechanical coupling to highly damped modes.

7.3.6 Realization of Quartz Temperature Sensors

7.3.6.1 General Properties

Measuring System

A general measuring system is shown in Figure 7-26 [73–76]. For the realization of a quartz temperature sensor a quartz resonator and a quartz oscillator are required. The temperature-dependent frequency is measured and the temperature is calculated from the frequency signal or a transformed time-coded signal. In order to obtain this temperature information, different solutions are possible, and they are described in Sections 7.3.6.2 and 7.3.6.3.

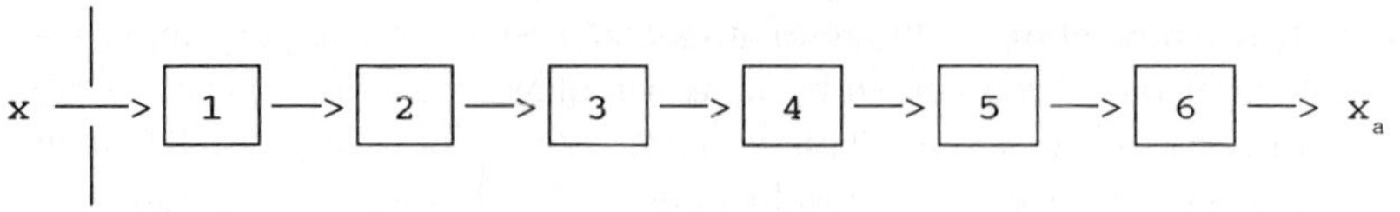

```
(1) measuring value registration
(2) value transformation and amplification
(3) signal transport
(4) calculating operation and comparison with reference
(5) measuring value transport
(6) measuring value output
```

Figure 7-26. Block circuit diagram of a measuring system [73].

Production Spread and Accuracy

Quartz resonators require highly pure single crystals. Therefore, synthetic quartz crystals are grown in autoclaves at about 400 °C and 100 MPa. The growth direction is defined by an oriented germ crystal.

The oriented grown quartz crystal bar is sawn into thin quartz plates. The resonance frequency of the quartz plate is primarily defined by the thickness and the angle of cut. To ensure high quality, accurate parallelism of the surface is required.

Most mechanical processes are limited in absolute accuracy to about 0.1 μm. A frequency accuracy of $\pm 5 \cdot 10^{-6}$ Hz, corresponding to an accuracy of temperature of ± 50 mK, requires a thickness accuracy of only a few atomic layers. For typical frequencies of 20 MHz the thickness required is of the order of 100 μm. This would result in a frequency error of the order of 1000 ppm. With a temperature coefficient of the order of 100 ppm/K, this corresponds to a temperature offset up to 10 K.

Additional fine-tuning techniques using polishing (subtractive) or metal deposition on the electrodes (additive), as used in fixed-frequency quartz production, often will not work with temperature sensor quartzes since the absolute temperature stability required of better than 50 mK during the processing is not achievable.

Note that an incorrect offset due to a different f_0 in Equation (7-25) will not affect the slope of the frequency temperature curve if the relative frequency shift $\Delta f / f_0$ is used instead of the absolute frequency shift.

The frequency-mass loading dependence due to metal electrodes such as silver on the surfaces of the quartz plate has been described [77, 78].

Aging, Relaxation, and Hysteresis

Aging of a quartz resonant sensor means a permanent drift of the resonant frequency with time. Data about aging of AT-cut quartz resonators at different temperatures have been measured for over 20 years. For temperature cut quartz resonators only the cut angle is different from the AT-cut quartz resonators. All other fabrication steps are the same as before

[18]. Hence, even the first temperature quartz resonators could rely on the high quality of an established mass production and effects caused by aging, eg, quartz impurities, purity of protection gas, tightness of seal, are well known. For an average temperature of about 100 °C the aging effects for quartz resonators from mass production are smaller then $5 \cdot 10^{-6}$ during the first year. This is equivalent to a temperature drift of less than 50 mK [18, 41].

After a temperature excursion followed by restoration of the starting temperature, the resonant frequency of a quartz resonator can differ from the initial resonant frequency.

If this offset is constant with time and accumulative after several prolonged high-temperature periods, it is due to high-temperature aging. If this offset is reversible with each temperature cycle, it is called hysteresis. If this offset slowly disappears with time, it is called relaxation. The reason for hysteresis and relaxation can be thermal stress mainly caused by a nonhomogeneous temperature distribution inside the quartz. Elastic crystal distributions through lattice structural faults or mechanical stress caused by the mounting support can also have such effects [79].

With quartz temperature sensors the greatest effect is usually relaxation. After periods at high temperature (eg, 300 °C) and rapid cooling to the reference ice point, it can initially be as high as 100 mK. This decreases gradually within a few days to typically less than 20 mK.

Packaging

The quartz resonator is mounted in a case of high-grade steel or glass. This case must be hermetically sealed from the ambient environment. Inside the case the quartz plate is located under a protective gas without atmospheric humidity. If the gas contains water it precipitates at the quartz plate surfaces at low temperatures and affects the resonance frequency. The gas also improves the thermal conductivity from the case to quartz.

The quartz plate is held in a mounting support with low damping. For this purpose the plate is held at points with a small vibration amplitude.

On the surfaces of the plate metal electrodes are evaporated. The signal connection is done through the metal springs and the metal electrodes. Figure 7-27 shows a quartz plate support.

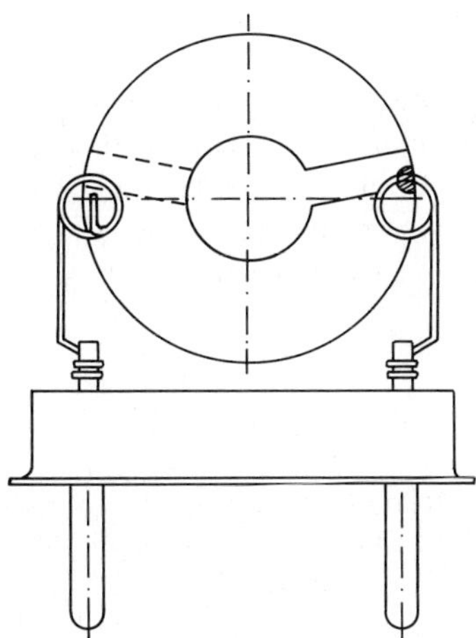

Figure 7-27.
Mounting of a quartz plate in a support with low damping [80].

The oscillator and sensor electronics are located in a plastic case at the end of the steel case tube. The high-frequency signal is conducted through a coaxial cable.

7.3.6.2 Frequency Sensor

Sensor Conception

The temperature-dependent frequency of an LC-cut quartz resonator is comparable to the frequency of a quartz resonator with minimum temperature dependence. The beat frequency is measured by a frequency counter and then the temperature information is calculated. Figure 7-28 shows schematically the arrangement of the sensor system.

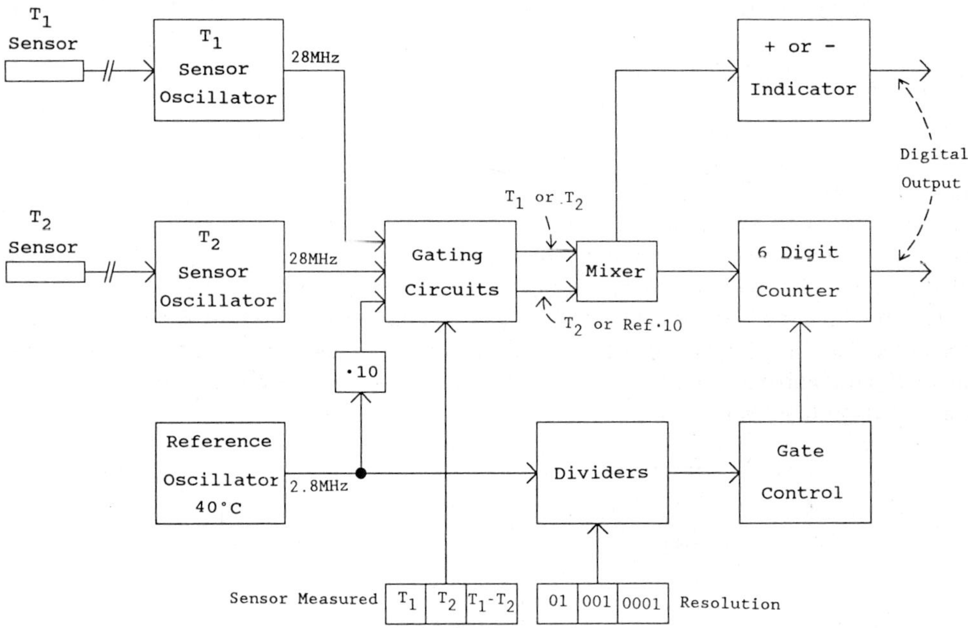

Figure 7-28. Arrangement of a two-channel thermometer [12].

The quartz oscillator must be calibrated in order to correct for the influence of the length of the cable to the quartz sensor. The connection cable between the quartz resonator and oscillator and also between the oscillator and frequency counter must be a high-frequency coaxial cable.

Analog Calibration

For the LC-cut the frequency depends linearly on temperature:

$$\frac{f(T) - f_0}{f_0} = \alpha \cdot (T - T_0) \; . \tag{7-35}$$

The first-order temperature coefficient is about $\alpha \approx 35$ ppm/K [12]. In order to obtain a slope of 1000 Hz/°C the resonance frequency must be

$$f_0 \cdot \alpha = 1000 \text{ Hz/°C} \rightarrow f_0 \approx 28 \text{ MHz}$$

for a reference temperature of 0 °C. The deviation from the linear characteristic curve is of the order of 0.05% [53, 73]. For each sensor a reference oscillator must be calibrated individually. When the quartz sensor is exchanged the module containing the sensor coefficients must also be exchanged.

7.3.6.3 Time Interval Sensor

Time Interval Calibration

The electric field of a temperature quartz resonator changes polarity at a frequency of about 16 MHz. The coaxial cable must be reloaded corresponding to the high frequency. Depending on the length of the dielectrically coaxial cable the power increases, the power being proportional to the length of the cable and the frequency. A high power leads to a greater self-heating of the quartz plate resonator, which influences the temperature measurement.

To solve this problem, a short coaxial cable is necessary. In the electronics near the quartz resonator plate, at a distance of about 0.5 m, the high frequency is divided down to 2 Hz. This signal is transferable by inexpensive wiring of arbitrary length. Figure 7-29 shows the distributions of these functions.

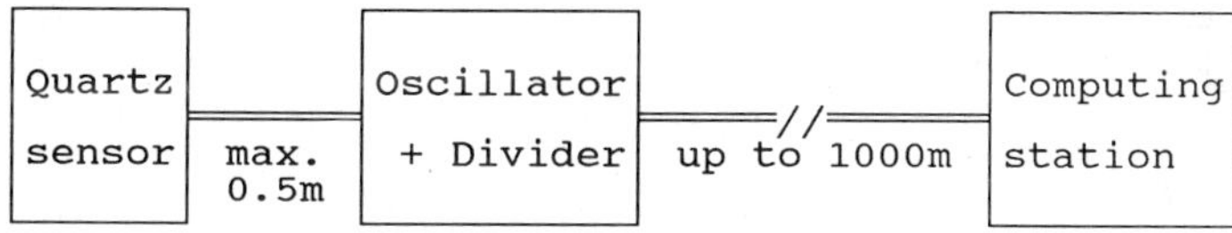

Figure 7-29. Components of a quartz temperature sensor system [81].

Further, the frequency signal can be transferred to a time interval signal. A pulse-forming module generates from the 2-Hz frequency short current pulses in time intervals of 0.5 s. The entire temperature information is contained in this time interval. Transmitting temperature information via the time interval between pulses has several other advantages:

- any ordinary pair of wires can be used instead of a coaxial cable;
- there is only minimum high-frequency radiation;
- time intervals can be measured with higher resolution than frequencies;
- resolution is independent of sensor frequency and depends only on the computing station; and
- the power supply for the oscillator and divider can use the same pair of wires.

Slope Calibration and Accuracy

The high resonance frequency of a quartz resonator is divided into a low-frequency signal and transformed in a time-interval signal. The frequency-temperature behavior is given by Equation (7-25). It follows that the time interval t is given by [81]

$$t = \frac{N}{f_0 \left[1 + \sum_{n=1}^{3} T_f^{(n)} \cdot (T - T_0)^n \right]} \tag{7-36}$$

where

t = time interval
N = divisor
f_0 = resonance frequency at temperature T_0
$T_f^{(n)}$ = temperature coefficients of frequency of order n
T = temperature
T_0 = reference temperature

An approximation for Equation (7-36) with reference temperature $T_0 = 0\,°\mathrm{C}$ is

$$t \approx t_0 \left[1 - \alpha T - (\beta - \alpha^2)\, T^2 - (\gamma + \alpha^3 - 2\alpha\beta)\, T^3\right] \tag{7-37}$$

where

t = time interval at temperature $T_0 = 0\,°\mathrm{C}$
α, β, γ = temperature coefficients of frequency
T = temperature

The term in brackets is independent of the frequency of the resonator and depends on the quartz cut angles only. The slope of the frequency-temperature characteristic curve is adjustable by the divisor N. It can be chosen so that the time interval acquires a defined time at the reference temperature T_0.

Additionally, since $\beta > 0$, the quadratic term in Equation (7-37) is smaller than the quadratic term in Equation (7-25).

Remote Correction

To transmit small corrections from the quartz sensor to the computing station, the length of the current pulses can be used for additional information transfer. Thus, small variations of the slope could be transmitted by coding it in the width of the pulses. Then the computing station can correct the temperature information transmitted via the time interval between two current pulses. This leads to a fully interchangeable sensor.

ILTIS Bus

The time coding of the temperature information via short current pulses allows for the realization of a bus system, called ***I**nter**L**eaved **T**ime**I**nterval **S**ensor* (ILTIS) bus. It is possible to connect up to 16 sensors on an ordinary pair of wires to a computing station. The time signals are interleaved. Every sensor obtains a different code number between 1 and 16. Figure 7-30 shows the time interval scheme of the ILTIS bus. The synchronization occurs by a voltage pulse on the same wire. A complete bus cycle takes about 1 s.

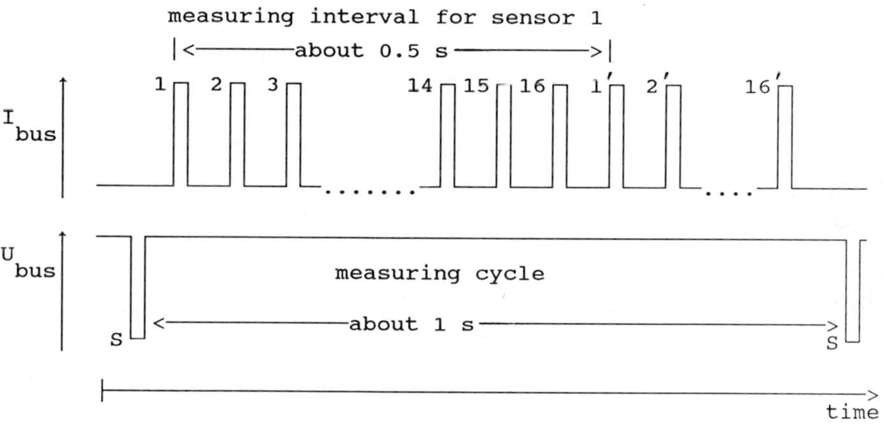

Figure 7-30. Time interval scheme of the ILTIS bus [58].
1 = First signal of sensor 1;
1′ = second signal of sensor 1;
S = synchronization signal.

7.4 References

[1] Moore, G., "Acoustic Thermometry: a Sound Way to Measure Temperature", *Electronics and Power* **9**, No. 30 (1984) 675–677.

[2] Henning, E., Moser, H., *Temperaturmessung, Akustisches Thermometer;* Heidelberg: Springer, 1977.

[3] Hedrich, A. L., Pardue, D. R., "Sound Velocity as a Measurement of Gas Temperature" in: *Temperature: Its Measurement and Control in Science and Industry* Vol. II; New York: Reinhold, 1955, p. 383.

[4] Pierce, A. D., *Acoustics – An Introcuction to its Physical Principles and Applications;* New York: McGraw-Hill, 1981.

[5] Wilson, W. D., *J. Acoust. Soc. Am.* **31** (1959) 1070.

[6] Trendelenburg, F., *Akustik;* Berlin: Springer, 1961.

[7] *J. Acoust. Soc. Am.* **19** (1947) 701.

[8] Fam, S. S., Lynnworth, L. C., Carnevale, E. H., "Ultrasonic Thermometry", *Instruments and Control Systems* **42** (1969) 107–110.

[9] Satyabala, S. P., Mallikarjun, R. S. P., Suryanarayana, M., "A Sing-Around Ultrasonic Velocity Meter Using Integrated Circuits", *Acoust. Lett.* **2** No. 29 (1979).

[10] Greenspan, M., Tschieg, C. E., "Sing-Around Ultrasonic Velocimeter for Liquids", *Rev. Sci. Inst.* **28** (1957) 897-901.

[11] Satyabala, S. P., Madhusudan-Rao, B., Sampath-Kumar, J., Mallikarjun, R. S. P. "Ultrasonic Digital Thermometer Using the Sing-Around Technique", *Acoust. Lett.* **3**, No. 10 (1980).

[12] Hammond, D. J., Benjaminson, A., "The Linear Quartz Thermometer – a New Tool for Measuring Absolute and Difference Temperatures", *Hewlett-Packard Journal* **16**, No. 7 (1965) 1-7.

[13] Shemlev, O. Y., Prokop'ev, V. I., "Instrum. & Quartz-Crystal-Thermometer", *Exp. Tech. (USA)* **28**, No. 5 (1985) 1220-1222.

[14] Priess, S., "Schwingquarze als hochauflösende Temperaturfühler", *Elektronik* **6** (1982) 49.

[15] Gorini, I., Sartori, S., "Quartz Thermometer", *Rev. Sci. Instrum.* **33** (1962) 883-884.

[16] Wade, W. H., Slutsky, L. J., "Quartz Crystal Thermometer", *Rev. Sci. Instrum.* **33** (1962) 212.

[17] Schöltzel, P., "Temperaturmessung mit Quarzsensoren", *VDI Zeitschrift* **112** (1970) 14.

[18] Ziegler, H., "Temperaturmessung mit Schwingquarzen", *Technisches Messen* **4** (1987) 124.

[19] Moldover, M. R., Mehl, J. B., "Spherical Acoustic Resonators: Promising Tools for Thermometry and Measurement of the Gas Constant", *Precision Measurement and Fundamental Constants, II Proc. 2nd Int. Conf. 8-12 June 1981, Gaithersburg, MD;* (NBS-SP-617) pp. 281-286.

[20] Moldover, M. R., Mehl, J. B., Greenspann, M., "Gas-Filled Spherical Resonators: Theory and Experiment", *J. Acoust. Soc. Am.* **79**, No. 2 (1986).

[21] Bancroft, D., "Measurement of Velocity of Sound in Gases", *Am. J. Phys.* **24** (1956) 355.

[22] Lynnworth, L. C., Papadakis, E. P., "Ultrasonic Thermometry", *Ultrasonics Symp. Proc.*, 1970, pp. 83-93.

[23] Gujral, R., "Acoustic Thermometry", *J. Pure Appl. Ultrason. (India)* **6**, No. 1 (1984).

[24] Dadd, M. W., "Acoustic Thermometry in gases Using Pulse Techniques", *The Transducer Tempcon Conf. Papers, London, June 14-16, 1983.*

[25] Apfel, J. H., "Acoustic Thermometry", *Rev. Sci. Instr.* **33**, No. 4 (1962) 428.

[26] Plumb, H. H., Cataland, G., "Acoustical Thermometer", *Science* **150** (1965) 155.

[27] Bell, J. P. W., "A Solid Acoustic Thermometer", *Ultrasonics* **6** (1968) 11-14.

[28] Tasman, H. A., Schmidt, H. E., "The Ultrasonic Thermometer-Construction; Application and Operating Experience", *High Temp. – High Pressures* **4** (1972) 477-481.

[29] Mobsby, E. G., "Practical Ultrasonic Thermometer", *Ultrasonics* **7**, No. 4 (1969) 39-42.

[30] Kneidel, K. E., "Advances in Multizone Ultrasonic Thermometry Used to Detect Critical Heat Flux", *IEEE Trans. on Sonics and Ultrasonics* **SU-29**, No. 3 (1982).

[31] Bell, J. F. W., "Ultrasonic Thermometry Using Resonance Techniques", *Science and Industry (ISA)* **4**, p. 709.

[32] Cataland, G., Plumb, H. H., Edlow, M. H., "The Determination of Absolute Temperature from Sound Velocity Measurements" in: *Symposium on Temperature – Its Measurement and Control in Science and Industry;* Columbus, OH: American Institute of Physics, 1961.

[33] Carneval, E. H., Poss, H. L., "Ultrasonic Temperature Determination in a Plasma", *Temperature: Its Measurement and Control in Science and Industry* **3** (1962); AVCO RAD-TR-61-13 (1961).

[34] Suits, C. G., "The Determination of Arc Temperature from Sound Velocity Measurements", *Physics* **6** (1935) Parts I and II.

[35] Tichy, J., Gautschi, G., *Piezo-elektrische Meßtechnik;* Berlin: Springer, 1980.

[36] Ullmann, *Encyklopädie der technischen Chemie* Vol. 15 1964, and Vol. 21 1982.

[37] Cady, W., *Piezoelectricity* Vol. 1 and Vol. 2; New York: Dover, 1964.

[38] Voigt, W., *Lehrbuch der Kristallphysik;* Stuttgart: Teubner, 1966.

[39] Bechmann, R., Hearmon, R. F. S., Kurtz, S. K., „Zahlenwerte und Funktionen aus Naturwissenschaft und Technik" in: *Landolt-Börnstein, Neue Serie,* Vol. III/1 (1966) and Vol. III/2 (1969) Hellwege, K. H., Hellwege, A. M. (eds.); Heidelberg: Springer.

[40] Kittel, Ch., *Einführung in die Festkörperphysik;* München: Oldenbourg, 1973.

[41] "Schwingquarze – Ein unverzichtbares Bauelement in der Elektronik", *Tagungs-Dokumentation des SVEI-Quarz-Symposiums 1985.* Frankfurt, Main.

[42] Sheludew, I. S., *Elektrische Kristalle,* WTB: Akademie-Verlag, 1975.

[43] "IRE Standards on Piezoelectric Crystals", *Proc. of the IRE* **33** (1945).

[44] "IRE Standards on Piezoelectric Crystals", *Proc. of the IRE* **37** (1949) 1378.

[45] Bechmann, R., Ballato, A. D., "Higher-Order Temperature Coefficients of the Elastic Stiffness of Alpha-Quartz", *Proc. of the IRE* **50** (1962) 1892.

[46] "IEEE Standard on Piezoelectricity", *IEEE STD 176–1978,* 1978, IEEE, New York.

[47] Cristoffel, E. W., "Über die Fortpflanzung von Stößen durch elastische feste Körper", *Ann. die Mat.* **II**, No. 8 (1877).

[48] Love, A. E., *Mathematical Theory of Elasticity;* New York: Dover, 1944.

[49] Schmitt, P., "Über das Nebenwellenverhalten und die Schwinggüte von planen Quarzschwingern im HF- und unteren VHF-Bereich", Ph. D. Thesis, 1978, Universität Stuttgart.

[50] Mason, W. P., "Low Temperature Coefficient Quartz Crystals", *Bell System Technical Journal* **19** (1940) 74.

[51] Hammond, D. A., Adams, C. A., "A Linear Quartz Crystal Temperature-Sensing Element", *ISA Trans.* **4** (1965) 349.

[52] Schaudel, D., Brendecke, H., Ziegler, H., "Schwingquarz-Thermometer für die Labor- und Prozeßtechnik", *Automatisierungstechnische Praxis* **5** (1988) 219.

[53] Schöltzel, P., "Temperaturmessung mit Quarzsensoren", *VDI Zeitschrift* **112** (1970) 14.

[54] Quint, H., "Physikalische Ursachen von Meßfehlern bei Quarztemperatursensoren", Diplomarbeit, 1989, Universität Paderborn.

[55] Ueda, T., Kohsaka, F., Iino, T., "Quartz Resonator for Measuring Temperature", *Japanese Patent Laid Open No. 61–111010,* May 29, 1986.

[56] EerNisse, E. P., Wiggins, R. B., "A Resonator Temperature Transducer with no Activity Dipth", *40th Annual Frequency Control Symposium,* 1986, pp. 216.

[57] Ueda, T., "Quartz Thermometer", *Japanese Patent Laid Open No. 58–206936,* December 2, 1983.

[58] Ueda, T., Kohsaka, F., Iino, T., "Quartz Thermometer", *Japanese Patent Laid Open No. 59–128422,* July 24, 1984.

[59] Ueda, T., Kohsaka, F., Iino, T., "Quartz Thermometer", *Japanese Patent Laid Open No. 60–196634,* October 5, 1985.

[60] Ueda, T., Kohsaka, F., Iino, T., "Quartz Thermometer", *Japanese Patent Laid Open No. 60–152946,* October 11, 1985.

[61] Ueda, T., Kohsaka, F., Iino, T., "Quartz Thermometer", *Japanese Utility Model Laid Open No. 60–154831,* October 15, 1985.

[62] Ueda, T., Kohsaka, F., Iino, T., "Measuring System", *Japanese Utility Model Laid Open No. 60–158300,* October 21, 1985.

[63] Ueda, T., Kohsaka, F., Iino, T., "Quartz Thermometer", *Japanese Patent Laid Open No. 61–138131,* June 25, 1986.

[64] Ueda, T., Kohsaka, F., Iino, T., "Quartz Thermometer", *Japanese Patent Laid Open No. 61–288132,* December 18, 1986.

[65] Ueda, T., Kohsaka, F., Iino, T., "Quartz Resonator Thermometer", *UK Patent Application GB 2176892A,* January 7, 1987.

[66] "Quarze als Sensoren zur Temperaturmessung", *Elektronik Industrie* **3** (1987) 26.

[67] Tomikawa, Y., Oyama, S., Konno, M., "A Quartz Crystal Tuning Fork with Modified Basewidth for a High Quality Factor: Finite Element Analysis and Experiments", *IEEE Trans. on Sonics and Ultrasonics* **Su29**, No. 6 (1982) 217.

[68] Herzog, W., *Oszillatoren mit Schwingkristallen;* Berlin: Springer, 1958.

[69] Omlin, L., "Analyse der Dimensionierung von Quarzoszillatoren", *Elektroniker* **6** (1977) 30–36; and Elektroniker 9 (1977) 8–14.

[70] *Schwingquarze, Quarzoszillatoren, Quarzfilter,* Valvo-Handbuch 1979.

[71] Nührmann, D., *Das große Werkbuch Elektronik* Teil B; Franzis-Verlag, 1989.

[72] Tietze, U., Schenk, Ch., *Halbleiter-Schaltungstechnik;* Berlin: Springer, 1986.

[73] Profos, P., *Handbuch der industriellen Meßtechnik;* Essen: Vulkan, 1974.

[74] Ziegler, H., "Digitale Sensoren", *Hard and Soft* **2** (1987), Fachbeilage Mikroperipherik.

[75] Brendecke, H., "Digitale Temperatursensoren", *Sensor Magazin* **1** (1986) 22.

[76] Brendecke, H., "Digitale Temperaturmessung", *Hard and Soft* **2** (1987), Fachbeilage Mikroperipherik.

[77] Sauerbrey, G. Z., "Verwendung von Schwingquarzen zur Wägung dünner Schichten und zur Mikrowägung", *Z. Phys.* **155** (1959) 206.

[78] Benes, E., "Auto-Z-Match – Eine neue Generation von Schwingquarz-Schichtdickenmeßgeräten", *Optik Symposium, Moskow, May 25–27, 1987.*
[79] Wieczorkowski, A., "Hysterese und Tieftemperaturverhalten von Quarztemperatursensoren", Diplomarbeit, 1984, Universität Paderborn.
[80] "Quarzschnitte und Ersatzdaten", *Markt und Technik* **43** (1981) 86.
[81] Ziegler, H., "A Low-cost Digital Sensor System", *Sens. Actuators* **5** (1984) 169.

8 Heat Flux Sensors

F. van der Graaf, TNO Institute of Applied Physics, Delft, The Netherlands

Contents

8.1 Introduction

In the literature and commercial documentation different expressions are used for the same type of sensor, namely heat flow meter, heat flow sensor, heat flux sensor, and heat flux transducer. In this chapter we shall use the abbreviation HFS for "heat flux/flow sensor" but shall distinguish between "flow" and "flux".

In the main field of application, viz., the measurement of heat transmission through walls, the purpose is to measure the "density of an energy current" with the dimensions W/m^{-2}. We shall call the sensor used for this purpose a *heat flux sensor.*

The field of thermal analysis represents a different application of the HFS. Here the purpose is to measure all types of caloric effects, such as specific heat, heat of melting, heat of solidification, and heat of hydration, mainly on small quantities of sample materials. The aim is to determine the amount of heat (W) exchanged by the sample during exposure to a certain temperature program.

Another useful application of HFS is the measurement of the energy levels of laser beams or proton beams. Here also the purpose is to measure heat (W).

It is clear that for a sensor used to measure heat (W), the term *heat flow sensor* is more appropriate.

In fact, there is no basic difference between the sensors intended for the above purposes. Only the construction, the dimensions, and the calibration method differ.

8.2 Heat Flux Sensors

8.2.1 Principle of heat flux measurement

For a stabilized one-dimensional heat flow through a homogeneous material we can write

$$q = -\lambda \cdot \frac{\partial T}{\partial x}, \tag{8-1}$$

where λ is the thermal conductivity, and T the temperature.

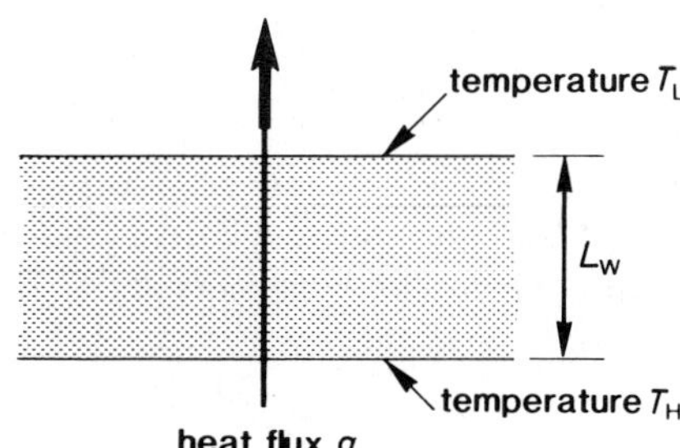

Figure 8-1.
Heat flux q passing through a wall of thickness L_w and thermal conductivity λ_w.

If the material forms a building wall with thickness L_W and surface temperatures T_H and T_L as shown in Figure 8-1, according to Equation (8-1) the heat flux through the wall then is given by

$$q = \frac{\lambda_W}{L_W} \cdot (T_H - T_L) . \tag{8-2}$$

If a thin layer of material is added to the wall surface (see Figure 8-2) and we assume that there is no change in heat flux, the heat flux through the layer is given by

$$q = \frac{\lambda_s}{L_s} \cdot \Delta T , \tag{8-3}$$

where ΔT is the temperature difference across the layer.

If we measure ΔT and know λ_s and L_s, q can be calculated.

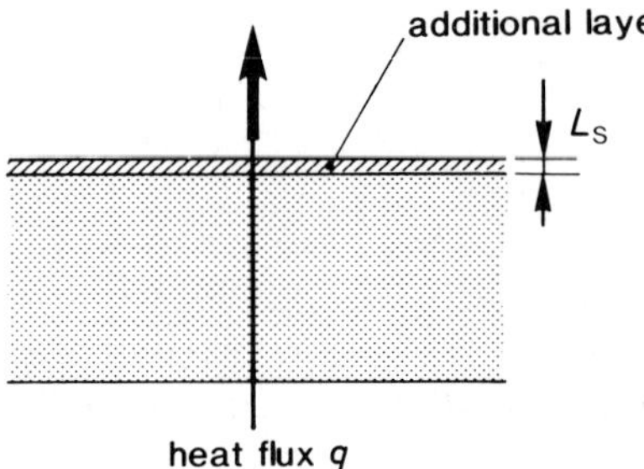

Figure 8-2.
Wall supplied with an additional layer of thickness L_s and thermal conductivity λ_s.

In practice, the heat flux through a wall depends on the differences between the inside and outside air temperatures and the thermal resistance of the wall. An additional layer mounted on the wall, for measuring the heat flux, will increase the thermal resistance and thus cause a decrease of heat flux. Consequently, a decreased heat flux will be measured. In Sections 8.2.3.1 and 8.2.3.2 this influence is dealt with and is called the one-dimensional error. It is clear that the aim of measuring the undisturbed heat flux can be approached by using a thin additional layer (low value of L_s) and a high thermal conductivity (λ_s).

From Equation (8-3), this leads to low values for ΔT. In practice, ΔT can only be measured with a large number of serially connected differential thermocouples (thermopile).

The output E of the thermopile is given by

$$E = n S \Delta T \tag{8-4}$$

where n and S are the number of differential thermocouples, and the thermocouple-sensitivity, respectively (in fact the difference between the Seebeck-coefficients of the thermocouple materials). By combining Equations (8-3) and (8-4) we obtain

$$q = \frac{\lambda_s}{L_s} \cdot \frac{1}{n S} \cdot E \tag{8-5}$$

or

$$q = CE. \tag{8-6}$$

In principle the calibration constant C can be derived from λ_s, L_s, n, and S, but it is generally more practical and accurate to measure C directly (see Section 8.2.4). Generally, disc, square or right-angled shapes are used for the additional layers and do not cover the wall surface completely.

Figure 8-3 shows the schematic set-up of the HFS.

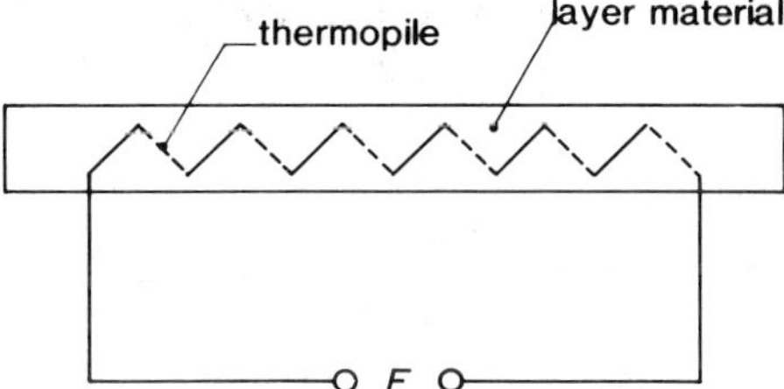

Figure 8-3. Schematic set-up of the HFS, the layer in Figure 8-2 being supplied with an embedded thermopile.

8.2.2 Types and Construction of Heat Flux Sensors

All known HFS consist basically of a thin layer of material (rubber, plastic) in which a thermopile is encapsulated. Several techniques can be used to manufacture the thermopile. One of the oldest is the "half-plated wound wire" technique [1, 2]. Constructional details are shown in Figure 8-4.

A constantan wire is wound around a flexible band of plastic and one half of each winding is plated with a thin layer of pure copper or pure silver. The junctions between the bare and the plated wires act as junctions of a thermocouple.

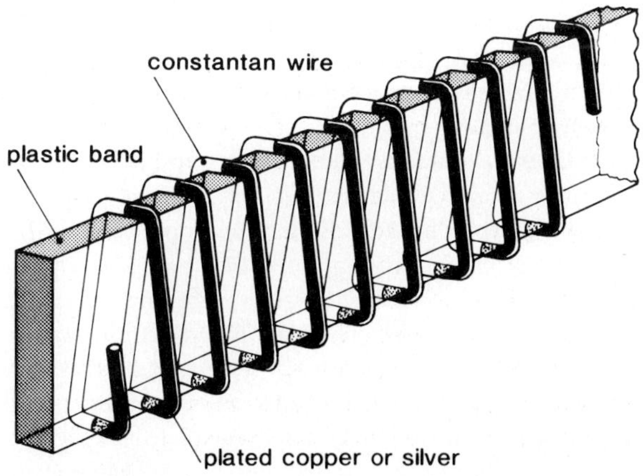

Figure 8-4. Constantan wire wound around a flexible band of plastic. Half of each winding is plated with a thin layer of pure copper or silver. The result is a thermopile.

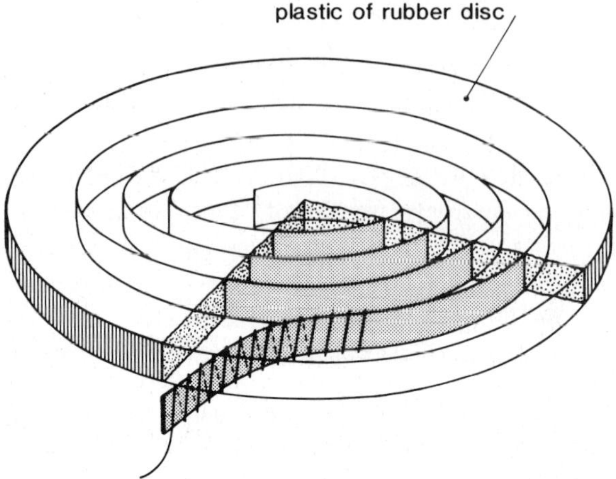

Figure 8-5. Thermopile embedded in plastic or rubber, resulting in an HFS.

Figure 8-5 shows how the thermopile is located in a plastic or rubber disc (core or substrate material). To achieve this configuration, molding techniques are applicd. Typical shapes and sizes for this type of HFS are discs with a diameter of 1–30 cm and a thickness of 2–5 mm. The number of differential thermocouples is between 100 and 2000.

Other techniques for obtaining thermopiles for HFSs are the following:

- Combination of *plating and photo-etching* [3]. Using this technique, very thin, large- and small-sized HFSs are obtained.
- Technique of *printed-circuit and plating through holes* of dissimilar metals [4].
- *Strain-gage techniques* applied to very thin etched metal foils or metallized layers in plastic films [5].
- *Thick-film technology* on ceramic substrates [6].

The combinations copper-constantan, copper-silver, and copper-nickel are the most popular thermocouple material.

The *core or substrate materials* are epoxy, epoxy-glass laminates, silicon elastomers, polyurethanes, and ceramic materials.

The surface of an HFS is sometimes provided with a laycr (eg, aluminium foil) to achieve the desired radiation characteristics (eg, low or high emissivity for infrared radiation). Moreover, high-conductivity metal plates are sometimes mounted on one or both sides of the HFS to eliminate lateral temperature gradients. Furthermore the HFS can be rigid or flexible.

To decrease measuring errors (see Sections 8.2.3.1 and 8.2.3.2), HFSs can be provided with an "integrated guard". The edge of, for instance, a disc-shaped HFS in most applications causes heat flux deflection, which leads to measuring errors. To bring this deflection out of the sensitive region, a "guard" has to be applied around the transducer. This guard consists of a layer around the HFS made of a material with an identical thermal resistance. The heat flux sensor will thus be a disc partly filled with a thermopile, as shown in Figures 8-6 and 8-7.

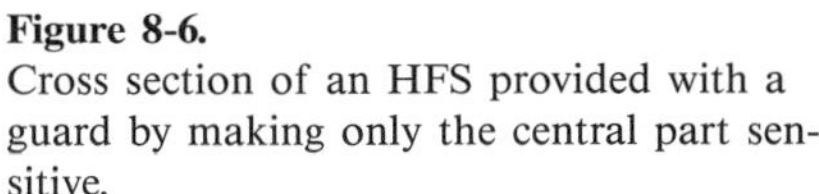

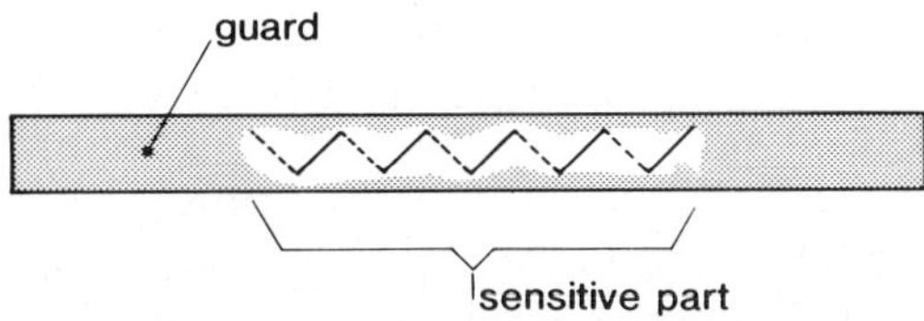

Figure 8-6.
Cross section of an HFS provided with a guard by making only the central part sensitive.

Figure 8-7. Photograph of the range of heat flux sensors which can be supplied by TPD-TNO-TH, The Netherlands. Diameter from 25 to 150 mm.

Some ranges of specifications for existing HFSs are as follows:

- Sensitivity range: from ca. 0.005 to 1000 $\mu V\ m^2\ W^{-1}$.
- Temperature range: from ca. -150 to 800 °C.
- Range of outer dimensions (inclusive guard): from 0.8×1.2 to 50×50 cm (or possibly larger).
- Range of thickness: from 0.07 mm to 6 mm.

More details of the construction and performance of commercial HFSs available in the USA can be found in [7].

8.2.3 Applications and Measuring Errors

Research into energy-saving measures in buildings industrial plants, and energy production from alternative sources such as solar energy often requires the determination of energy balances. Experiments are carried out under controlled or natural climatological conditions. Measurements of heat losses, heat transmission, and heat gain through walls, floors, roofs, and soils are essential. Further, the need to measure material properties such as thermal conductivity is closely related to this type of research.

In the following, the use of HFSs in the above applications is described and specific kinds of measuring errors are considered.

8.2.3.1 Measurement of Heat Losses Through Walls, Floors, and Roofs with a Built-in Heat Flux Sensor

In experimental buildings for energy-saving tests it is often possible to build in the necessary devices for heat-balance determinations. Temperature sensors, HFSs and the routes of measuring wires from the sensors to data-acquisition systems can be incorporated at the architectural design stage.

The main reason for preferring this approach is that sensors are well protected from atmospheric contaminants and uncontrolled infrared and/or solar radiation. It will be seen that some of these items can lead to lower measuring errors. However, the disadvantages are as follows:

- it is difficult to check correct locations and attachments;
- it is not easy to repair defective components;
- usually the sensor is not reusable.

The HFS can be mounted mid-way within the insulation or a concrete wall, or between the concrete layer (or wooden planking) and the insulation.

Figures 8-8 a and b show cross sections of a disc-form HFS incorporated in different wall materials (a, $\lambda_W > \lambda_s$; b, $\lambda_W < \lambda_s$). The heat flux direction lines and plots of the heat flux against the distance r from the center are shown.

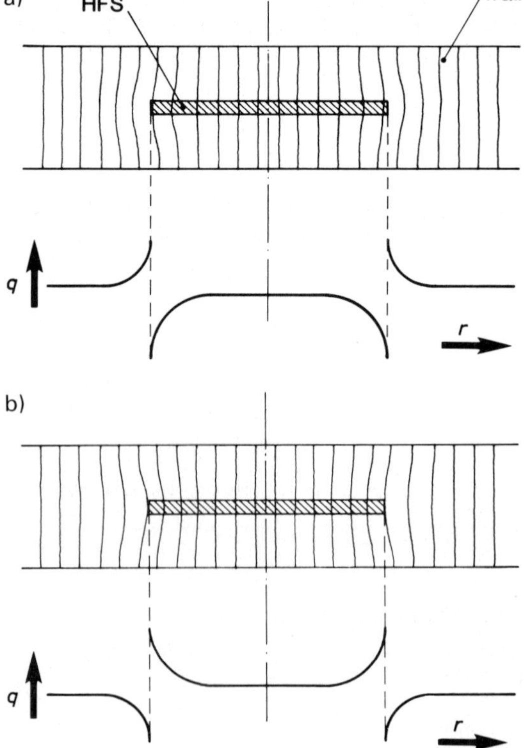

Figure 8-8.
Heat flux lines and profiles for HFSs incorporated in wall material with (a) $\lambda_w > \lambda_s$ and (b) $\lambda_w < \lambda_s$.

The heat flux "seen" by the HFS deviates from the original heat flux and shows also strong gradients at the HFS edge. The distortion can be split into two components: a one-dimensional and a three-dimensional distortion. Figure 8-9 illustrates the separation of these two components.

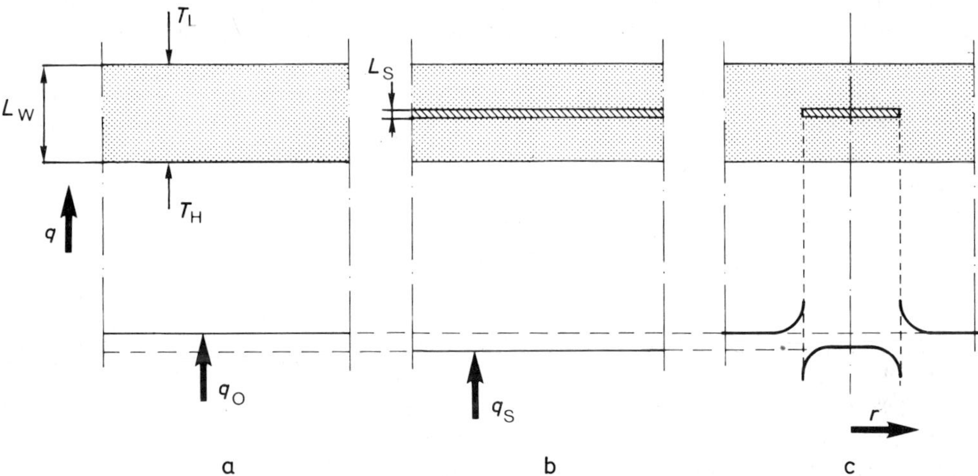

Figure 8-9. Explanation of separation of one-dimensional and three-dimensional (deflection) distortion. (a) Undistorted heat flux level q_0. (b) One-dimensional distortion caused by an infinite HFS. The heat flux changed from q_0 to q_s. (c) Heat flux profile for a finite HFS. The heat flux level in the center is equal to q_s. The heat flux at the edge shows strong gradients.

One-dimensional distortion

Figure 8-9 shows (a) the undistorted heat flux q_0, (b) shows the heat flux q_s if the sensor had an infinitely large diameter and $\lambda_s < \lambda_W$, and (c) both the one and three-dimensional distortions.

From Figure 8-9b, for the one-dimensional heat flux ratio q_s/q_0 we can write

$$\left(\frac{q_s}{q_0}\right)_{1-\text{dim.}} = \frac{L_W}{L_W - L_s + \dfrac{\lambda_W}{\lambda_s} \cdot L_s} . \tag{8-7}$$

In practice it is easy to estimate roughly the one-dimensional heat flux ratio, and hence the one-dimensional error.

Example:

A heat flux sensor has been incorporated in the insulation of a wall.

Insulation: thickness $L_W = 60$ mm
thermal conductivity $\lambda_W = 0.03 \text{ W m}^{-1} \text{ K}^{-1}$

HFS: thickness $L_s = 3$ mm
thermal conductivity $\lambda_s = 0.3 \text{ W m}^{-1} \text{ K}^{-1}$

From Equation (8-7),

$$\left(\frac{q_s}{q_0}\right)_{1-\text{dim.}} = 1.047 \,.$$

Thus the HFS would indicate a heat flux level ca. 5% higher than the undistorted heat flux if only the one-dimensional distortion is considered.

Note that actually the thermal resistance of the wall construction is higher than that calculated for the insulation layer alone. Consequently, the calculated one-dimensional distortion of 5% will be lower in practice.

Three-dimensional distortion

This distortion is related to the strong heat flux gradients at the edge of the HFS as shown in Figures 8-8a, 8-8b, and 8-9c. The three-dimensional distortion is complicated and has been the subject of several studies and experiments [8-10].

The importance of the application of guards, as demonstrated by the application of computer-model calculations and experiments, can be explained as follows. In Figure 8-10 the distorted heat flux already illustrated in Figure 8-9 is shown. However, now only the center part of the HFS is sensitive to the heat flux. The so-called guard frame protects the sensor's sensitive area from the effects of the distorted field around the edge of the sensor. The result is that the heat flux through the center section will approach the one-dimensional distorted heat flux according to Equation (8-7). The required size of the guard, leading to an acceptable level of the error due to the three-dimensional distortion, is a function of λ_W, λ_s, L_W, L_s, D_s, and the acceptable residual three-dimensional error.

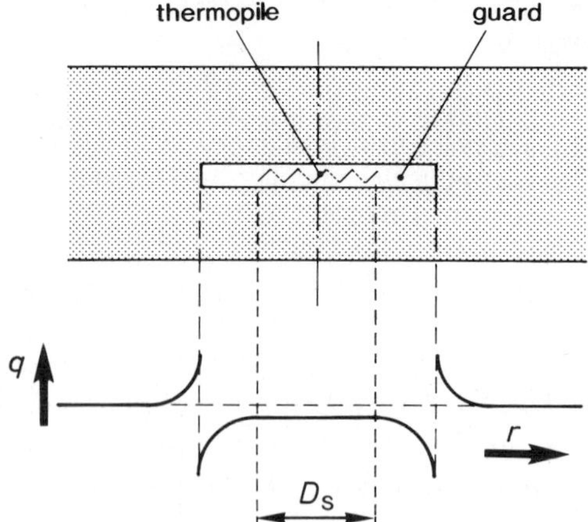

Figure 8-10.
If only the central part of the disc-shaped HFS is sensitive, the measured heat flux corresponds to q_s. The insensitive ring-shaped part of the HFS acts as a guard.

From an earlier publication [9] we can consider the following example, where

$\lambda_W = 0.03$
$\lambda_s = 0.6$
$L_W = 100$ mm
$L_s = 5$ mm
$D_s = 40$ mm

According to Equation (8-7), the one-dimensional heat flux ratio is

$$\left(\frac{q_s}{q_0}\right)_{1-\text{dim.}} = 1.05 \qquad (5\% \text{ error})$$

and the three-dimensional heat flux ratio is without guard:

$$\left(\frac{q_s}{q_0}\right)_{3-\text{dim.}} = 1.4 \qquad (40\% \text{ error})$$

with guard (guard diameter 120 mm):

$$\left(\frac{q_s}{q_0}\right)_{3-\text{dim.}} = 1.02 \qquad (2\% \text{ error})$$

The influence of the guard dimensions on the three-dimensional error as a function of several parameters has been considered [8, 9]. The three-dimensional error can be sufficiently decreased by using guards of large sizes.

Although not proved by calculations, it is generally assumed that the thermal conductivity of the guard material should match that of the material of the sensivitive part. It is uncertain how differences between the thermal conductivities of the guard and sensor influence the three-dimensional errors in the above "built-in" application. Some computer calculations on this aspect are required.

8.2.3.2 *Surface-mounted Heat Flux Sensors*

In Figure 8-11 a cross section of a wall surface-mounted HFS is shown. Both the wall surface and the HFS surface are exposed to radiation heat transfer from surrounding objects and convective heat transfer from air.

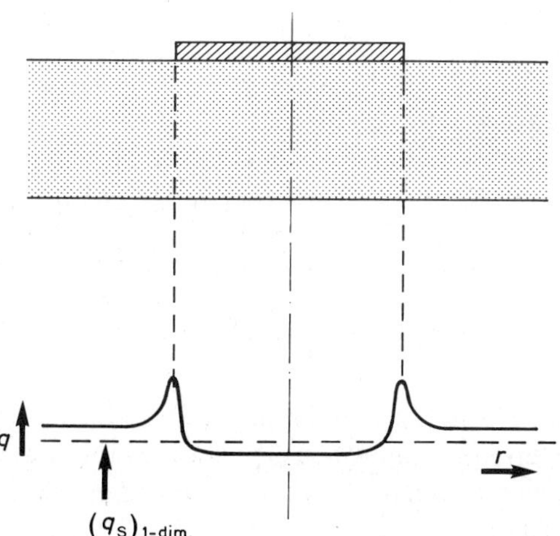

Figure 8-11.
Cross section of a surface-mounted HFS. The heat flux is plotted as a function of the distance from the center.

We exclude the case of direct exposure of the wall and HFS surface to solar radiation since the application of HFSs under these conditions is not to be recommended. In Figure 8-11 the heat flux is plotted against the distance from the center. The heat flux level $(q_s)_{1-\mathrm{dim.}}$ for the case when the HFS is of infinite size (one-dimensional distortion) is also indicated. It can be noted that:

- there are strong edge gradients (usually called the "edge spill");
- with the exception of the edge, the HFS "sees" a lower heat flux level than the undistorted heat flux;
- this heat flux level is lower than that of the one-dimensional distorted heat flux.

To quantify the measuring errors caused by the heat flux distortion, it is necessary to know the interactions between the following parameters:

- Wall construction:
  - thermal resistance;
  - lateral conductivity of the surface layer (on which the HFS is mounted).
- Heat transfer from the wall surface and the HFS surface to the surroundings:
  - radiation;
  - convection.
- Contact resistance between HFS and wall surface, and thermal resistance of the HFS.
- HFS dimensions: size and thickness.
- Size of the guard around the HFS.

Results of studies in this field have been published [10–14] and some conclusions can be drawn as follows.

Wall construction [11, 13]:

- A higher thermal resistance of the wall leads to lower measurement errors (especially the one-dimensional error).
- A highly conductive wall surface layer (eg, a metal layer or a metal wall) increases the error.

The extreme situation is when the temperature of the surface layer is not influenced by the presence of the HFS.

If the resistance due to heat transfer from the wall surface (and HFS) to the surroundings is R_T ($\mathrm{C\,m^2\,W^{-1}}$) and the thermal resistance of the HFS including contact resistance is R_{HFS}, the ratio of the measured heat flux q_s and undistorted heat flux q_0 is

$$\frac{q_s}{q_0} = \frac{R_T}{R_T + R_{HFS}} . \tag{8-8}$$

The practical importance of this ratio is that the highly conductive layer eliminates the influence of the wall construction on the error and makes the error more predictable [13]. Further, the error decreases if low values for R_{HFS} (thin HFS and low contact resistance) are chosen.

Heat transfer to surroundings

Errors increase when the heat transfer from both the wall and HFS increase [11–13]. The radiation characteristics of the wall and HFS surface have to be matched if the above-mentioned measurement on a highly conductive wall surface is applied [14].

Contact resistance between HFS and wall surface, and thermal resistance of the HFS

In building applications the contact resistance can often be high, which makes values of the HFS resistance of ca. 0.01 °C m^2 W^{-1} sufficient (lower values give little advantage [12]). If low values for the contact resistance can be achieved for other applications, or by the use of paste (eg, gel toothpaste [10]), low-resistance HFSs are sometimes preferred but are not essential (see HFS guards, Section 8.2.2).

An important practical aspect is the *homogeneity of the contact resistance*. If the contact material is missing locally (enclosures of air bubbles caused during placement of the HFS), the heat flux passing the HFS is inhomogeneous. This leads to additional errors, especially if the HFS has an inhomogeneous sensitivity.

HFS size and thickness

The surface area of an HFS should be much larger than the area of the edge, so the thickness/diameter ratio should be very small. The thickness and the thermal conductivity of the HFS material determine the thermal resistance. Related to Equation (8-3), the thermal resistance is

$$R_{\text{HFS}} = L_s/\lambda_s \,. \tag{8-9}$$

As can be seen, the value of the contact resistance indicates the advisable maximum value of L_s. For building applications a thickness of 1–3 mm (and a thermal conductivity $\lambda_s \approx 0.3$ W m^{-1} K^{-1}) will be adequate in most instances if the size is 75–150 mm square, or 75–150 mm in diameter.

For the measurement of the heat flux through a low-resistance wall construction where the HFS can be mounted on a smooth wall surface, an HFS with a low thermal resistance should be utilized. However, careful fixing with a thin layer of paste between the HFS and the surface is required. The size of this type of HFS could be smaller (25–50 mm).

Size of the guard around the HFS

The necessity for guards, especially for building applications, has been outlined [11] and is advised for diameters of 50–200 mm. This should bring the error near to the level of the one-dimensional error.

Other results [13], however, lead to the conclusion that only a small guard (ca. 10 mm) is required, based on the knowledge that the edge spill is concentrated on a small edge region. With the exception of this region, the heat flux through the HFS is nearly homogeneous, but the level is a function of the HFS size (diameter). This is illustrated in Figure 8-12. It can be seen that (according to [13]) the error is caused by a global rather than local distortion.

From Figure 8-12 one can also expect that a large HFS (large diameter) will bring the level of the resulting error to that of the one-dimensional case. The two publications [11, 13] do not conflict in that it is concluded that for the total area of the sensor, including the guard, large values (75–150 mm square or diameter) should be chosen.

From [11], however, we conclude that an HFS with a diameter of 150 mm should contain a central sensitive area less than 50 mm in diameter, thus having a guard width of 50 mm. From [13], we conclude that for an HFS of 150 mm diameter the diameter of the sensitive area could be 130 mm or less.

It is important to be aware that the discussed dimensions of the HFS and guard in relation to measuring errors are strongly dependent on the thermal wall parameters and on heat transfer values from the surface to air. Therefore, a comparison of the above two sets of results [11, 13] is questionable in some respects.

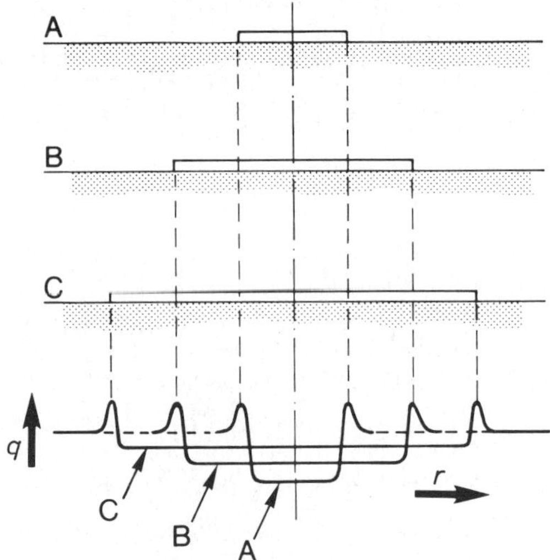

Figure 8-12.
Influence of the size of the HFS on the heat flux, indicating that a small guard and a large size of the HFS decrease errors.

8.2.3.3 *Application of Heat Flux Sensors to the Measurement of Heat Losses from Insulated Pipes*

This application involves distortion effects similar to those in Section 8.2.3.2. It is necessary to apply flexible HFSs. An HFS in belt form can be wrapped around a pipe to avoid errors caused by differences in heat losses above or below the insulated pipe. Different types of HFSs and some measurement results have been described [7].

8.2.3.4 *Application of Heat Flux Sensors to the Determination of the U Value of a Wall under Field Circumstances*

The U value under steady-state conditions is

$$U = q/\Delta T_{\text{air}} \tag{8-10}$$

where ΔT_{air} is the difference between the air-temperatures inside and outside the wall.

Thus, by measuring this temperature difference and the heat flux with an HFS (mounted on the inside of the wall), the U value can be calculated if steady-state conditions are reached. Under field circumstances, it is virtually impossible to attain true steady-state conditions owing to fluctuations in the air temperature and the thermal mass of the wall.

In order to determine the U value, it is necessary to integrate the temperature difference and the heat flux:

$$U = \frac{\int_0^t q(t)\,\mathrm{d}t}{\int_0^t \Delta T(t)\,\mathrm{d}t}. \tag{8-11}$$

If the errors dealt with in Section 8.2.3.2 are minimized with the application of a well designed HFS, the error in determining U now depends only on the thermal mass of the wall and the integration time t (= measurement time). In general, for most building walls a measurement time of 24 h is a minimum. If weather fluctuations are unfavorable, a multiple of a 24 h measuring time is advised [15].

8.2.3.5 *Application of Heat Flux Sensors in Instruments to Measure the Thermal Conductivity of Materials*

Several methods have been developed for the measurement of the thermal conductivity of materials, such as the guarded hot-plate method, Poensgen method, heat flow meter apparatus and Hamaker method. The first two methods, and in most instances also the third, are called "absolute" methods [16]. A known heat flux is created by electrical heating and is led through the sample material. In the last method and sometimes the third a generated heat flux is measured by a calibrated HFS.

In Figure 8-13 some configurations of methods involving HFSs are shown.

The general operating principles are as follows. The sample material (thickness) L_M and the layer(s) containing the HFS(s) are placed between a heating and a cooling plate. The plate temperatures are uniform over the plate area. The temperature difference ΔT over the sample

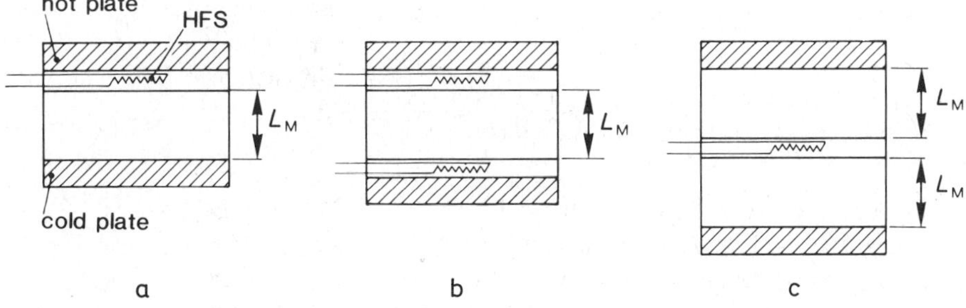

Figure 8-13. Different types of heat flow meter apparatus for measuring the thermal conductivity of building materials. Basically the instrument consists of a cold and a hot plate between which the material is placed together with one or more HFSs.

material is measured in the central area. In a steady-state situation, from ΔT, the measured heat flux q, and L_M the thermal conductivity can be calculated according to Equation (8-1):

$$\lambda = \frac{L_M}{\Delta T} \cdot q \; (\text{W m}^{-1} \text{K}^{-1}) . \tag{8-12}$$

A detailed description of the dimensioning of this type of measurement can be found in [17]. In [18] these methods are compared with the "absolute" method. In [19] an analysis of the measuring accuracy was carried out for the set-up in Figure 8-13 b and the method was compared with the Poensgen method. In [20] a review is given of commercially available instruments for measuring thermal conductivity according to the described principles.

8.2.4 Calibration of Heat Flux Sensors

8.2.4.1 *Intrinsic Calibration Value*

If the HFS is part of a field with a homogeneous heat flux q in the direction perpendicular to the HFS and a signal E is produced, the intrinsic calibration value C is defined as

$$EC = q . \tag{8-13}$$

The most common technique for measuring C is based on the use of a heat flow meter apparatus. This method requires an insulating material of known thermal conductivity as a reference. We therefore call it a comparative measurement. The principle is shown in Figure 8-14.

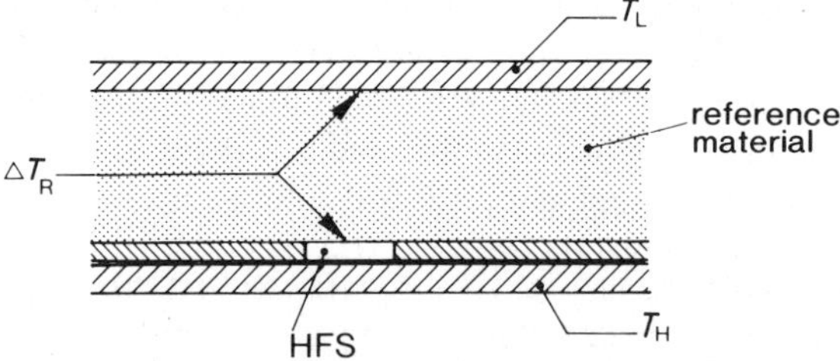

Figure 8-14. Cross section of an apparatus for calibrating an HFS using a reference material of known thermal conductivity.

Two layers are placed between the cold and the hot plate. One is the insulating material of known thermal conductivity λ_R and thickness L_R (the reference material). The other consists of the HFS and a material matched with the HFS for thickness and thermal conductivity (if the HFSs are square, a number of them can be used as surrounding material).

The temperature difference over the insulating layer (ΔT_R) is measured. The calibration value is obtained from

$$C = \frac{\lambda_R \Delta T_R}{E L_R} . \tag{8-14}$$

Details of the dimensions, temperatures, and measurements can be found in [17]. The same method, but in a special practical application, is described in [21].

Another technique for measuring the intrinsic calibration value is an absolute method based on electric heating. An instrument based on this principle has been described [9] and is shown in Figure 8-15.

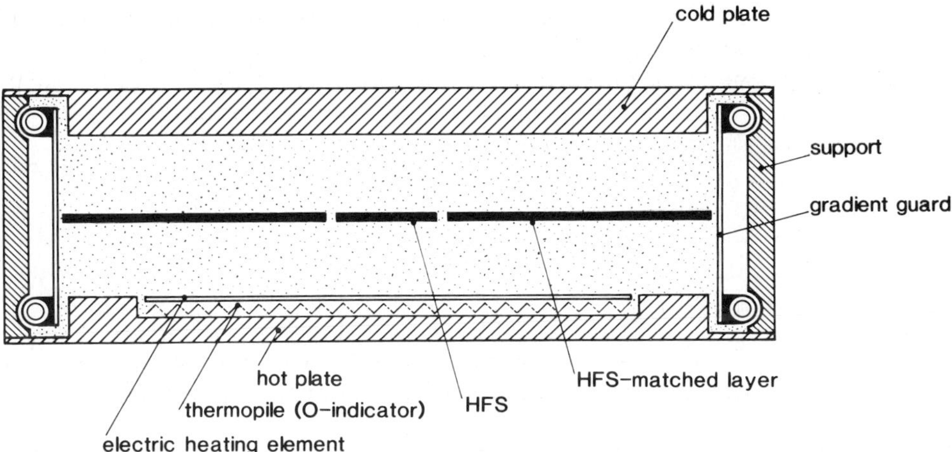

Figure 8-15. Cross section of an absolute HFS calibration apparatus as developed at TPD-TNO-TH in The Netherlands. The application of the gradient guard eliminates the "edge-loss" problem.

The main parts are a cold plate, a hot plate, and an electric heating element located on a layer of insulating material in a space on the hot plate. The surface of the heating element lies at exactly the same level the surface of the hot plate surrounding it. The layer of insulating material between the heater and hot plate incorporates a thermopile (about 15 000 copper-constantan junctions) acting as a "zero-q" indicator.

Insulating material is placed between the hot and the cold plates, and a layer of HFS-matched material and the HFS itself are located half-way between. A metal wall around the insulation supplied separately with hot and cold water provides a linear temperature gradient, corresponding to that through the insulating material ("gradient guard"). Hence, edge loss is eliminated whatever the surrounding temperature.

If the temperatures of the hot and cold plates (and the gradient guard) are stabilized, the dissipation of the electric heater is controlled so as to make the heat flux from the heater to the hot plate equal to zero, as indicated by the "zero-q" indicator. From the heater area and the electric power to the heater, the heat flux through the HFS and the calibration value can be calculated.

It should be noted that this instrument also offers attractive possibilities for measuring the thermal conductivity of materials owing to its capability to create a well-known homogeneous heat flux also through sample materials with extreme thickness/area ratios.

Another comparable absolute calibration instrument has been described [19].

8.2.4.2 *In Situ Calibration*

For some applications measuring errors as described in Sections 8.2.3.1 and 8.2.3.2 may be unavoidable. It is also possible that a more complicated wall, roof, or floor construction may lead to inhomogeneous transmission of heat.

If the measurement of the (total) heat flux is still desired, a solution can be found in the relationship between the signal of an installed HFS and an in situ generated known heat flux through the considered object. If the relationship is quantified, the HFS signal can be translated to the heat flux to be measured.

8.2.5 Influence of Internal Heat Flux Sensor Construction on Calibration and Application Errors

As already mentioned, all or maybe almost all HFSs are constructed as a thermopile with hot and cold junctions positioned near the two opposite surfaces of the HFS. Figure 8-16 shows an example of the construction in which the thermopile is embedded in a filling material of low thermal conductivitiy. The thermocouples of the thermopile consist of metals (eg, copper/constantan) of moderate to thigh thermal conductivity. They act as "cold bridges" in the filling material.

Case 1

Let us consider this HFS incorporated in an insulating layer through which a heat flux is passing. Temperature profiles exist on both surfaces (planes 1 and 4 in Figure 8-16) of the HFS owing to "cold bridges" and also to the low thermal conductivitiy of the surrounding in-

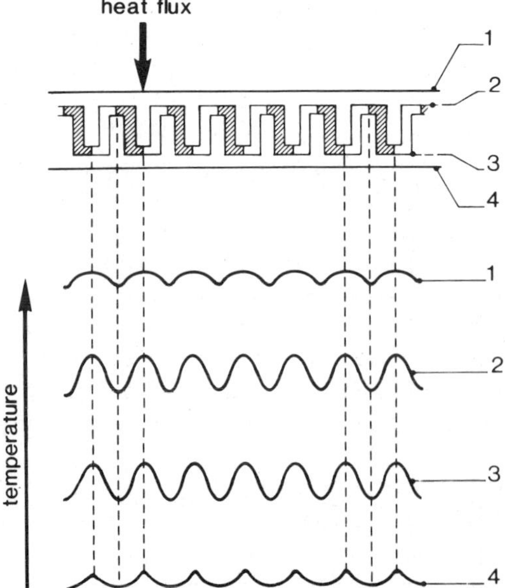

Figure 8-16.
Detailed cross section of the HFS with the thermopile embedded in the filling material. The thermocouples act as cold bridges and cause a pattern of hot-cold spots corresponding to the thermocouple pattern.

sulating material. Similar temperature profiles exist under the HFS surfaces at the junction planes (planes 2 and 3). These gradients produce a certain temperature difference between the junctions of every thermocouple and consequently a voltage signal is generated. The temperature profiles in the planes of the thermocouple junctions correspond to a similar heat flux distortion for every thermocouple.

The effects that occur on a macro-scale for an HFS incorporated in a material (see Section 8.2.3.1) are also found on a micro-scale for every thermocouple in the filling material. Therefore, we call this distortion the "micro-effect".

Case 2

Consider an HFS incorporated in a good conductive layer (eg, metal) through which the same heat flux is passing as in case 1. At the surfaces (planes 1 and 2 in Figure 8-17) of the HFS the temperature gradients are now more or less straightened and this influences the temperature profiles at the junction planes. The result is a change in the temperature difference over the thermocouples as compared with case 1, and consequently a change in the signal. It can be proven that the temperature-difference between hot and cold junctions for case 2 is higher that that for case 1.

Two conculsions can be derived from the above:

1. The calibration value of an HFS depends on the thermal conductivity of the surrounding material.
2. A measured calibration value is applicable only when an HFS is operated with similar surroundings to those during calibration.

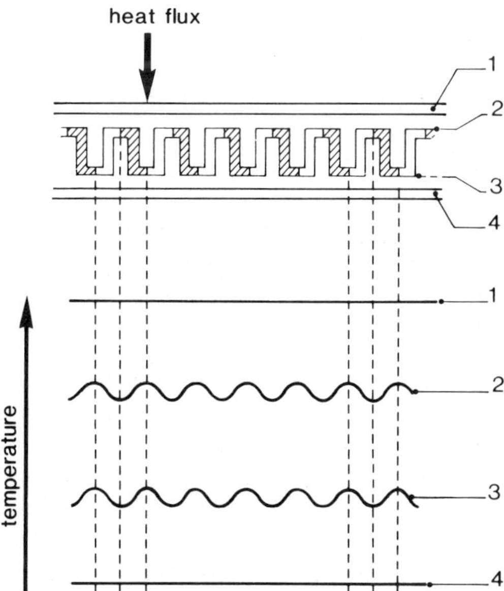

Figure 8-17.
Detailed cross-section of the HFS with isothermal planes on both surfaces. The temperatures of the thermocouple junctions differ from those in Figure 8-16.

Decrease in the influence of the micro-effect

An increase in the distance between the HFS surface and the nearest thermocouple junction plane (see Figures 8-16 and 8-17) leads to a decrease in the influence of the micro-effect.

Published calculation results [13] show that for a certain HFS configuration a distance of 0.1 mm or more decreases the influence of the micro-effect to a level of less than 3%. For users of HFSs it is important to know whether the HFS being applied is free from micro-effect influences or not. The HFSs shown in Figure 8-7, for example, have junctions which are positioned about 0.5 mm below the disc surface. Therefore, they are considered [13] to be free from micro-effect influences.

8.3 Heat Flow Sensors (to Measure Transported Heat)

In this section different kinds of measurements of transported heat (W) with HFSs are considered. There are no basic differences between these HFSs and those described in the previous sections. Only the aim of the application differs substantially and hence the performance and calibration.

8.3.1 Heat Flow Sensors Applied in Differential Thermal Analysis or Differential Scanning Calorimetric Instruments

In the field of thermal analysis a sample material is exposed to a temperature program and the reaction, eg, the amount of heat absorbed or produced, is measured using the techniques of differential thermal analysis (DTA) and differential scanning calorimetry (DSC). Many instruments are commercially available for this purpose.

Basically the principle is to measure small temperature differences and to convert them into heat quantities by calibration. In some cases HFSs are applied, and Figure 8-18 shows the principle of DSC with application of an HFS. The heat-sink temperature increases with time with a constant heating rate, $\partial T/\partial t$.

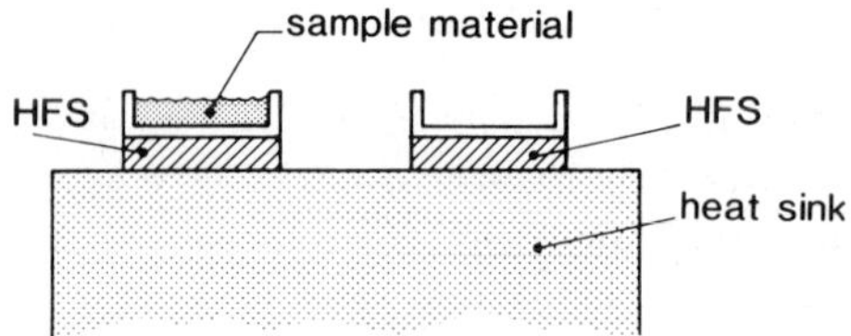

Figure 8-18.
Principle of the heat flux sensor DSC. Two HFSs are mounted on one heat sink. Both HFSs are supplied with sample containers, one being empty and the other filled with sample material.

On each of the two HFSs a container is placed, one empty (with heat capacity C_c) and the other filled with a known weight (M) of the sample material (with unknown specific heat c_p). After a short starting period the containers and sample material will increase in temperature with the heating rate $\partial T/\partial t$.. The heat passing through the HFSs is given by

$$q_c = C_c \frac{\partial T}{\partial t} \tag{8-15}$$

for the empty container and

$$q_{c+s} = \left\{C_c + Mc_p\right\} \cdot \frac{\partial T}{\partial t} \tag{8-16}$$

for the filled container. If we assume that the sensitivities (S) of the two HFSs are equal, for the difference E between the two signals we can write:

$$E = S \cdot M \cdot c_p \cdot \frac{\partial T}{\partial t} \,. \tag{8-17}$$

It is obvious that Mc_p can be directly measured as a function of the temperature. If the sample material melts at a temperature within the operating range the process is more complicated and Equation (8-17) is no longer valid, because the temperature increase of the sample material (and its container) lags behind the original $\partial T/\partial t$.

This effect is shown in Figure 8-19. The signal E and the temperature of the empty container are recorded as a function of time. We assume that the material has different values for the specific heat before and after the melting and there is a sharp melting point. The heat of melting can be found by integrating the shaded area. To a good approximation (if $C_c \ll Mc_p$) the melting point T_M can be found at the time corresponding to the intersection of the extrapolated melting curve (constant slope) with the time axis.

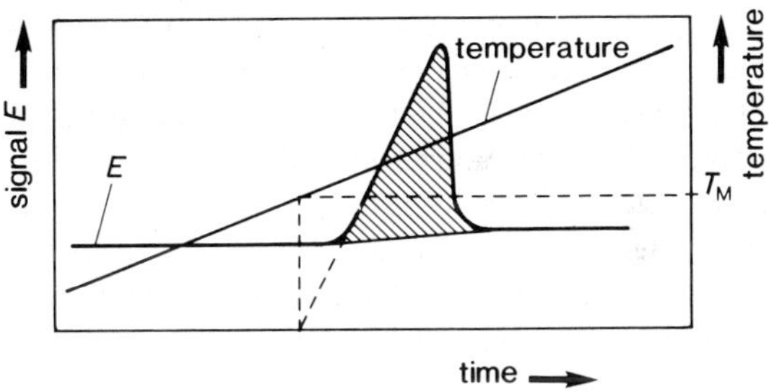

Figure 8-19.
Plot of an HFS DSC run. The difference in HFS signals E and the temperature are plotted against time. The shaded area corresponds to the heat of melting of the sample material.

It must be said that, in general, the interpretation of DSC curves is more complicated, especially if sample materials have melting ranges and/or more than one melting peak.

Relevant publications on this topic can be found in [22] and [23].

Concerning *HFS design* for these applications, the following can be specified:

- small dimensions (diameter 10–25 mm);
- thickness not critical (ca. 2 mm);
- no guard required;
- low thermal resistance;
- high sensitivity (eg, 500 mV W^{-1}).

Figure 8-20 shows a DSC that includes three sample units and one compensation unit.

Figure 8-20.
Heat flux sensor DSC consisting of three measuring units and one compensating unit.

For the calibration of each individual HFS two different methods can be applied:

1. An electrical heater is placed in the sample container in good thermal contact with the bottom of the container. A constant known quantity of heat is dissipated by the heater and the heat sink is maintained at a constant temperature. From the measured HFS signal and the dissipated energy the sensitivity can be calculated.
2. The sample container is filled with a material with a known Mc_p and exposed to a temperature increase ("run") with a known heating rate $\partial T/\partial t$. An identical run is then made with an empty container. The difference in the signals between the two runs corresponds to Mc_p. The calibration value can then be calculated using Equation (8-17).

8.3.2 Heat Flux Sensors Applied in Other Thermal Analysis Instruments

8.3.2.1 Cement Calorimeter

The production of heat of hydration during the curing process of cement is a useful parameter with regard to some of its characteristics. It can be measured using the instrument shown in Figure 8-21. The heat sink is maintained at a constant temperature. The sample container placed on the HFS is filled with a mixture of dry cement, mortar, and sand. When thermal equilibrium has been attained, a known quantity of water is injected into the container

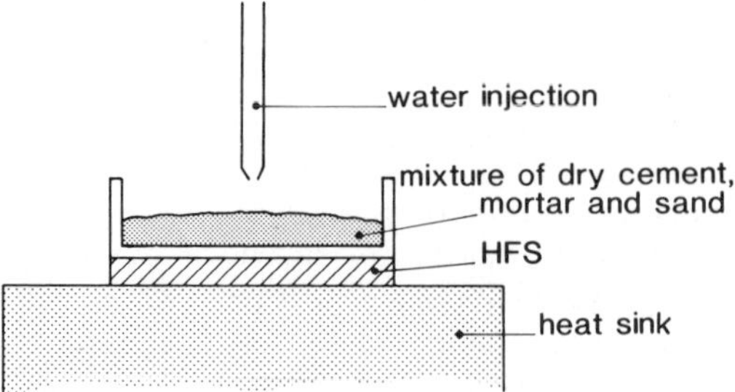

Figure 8-21.
Principle of the measurement of the heat of hydration using an HFS.

to start the hydration process and the HFS signal is recorded. This instrument can be calibrated using method 1 described in Section 8.3.1.

The requirements for heat flux sensor design are as follows:

- diameter ca. 50 mm;
- thickness not critical;
- no guard required;
- high sensitivity.

A complete calorimeter can contain up to six HFS units, five for the measurement and one for compensation of temperature fluctuations. An example is shown in Figure 8-22. A standardized measurement method has been described [24].

Figure 8-22. Cement calorimeter consisting of five measuring units and one compensating unit.

8.3.2.2 Mixing Heat Calorimeter

This instrument is similar in principle to the cement calorimeter. To stimulate the mixing, stirrers can be applied.

8.3.2.3 Measurement of Fluid Evaporation

The set-up for this application is shown in Figure 8-23. The HFS is mounted on the temperature-controlled heat sink. A layer of evaporating liquid is maintained on the top of the HFS by a fluid supply. If the specific heat of evaporation for the liquid is known, from the HFS signal the amount of liquid evaporated can be calculated.

Calibration is performed using method 1 described in Section 8.3.1.

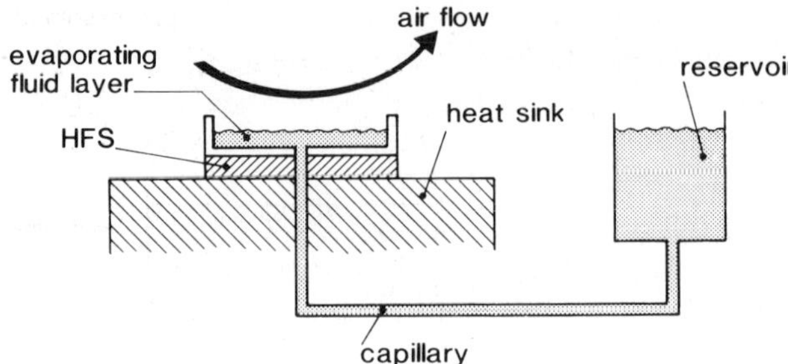

Figure 8-23.
Principle of fluid evaporation measurement.

8.4 Heat Flow/Flux Sensors Applied in Instruments for Radiation Measurement

In general, these instruments are constructed from the following components:

1. An interface to achieve a well defined absorption of radiation, either depending on spectral wavelength or not. The absorbed radiation is converted into measurable heat.
2. The HFS.
3. The heat sink on which the HFS is mounted. Often a good thermal contact between this heat sink and the surrounding air is required.
4. Sometimes measures are necessary to insulate the absorber from its surroundings. One or more layers of convection shields (eg, glass domes for solarimeters) are then applied.

For practical application there are two different kinds of construction. In one the plane of absorption is perpendicular to the direction of the heat flux in the HFS (transversal-constructed sensor). In the other the plane of absorption is parallel to with the direction of the heat flux in the HFS (lateral-constructed sensor).

8.4.1 Transversal-Constructed Heat Flux Sensors for Radiation Measurement

8.4.1.1 Solarimeters

Solarimeters are used to measure the intensity of solar radiation (W m^{-2}). A cross section of the instrument is shown in Figure 8-24. Two glass domes transmit the solar radiation. They act as a thermal resistance for convection and infrared heat transfer between the absorbing surface and the surroundings. The absorbing layer consists of a material (eg, paint) which has a high and very stable absorption for the solar-radiation wavelengths, independent of the angle of incidence. The heat sink is in good thermal contact with the surrounding air. However, optimization is necessary to eliminate signals caused by the dynamic response to sudden changes in the air temperature.

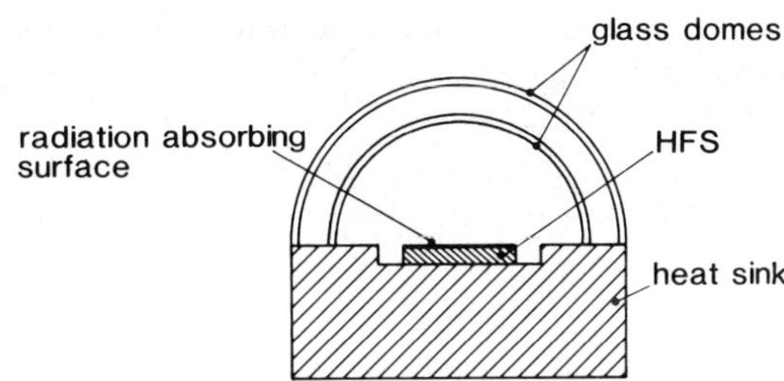

Figure 8-24. Cross section of a solarimeter.

8.4.1.2 Measurement of High Peak Power Laser Beams

The average energy level of short pulses of high-energy laser light can be measured by absorption of the pulses in an aluminium layer. If this layer is fitted on an HFS mounted on a heat sink, the laser pulses are converted into an average heat flux, and thus can be measured.

8.4.1.3 Measurement of Proton Beams

The method is similar to that for the laser beam.

8.4.1.4 Sensor for Infrared (IR) Detection

HFSs are in general applicable for IR detection (see also Section 8.4.2), and a successful application is the "Infrared Astronomical Satellite" (IRAS). To control the satellite's position with respect to Earth, HFSs were applied to detect the IR radiation from the Earth [25].

8.4.1.5 Design Criteria

For the applications described in Sections 8.4.1.1–8.4.1.4, similar design criteria to those in Section 8.3.1 apply. The differences are mainly in the type of absorption layer.

8.4.2 Lateral-Constructed Heat Flux Sensors

Modern techniques such as thin-film technology (metal vapor deposition) and silicon IC technology offer attractive possibilities for the manufacture of thermopiles/HFSs. Figure 8-25 shows the principle of the construction of a thin-film thermopile on Mylar foil and strained between the legs of a heat sink.

The middle part of the foil (on the "hot junctions") is coated with an IR-absorbing material. The "cold junctions" are in contact with the heat sink legs, to which the absorbed heat is transported through the thin Mylar layer. Thermocouple materials commonly used are Bi-Sb and Bi-Te.

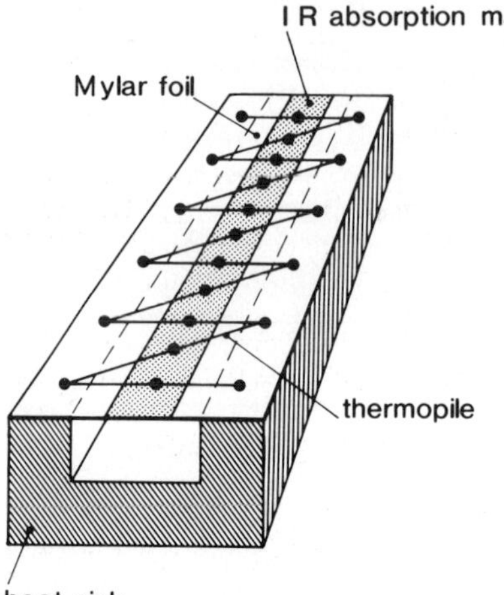

Figure 8-25. Thin-film HFS for measuring infrared radiation.

Some characteristic values are as follows:

- dimensions: $10-100\ mm^2$;
- sensitivity: $10-100\ V\ W^{-1}$;
- thermal resistance: up to $100000\ K\ W^{-1}$;
- time constant: 0.1–0.5 s.

The application of some of these types of sensors in space research has been described [26].

Figure 8-26 shows an IR detector manufactured according to silicon IC technology [27]. The absorbing layer and the thermopile are located on a cantilever beam (silicon of thickness 10 μm). The hot junctions are positioned near the IR-absorbing area and the cold junctions near the supporting rim (heat sink). The thermocouple materials are aluminium/p-type silicon. The thin silicon layer is obtained by an etching process.

Some characteristics values are as follows:

- dimensions of the cantilever beam: ca 4×4 mm;
- sensitivity: ca $10\ V\ W^{-1}$;
- thermal resistance: $300-400\ K\ W^{-1}$;
- time constant: 0.1–0.2 s.

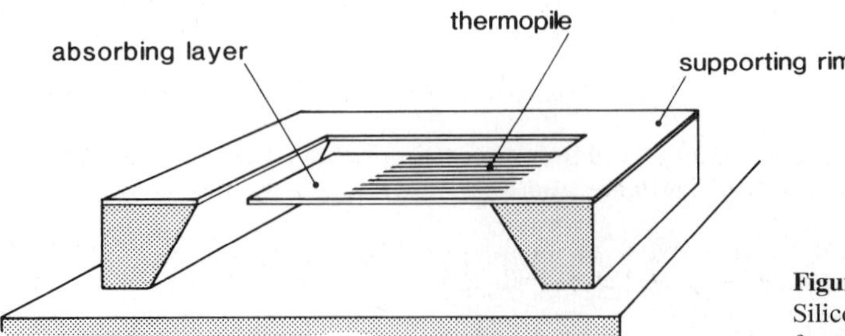

Figure 8-26. Silicon IC HFS for measuring infrared radiation.

8.5 References

[1] De Jong, J., Marquenie, L., *Instrum. Pract.* January (1962) 45–51.

[2] Van der Graaf, F., *Proc. 3 rd International Conference 'Temp/MEKO 87 Thermal and Temperature Measurement in Science and Industry', IMEKO Sheffield, September 15–17, 1987,* pp. 171–182.

[3] Miyake, Y., Eguchi, U., in: *Guarded Hot Plate and Heat Flow Meter Methodology, ASTM STP 879,* Shirtliffe C. J., Type, R. P., (eds.); Philadelphia: American Society for Testing and Materials, 1985, pp. 171–179.

[4] Degenne, M., Klarsfeld, S., in: *Building Applications of Heat Flux Transducers, ASTM STP 885,* Bales, E., Bomberg, M., Gourville, G. E. (eds.); Philadelphia: American Society for Testing and Materials, 1985, pp. 163–171.

[5] *Heat Flow Sensors, Bulletin 36614,* RdF Corporation. Hudson, New Hampshire, USA.

[6] Van Dort, A. C., Van der Graaf, F., Steenvoorden, G. K., *Sens. Acutators* **4** (1983) 323–331.

[7] Hauser, R. L., in: *Building Applications of Heat Flux Transducers, ASTM STP 885,* Bales, E., Bomberg, M., Courville, G. E., (eds.); Philadelphia: American Society for Testing and Materials, 1985, pp. 172–183.

[8] Apthorp, D. M., Bligh, T. P., in: *Building Applications of Heat Flux Transducers, ASTM STP 885,* Bales, E., Bomberg, M., Courville, G. E., (eds.); Philadelphia: American Society for Testing and Materials, 1985, pp. 45–64.

[9] Van der Graaf, F., in: *Building Applications of Heat Flux Transducers, ASTM STP 885,* Bales, E., Bomberg, M., Courville, G. E., (eds.); Philadelphia: American Society for Testing and Materials, 1985, pp. 79–96.

[10] Flanders, S. N., in: *Building Applications of Heat Flux Transducers, ASTM STP 885,* Bales, E., Bomberg, M., Courville, G. E. (eds.); Philadelphia: American Society for Testing and Materials, 1985, pp. 140–159.

[11] Trethowen, H. A., *Build. Environ.* **21**, No. 1 (1986) 41–56.

[12] Trethowen, H. A., in: *Building Applications of Heat Flux Transducers, ASTM STP 885,* Bales, E., Bomberg, M., Courville, G. E., (eds.); Philadalphia: American Society for Testing and Materials, 1985, pp. 9–24.

[13] Standeart, P., *Numerical Analysis of Operational Errors with Surface-mounted Heat Flux Sensors,* Report of Research by the Direction of Prime Minister's Services, Policy of Science National Program of Energy, Belgium, 1987.

[14] Baba, T., Ono, A., Hattori, S., *Rev. Sci. Instrum.* **56**, No. 7 (1985) 1399–1401.

[15] Modera, M. P., Sherman, M. H., Sounderegger, R. C., in: *Building Applications of Heat Flux Transducers, ASTM STP 885,* Bales, E., Bomberg, M., Courville, G. E., (eds.); Philadelphia: American Society for Testing and Materials, 1985, pp. 203–219.

[16] *Standard Test Method for "Steady State Heat Flux Measurements and Thermal Transmission Properties by Means of the Heat Flow Meter Apparatus C 518–85;* Philadelphia: American Society for Testing and Materials, 1985.

[17] *Steady-State Heat Flux Measurement and Thermal Transmission Properties by Means of the Heat Flow Meter Apparatus, ASTM C518.85;* Philadelphia: American Society for Testing and Materials, 1985.

[18] De Ponte, F., in: *Guarded Hot Plate and Heat Flow Meter Technology, ASTM STP 879,* Shirtliffe, C. J., Tye, R. P., (eds.); Philadelphia: American Society for Testing and Materials, 1985, pp. 101–120.

[19] Hamaker, H. J., *Technische Warmte-geleidingsmetingen, de Bepaling van de Warmtegeleidingscoëfficiënt van Bouw- en Isolatiematerialen,* Thesis, RU-Utrecht, 1939.

[20] Coumou, K. G., in: *Guarded Hot Plate and Heat Flow Meter Methodology, ASTM STP 879,* Shirtliffe, C. J., Tye, R. P. (eds.); Philadelphia: American Society for Testing and Materials, 1985, pp. 161–170.

[21] Bligh, T. P., Apthorp, D. M., in: *Building Applications of Heat Flux Transducers, ASTM STP 885,* Bales, E., Bomberg, M., Courville, G. H. (eds.); Philadelphia: American Society for Testing and Materials, 1985, pp. 25–44.

[22] Chen, R., Kirsch, Y., "Analysis of Thermally Stimulated Processes" in: *International Series on the Science of the Solid State* Vol. 15; Oxford: Pergamon Press, 1981 pp 97–115.
[23] Wendlandt, W. M., in: *Thermal Analysis,* Elving, P. J., Winefordner, J. D., Kolthoff, I. M. (eds.); New York: Wiley, 1985 pp 213–460.
[24] "Ciments, Determination de la Chaleur d'Hydratation (Méthode par Conduction)", *Norme Belge NBN B 12-213,* 1 e éd., September 1975.
[25] de Boom, C. W., *Twee Sensoren van het IRAS Standregelsysteem, TNO Project No. 10,* October 1977, pp. 355–360.
[26] Rolls, W., Kale, B. M., Cirino, P., *State-of-the-Art Thermopiles – Design and Applications;* Stanford, CT: Barnes Engineering.
[27] Sarro, P. M., *Integrated Silicon Thermopile Infrared Detectors,* Thesis, Delft University, 1987.

9 Thermal Mass-Flow Meters

Martin Hohenstatt, SENSYCON GmbH, Hanau, FRG

Contents

9.1 Introduction

The mass of a substance and the number of its constituents, atoms or molecules, N, are directly correlated. Therefore, in many technical applications the measurement of flow in terms of mass units is of increasing interest. In process control of chemical reactions, for example, the quantity of mass flow is needed in order to establish the exact stoichiometric relationship of the reactants.

Classical flow meters determine the quantity of volume flow $\dot{V}$, where the density ρ of the fluid has to be known or measured to obtain mass flow $\dot{M}$:

$$\rho \dot{V} = \dot{M} \sim N . \tag{9-1}$$

Especially in the case of gases, density varies strongly with temperature and pressure. In order to obtain a mass flow, the measurement of volume flow has to be supplemented by an additional measurement and computational analysis of these parameters. The *direct* measurement of mass flow therefore decreases the number of sensors and the expenditure of installation with obvious consequences regarding cost, accuracy and reliability. It especially provides for very simple handling of the measurements.

Three different approaches to direct mass flow measurement are known to which the following physical principles apply:

- moment of inertia of the fluid;
- modulation of the flow velocity;
- heat transfer from a heated body to the fluid.

Coriolis mass-flow meters [1] are representatives of the first type of instruments. They find applications particularly in the measurement of liquids and slurries. In the case of gases they can only be used at elevated pressure.

The second kind of mass-flow meter [2] is designed to measure gases at flows less than 100 kg/h. The method is based on the modulation of the flow velocity in a pipe by means of a mechanical oscillator which is damped by the mass flow.

This chapter deals with two different kinds of thermal mass-flow meters:

- calorimetric flow meters;
- boundary-layer thermal flow meters.

Their main field of application is the measurement of gases, and in some cases also of liquids. After a short discussion of the principles of measurement, the set-up of specific instruments with special emphasis on sensor elements is described. Finally, examples of commercial instruments for typical industrial applications are described.

9.2 Principle of Measurement

9.2.1 Basic Principles

A linear dependence exists between the mass M of a fluid and the amount of energy Q which is needed to raise its temperature by ΔT. The proportionality factor is the inverse of c_p, the specific heat of the medium at constant pressure. The pure *calorimetric* approach to mass-flow measurement makes use of this principle by measuring the ratio of the power $\dot{Q}$ transferred from a heater element to the fluid and the resulting temperature change of the fluid to obtain mass flow:

$$\dot{M} = \frac{1}{c_p} \frac{\dot{Q}}{\Delta T} . \tag{9-2}$$

The relevant ΔT is the mean temperature change of the whole portion of fluid per time invertal which is under consideration.

Thermal mass-flow meters of the *boundary-layer* type evaluate the flow-dependent power dissipation of a heater element rather than the temperature change of the fluid. The heat transfer to the moving fluid, also called forced convection, is described by the Navier-Stokes equations. Analytical solutions exist only for few geometrical set-ups. Therefore, the view of a simple model gives a better understanding than tedious calculations: a boundary layer forms across the surface of a body located within a moving fluid. Therein the flow velocity v decreases from the velocity of the free or undisturbed flow to zero at the surface of the body.

The heat which is transported away by the fluid has to be transferred across that boundary layer, the thickness of which limits the amount of heat flux.

Two counteracting forces taking effect on the fluid particles determine the thickness of the boundary layer:

- friction causes a viscous drag ($\sim \eta\, v/l$) and is responsible for the formation of the boundary layer; η is the dynamic viscosity of the fluid and l a characteristic length of the heated body. For example, in the case of a thin plate, l is its extension in the direction of flow;
- the moment of inertia ($\sim \rho\, v^2$) works against friction forces and limits the thickness of the boundary layer.

The ratio of these two quantities is the Reynolds number, *Re*, a very important parameter in fluid dynamics. For a given flow profile in a duct of cross-sectional area A_d, it is proportional to the mass flow $\dot{M}$:

$$Re = \frac{\rho\, v^2}{\eta\, v/l} = \frac{\rho\, v l}{\eta} = \frac{\dot{M}\, l}{A_d\, \eta} . \tag{9-3}$$

For the dissipated power P_d in many practical cases an exponential function of the Reynolds number gives satisfactory results [3]:

$$P_d \sim \lambda\, Pr^m\, Re^n\, A_b\, \Delta T_b \sim M^n \tag{9-4}$$

$$Pr = \frac{\eta\, c_p}{\lambda} . \tag{9-5}$$

λ is the thermal conductivity and represents, together with the Prandtl number *Pr*, the thermal behavior of the fluid; A_b is the surface area of the heated body and ΔT_b the difference between its temperature and the temperature of the fluid; *m* and *n* are constants which have to be determined experimentally.

9.2.2 Calorimetric Mass-Flow Meters

Flow meters which directly apply the calorimetric principle corresponding to Equation (9-2) [4] suffer from the problem that the electric power in the heater element is only partly transferred to the fluid. Even with a heater element mounted inside the duct, a set-up sometimes called a "heated-grid calorimetric flow meter", the heat conductance to the walls, insulation, etc., is not negligible and is often flow dependent. Consequently, most instruments apply a slightly different approach, the principle of which is illustrated schematically by Figure 9-1.

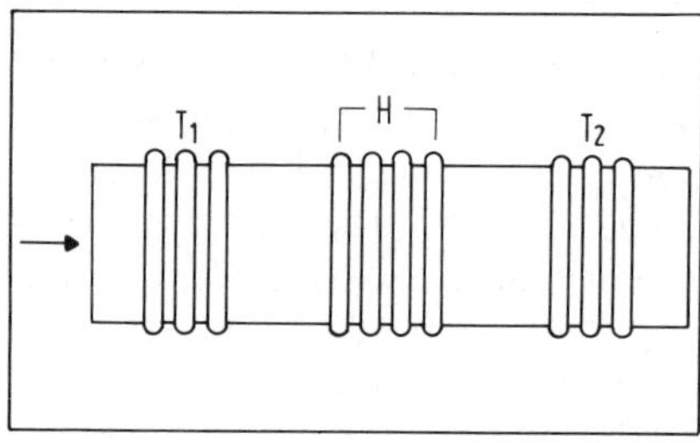

Figure 9-1.
Calorimetric mass-flow meter with external heater element (H) and external temperature sensors (T_1 and T_2) (from [19]).

A thin-walled capillary is heated at the mid-point by an external heater element (H) and the temperature of the wall is also measured externally by two sensor elements mounted symmetrically upstream (T_1) and downstream (T_2) of the heater element. With no flow a symmetrical temperature profile is produced along the capillary mainly by heat conduction inside the wall as shown in Figure 9-2. The temperature difference $T_{21} = T_2 - T_1$ between the two

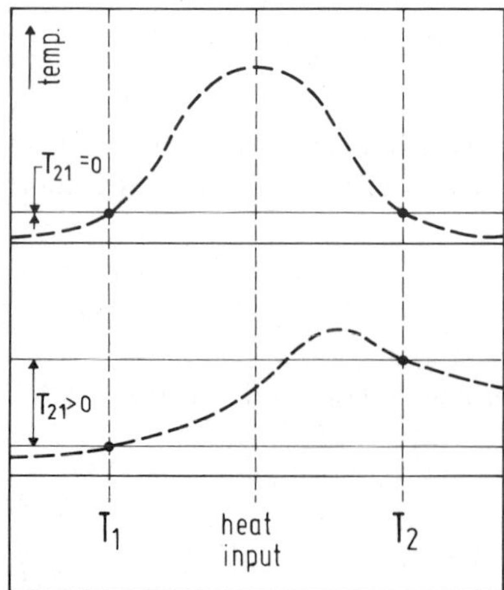

Figure 9-2.
Calorimetric mass-flow meter: temperature profile along the capillary for zero (upper) and non-zero (lower) gas velocity. T_1 and T_2 mark the position of the temperature sensors with the corresponding temperature difference T_{21} (after [19]).

temperature sensors is zero. However, with a fluid flowing, heat is transferred to the fluid and transported downstream. Superposed on this pure calorimetric effect is heat conduction in the wall. For the upstream temperature T_1 these two effects are of opposite sign and T_1 is reduced, whereas the downstream temperature T_2 is increased by both effects. This leads to a nonsymmetrical temperature profile as in Figure 9-2.

The proportionality between T_{21} and mass flow $\dot{M}$ can be understood by a simple model [5–7] which establishes a balance for the heat fluxes into and out of a short section of capillary of length dx, (see Figure 9-3):

$$\dot{q}\,\mathrm{d}x = \Delta\Phi_\mathrm{T} + \Delta\Phi_\mathrm{C} + \Phi_\mathrm{L}\,. \tag{9-6}$$

External heating $\dot{q}\,\mathrm{d}x$ with $\dot{q}$ being the heating power per unit length is compensated for by heat loss due to transportation by the fluid, $\Delta\Phi_\mathrm{T} = \Phi_4 - \Phi_3$, heat conduction in the wall, $\Delta\Phi_\mathrm{C} = \Phi_2 - \Phi_1$, and heat loss to the surroundings, Φ_L. On the basis of laminar flow and good thermal contact between the capillary and the fluid, the model presumes a uniform temperature $T(x)$ within a plane of cross-section including the wall.

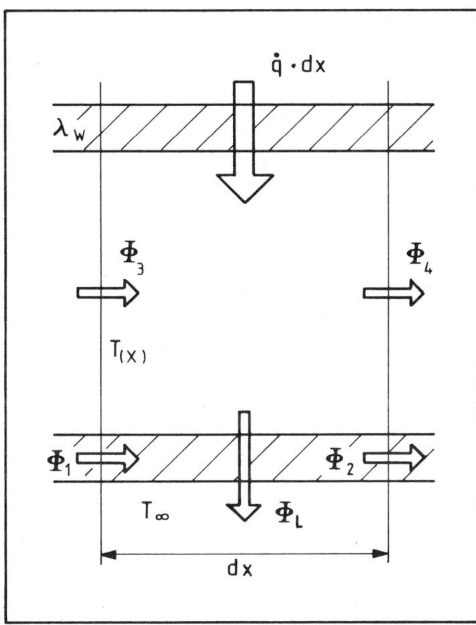

Figure 9-3.
Calorimetric mass-flow meter: balance for the heat fluxes into and out of a section of capillary (length dx) with external heating $\dot{q}\,\mathrm{d}x$, heat conduction in the wall Φ_1 and Φ_2, heat transportation by the fluid Φ_3 and Φ_4, and heat loss to the surroundings Φ_L (after [5]).

The essential term for the mass flow dependence is of course $\Delta\Phi_\mathrm{T}$, which represents the calorimetric principle:

$$\Delta\Phi_\mathrm{T} = \Phi_4 - \Phi_3 = c_p\dot{M}(T(x+\mathrm{d}x) - T(x)) = c_p\dot{M}\frac{\mathrm{d}T}{\mathrm{d}x}\mathrm{d}x\,. \tag{9-7}$$

Heat conduction and heat loss are represented by

$$\Delta\Phi_\mathrm{C} = \Phi_2 - \Phi_1 = -\lambda_\mathrm{w}A_\mathrm{w}\left(\frac{\mathrm{d}T(x+\mathrm{d}x)}{\mathrm{d}x} - \frac{\mathrm{d}T(x)}{\mathrm{d}x}\right) = \lambda_\mathrm{w}A_\mathrm{w}\frac{\mathrm{d}^2T}{\mathrm{d}x^2}\mathrm{d}x \tag{9-8}$$

and

$$\Delta \Phi_L = \mu (T(x) - T_\infty) \, dx \tag{9-9}$$

where A_w and λ_w are the cross-sectional area of the capillary wall and the thermal conductivity of its material, respectively, μ the heat loss coefficient per unit length, and T_∞ the temperature of the surroundings.

By inserting Equations (9-7), (9-8), and (9-9) into Equation (9-6), a differential equation for $T(x)$ is obtained:

$$\frac{d^2 T(x)}{dx^2} - \frac{c_p \dot{M}}{\lambda_w A_w} \frac{dT(x)}{dx} - \frac{\mu}{\lambda_w A_w} (T(x) - T_\infty) + \frac{\dot{q}}{\lambda_w A_w} = 0 \tag{9-10}$$

which can be solved for the special set-up in Figure 9-1. With the approximation of small mass flow, which implies that heat conduction in the wall and heat loss to the surroundings are the dominant effects, the temperature difference T_{21} is proportional to the mass flow:

$$T_{21} = \text{const.} \cdot c_p \, P_{el} \, \dot{M} \tag{9-11}$$

where P_{el} is the electrical power in the heater element and the constant includes the thermal and geometrical properties of the capillary set-up.

Figure 9-4 shows experimental results [7] obtained with nitrogen. The linear dependence is valid only for small amounts of fluid; at higher values it is even ambiguous.

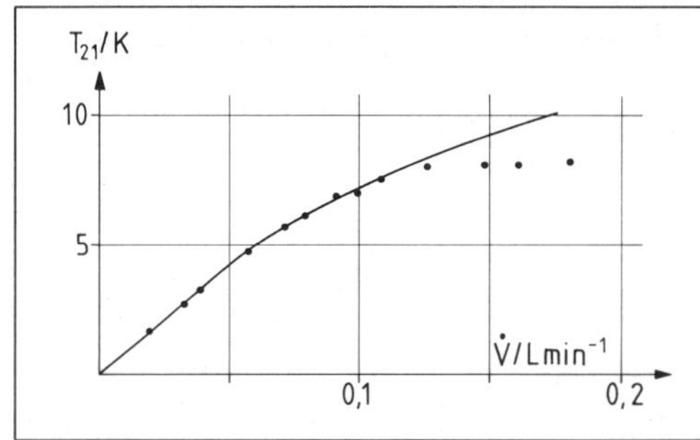

Figure 9-4.
Calorimetric mass-flow meter: temperature difference T_{21} vs. volume flow rate of nitrogen. The points mark measurements, the solid line a theoretical calculation (after [7]).

9.2.3 Thermal Mass-Flow Meters of the Boundary-Layer Type

This kind of instrument is often called a hot-wire or hot-film anemometer. They represent a very old (the first paper was published in 1914 [8]) but still very active field of research. Freymuth published a complete bibliography on the subject up to 1983 [9].

The most frequently used design is the so-called "constant-temperature anemometer", the set-up of which is shown schematically in Figure 9-5. Two temperature-dependent resistors R_S and R_T, in thermal contact with the flowing medium, are part of an electrical bridge circuit. The ratio of their resistance values is small, eg, of the order of 1 : 100, so that the major

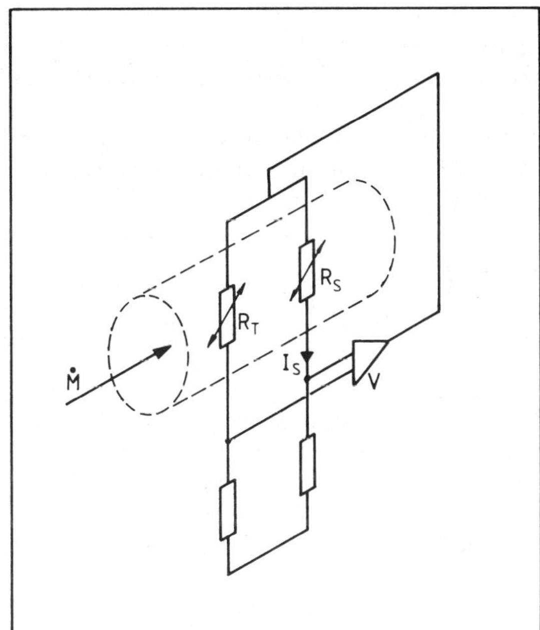

Figure 9-5.
Schematic diagram of a constant-temperature anemometer with temperature-dependent resistors R_S and R_T in contact with the fluid ($\dot{M}$) (from [19]).

part of the bridge current flows through the low-ohmic resistor R_S which is heated. R_T has the temperature of the fluid. The current flowing through the bridge is defined by an electronic control circuit (symbolized by amplifier V), so that the voltage at the bridge diagonal U_D is balanced:

$$U_D \sim \frac{R_1}{R_1 + R_S} - \frac{R_2}{R_2 + R_T} = 0\,. \tag{9-12}$$

R_1 and R_2 are resistors connected in series to R_S and R_T, respectively. The control circuit therefore establishes a constant relationship between the heated resistor R_S and the temperature sensor R_T, ie, the temperature of the fluid. The electrical power generated in resistor R_S is balanced by the thermal power which is transferred to the fluid from the heated resistor, according to Equation (9-4):

$$I_S^2\,R_S = K\,\lambda\,Pr^m\,Re^n\,A_S\,\Delta T_S \tag{9-13}$$

where I_S is the heating current through resistor R_S, K is a constant depending on material and geometrical parameters, and A_S and ΔT_S are the surface area of the heated resistor R_S and the difference between its temperature and the temperature of the fluid T_F, respectively.

Equation (9-13) can be transformed into the well known King equation [8]:

$$I_S \sim \sqrt{A + B\,Re^n} \tag{9-14}$$

where A represents flow-independent heat loss processes as, for example, free convection or heat flux to the supports and B summarizes the factors of Equation (9-13).

Figure 9-6 shows the characteristic curve of a constant-temperature anemometer with a heated resistor in the form of a thin plate oriented in parallel to the direction of flow. The solid line respresents a numerical approximation of the King equation (Equation (9-14)). The agreement with the experimental results (crosses) is better than 2%.

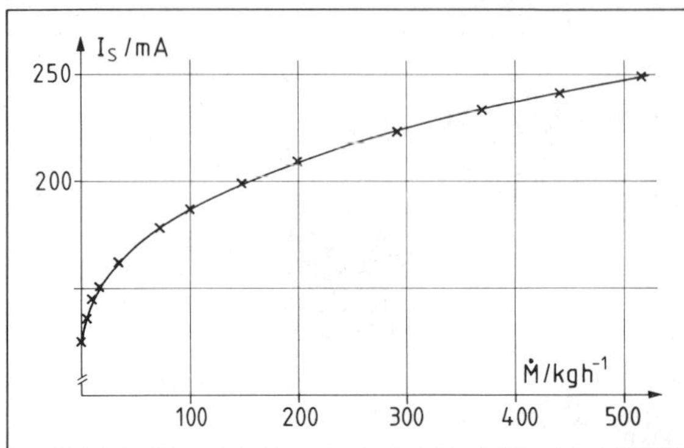

Figure 9-6. Characteristic curve of a constant-temperature anemometer (heated resistor in form of a thin plate oriented in parallel to the direction of flow). Crosses mark measurements, the solid line a numerical approximation with King's law (Equation (9-14)).

The non-linearity of the characteristic curve gives rise to another important feature of the boundary-layer thermal mass-flow meter: it exhibits a large turndown ratio of typically 40 : 1 with almost constant accuracy of reading because of the increasing resolution for decreasing mass flow.

A few other design concepts are known from the literature which will be mentioned here only briefly:

- *Constant-current anemometer* [10]. It works without active feedback and is therefore particularly simple. Because of the flow-dependent temperature of the heated sensor it is mainly used for the measurement of fluctuations in flow velocity.
- *Thermal mass-flow meter with two heated sensors* [11]. Both are heated to a constant temperature above ambient. If P_1 is the electrical power supplied to the first, which is in contact with the moving fluid, and P_2 is the power in the second, which is situated in a region without flow, then by calculating $(P_1 - P_2)/P_2$ one obtains a mass-flow-dependent signal.
- *Indirect heating of the sensor element* [12].This avoids high electrical current in the delicate sensor element but relies on good and stable thermal contact with the heating element.
- *Dynamic thermal mass-flow meters* [13]. In this concept only one sensor element is required which is periodically heated electrically and cooled by the fluid. The temporal temperature behavior is used as a measuring signal with obvious disadvantages concerning the response time.

9.3 Exemplary Description of Specific Instruments

9.3.1 Sensors, Sensor Elements

A wide variety of temperature sensors can be found in thermal flow meters. Most frequently Pt and Ni based resistive elements are used. However, there are also many examples where thermocouples, semiconductor devices or PTC (positive temperature coefficient) thermistors are applied.

In calorimetric mass-flow meters the sensor elements are often mounted outside the capillary. Frequently a wire of temperature-dependent resistance material is wound around it to detect the temperature of a certain section.

In the classical thermal mass-flow meter of the boundary-layer type, ie, the hot-wire or hot-film anemometer, thin Pt wires or thin films of Pt or Ni on small cylinders of glass or ceramic are used. The characteristic of these sensors is very much affected by the deposition of dirt particles on the surface. Their main field of application is therefore in the laboratory. A complete overview of different designs can be found, for example, in Reference 10.

There is, however, one important field, ie, the measurement of the intake-air of internal combustion engines, where the hot-wire sensor is found also outside the laboratory [14, 15]. Figure 9-7 shows the inside of a tube wherein a 70-μm Pt wire is mounted. Two holders serve

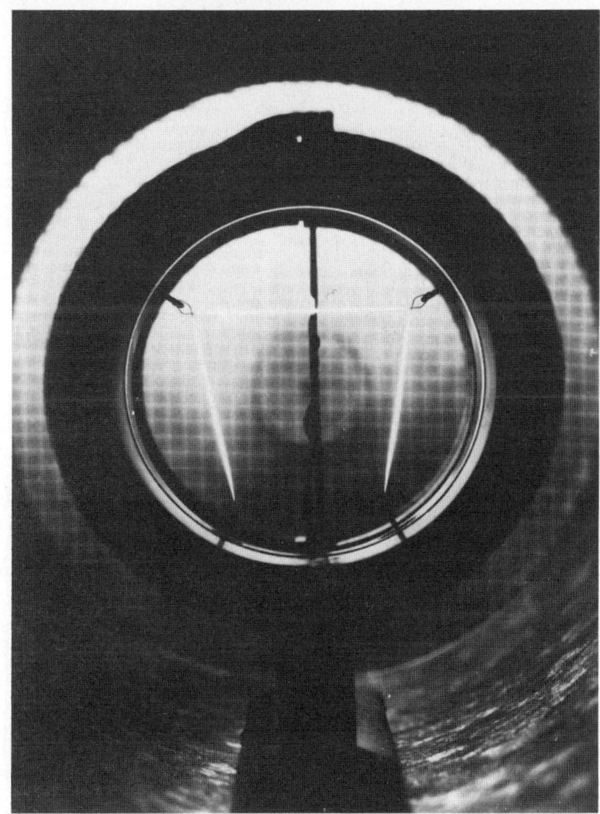

Figure 9-7.
Hot-wire anemometer: 70-μm heated Pt wire mounted inside a tube (Bosch Pressebild).

as electrical contacts; to the others the wire is fixed by loops so that no current is flowing through them. The hot wire exhibits a very short response time to changes in flow rate. The problem of dirt deposition is avoided by a special burn-off cycle, in which the wire is heated at a very high temperature for a short period.

Thin-film sensor elements offer the possibility of economic manufacture and are less delicate than wires. In most cases a thin film of Pt, Ni, etc., is applied on a ceramic substrate. The film is structured in a meandering-like resistor pattern. Usually the film is covered with an additional protective layer. The deposition of dirt particles is considerably reduced when a substrate in the form of a thin plate is used where the large surfaces are oriented in parallel to the direction of flow [16]. Figure 9-8 shows an assembly with two Pt thin-film resistors, one of them heated, mounted by point suspension on metallic clips to minimize thermal contact between the sensor elements and holders. The dimensions of the resistors in this sample are $10 \times 2 \times 0.3$ mm^3.

Figure 9-8.
Sensor assembly with two Pt thin-film resistors mounted by point suspension. The dimensions of the resistors are $10 \times 2 \times 0.3$ mm^3.

Figure 9-9 shows the heated resistor of Figure 9-8 in detail. The printed lines of the resistor pattern exhibit different widths in order to adjust spatially the distributions of the electric power and of the power transferred to the fluid; the printed line along the leading edge is the narrowest to compensate for the convective losses which are maximum there. This measure guarantees a short response time because it avoids rearrangement of thermal equilibrium on the substrate as a result of changing flow rate [16, 17].

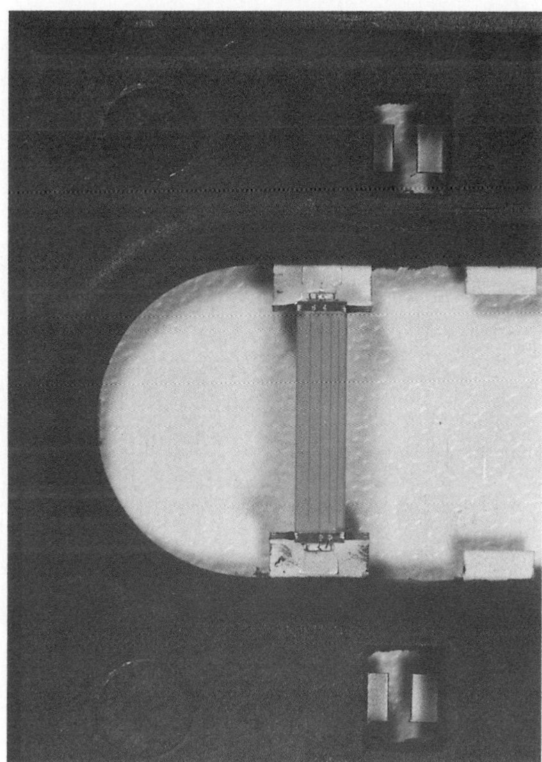

Figure 9-9.
Heated resistor of Figure 9-8 in detail.

An even more robust sensor assembly and its matching probe construction are shown in Figure 9-10. Pt thin-film resistors are fixed to a ceramic holder so that the sensor can be applied in chemically aggressive atmospheres [18, 19]. The dimensions of the resistors in this special case are $5 \times 4 \times 0.6\ mm^3$.

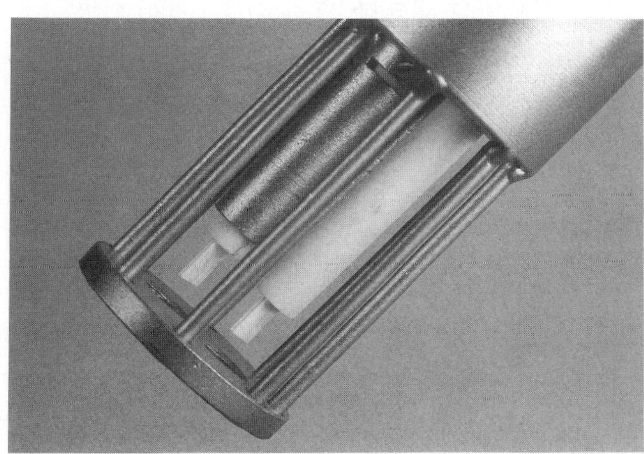

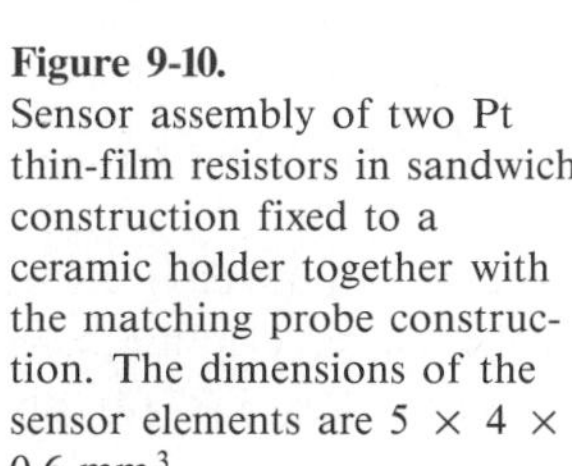

Figure 9-10.
Sensor assembly of two Pt thin-film resistors in sandwich construction fixed to a ceramic holder together with the matching probe construction. The dimensions of the sensor elements are $5 \times 4 \times 0.6\ mm^3$.

Integrated-circuit technology can be applied to the manufacture of monolithic mass-flow meters based on micromachined silicon for very low flow rates [20–22]. By nonisotropic etching processes flow channels of microscopic dimensions can be produced with high accuracy. Thin- and thick-film techniques are applied for the sensors and also for the electronics which can be integrated in the sensor. Figure 9-11 shows micrographs (different magnifications) of miniature sensor elements suspended like bridges over an etched flow channel.

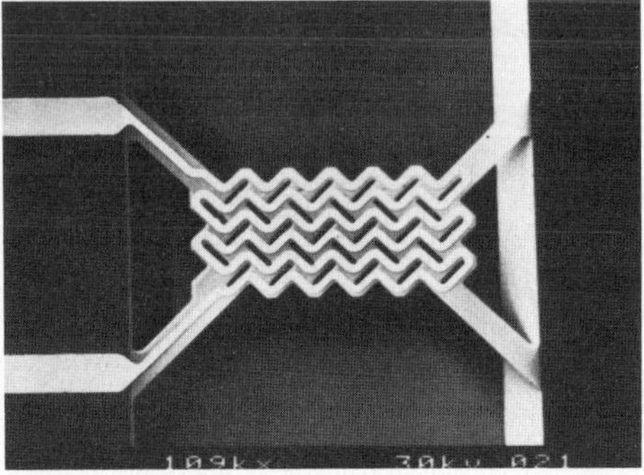

Figure 9-11. Micrographs (different magnifications) of miniature sensor elements suspended like bridges over an etched flow channel (reproduced with permission from Innovus, San Jose, CA, USA).

9.3.2 Mechanical Design

The calorimetric principle of measurement leads to a linear relationship between mass flow and resulting temperature difference only for very low flow rates (see Section 9.2.2.). Therefore, a by-pass design as in Figure 9-12 is often applied to cover a wider range of flow rates. Only a small but constant fraction of the flow (typically 1 : 100) is directed through the capillary carrying the sensor. This is achieved by a special restrictor which forces the flow in

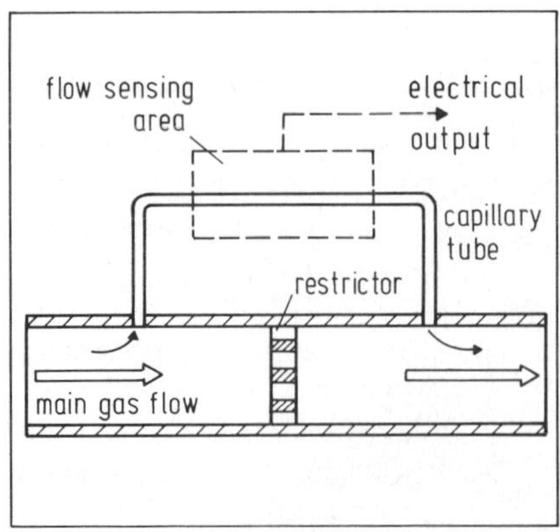

Figure 9-12.
By-pass design of a calorimetric flow meter (after [19]).

the main tube to remain in the laminar regime. The restrictor can be exchanged in some instruments to extend the flow range. Frequently these instruments are used in combination with regulating valves.

Figure 9-13 shows a cut-away model of an instrument of the boundary-layer type [19] which is designed for mass-flow measurements of air (or similar gases) at moderate temperatures and pressures. Two sensor elements, Pt thin-film resistors as in Figure 9-8, are mounted in the center of the tube, the plane of the films being parallel to the direction of flow. A combination of different honeycombs and meshes straightens the flow profile and avoids a dependence of the measurement on inlet conditions.

Figure 9-13. Cut-away model of a thin-film thermal mass-flow meter of the boundary-layer type.

The mass-flow meter shown in Figure 9-14 is designed for use in industrial process applications [18, 19]. It can withstand high pressures and temperatures and chemically aggressive gases. The instrument is constructed in a similar manner to an insertion-type probe. At its end it houses the sensor (see Figure 9-10) having direct thermal contact with the medium. The probe can be flanged to a piece of pipe of standardized length as in Figure 9-14, which is then installed between flanges in a line. For ducts with large diameters ($\gtrsim$200 mm) direct installation of the probe, eg, by welding, is preferable.

A device with the sensor elements protected by a cover of stainless steel has been described [23]. For measurements in large ducts an array of sensors can be assembled to eliminate uncertainties due to variable flow profiles.

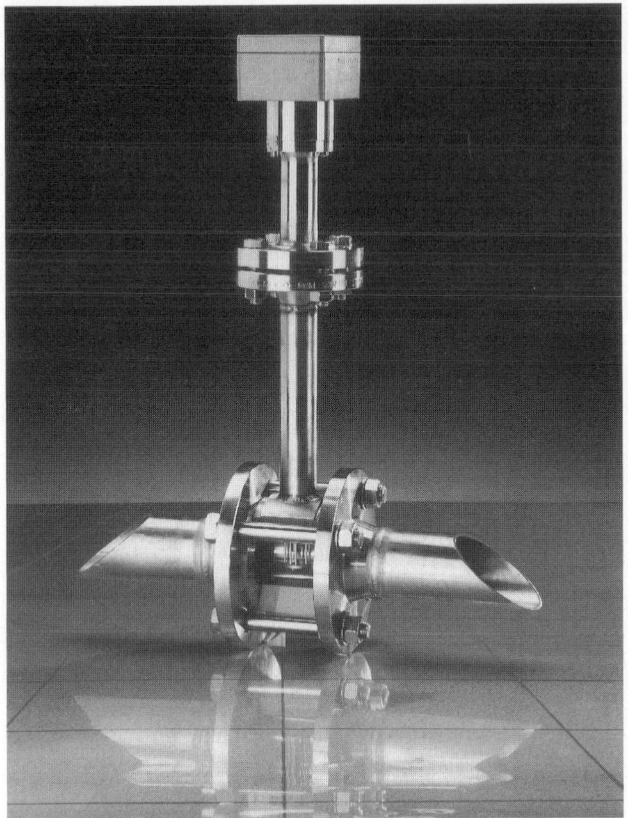

Figure 9-14. Insertion-type construction of a thin-film thermal mass-flow meter (cut-away model).

9.3.3 Electronics

In calorimetric flow meters the temperature difference between the inlet and outlet is measured by arranging the upstream and downstream temperature sensors in a classical Wheatstone bridge circuit. The power in the heating element is kept constant. Since the characteristic is inherently linear only a simple adaption to standardized output signals has to be performed.

Thermal mass-flow meters of the boundary-layer type normally also incorporate a Wheatstone bridge circuit but with an active feedback to define the temperature difference between the heated sensor and the temperature of the fluid, as indicated in Figure 9-5. The sensitivity of the circuit is optimum for the so-called equal-arm bridge. A fluctuation in mass flow rate results in a change in the resistivity of R_S and thus leads to an inbalance of the bridge diagonal voltage U_D:

$$dU_D = \frac{\partial U_D}{\partial R_S} dR_S \sim \frac{R_1 R_S}{(R_1 + R_S)^2} \frac{dR_S}{R_S} . \qquad (9\text{-}15)$$

$|dU_D|$ is a maximum for $R_1 = R_S$, a relationship which is also preferable in view of the signal-to-noise ratio and response time [24, 25].

Precise temperature compensation of the bridge can be guaranteed by a resistor network connected in series with the temperature sensor R_T to adjust its temperature coefficient relative to that of the heated resistor R_S [26]. In this way also a small temperature dependence of the fluid properties can be compensated for. As an example, Figure 9-15 shows the resulting small deviations after compensation with a single resistor for variable air temperature (0–60 °C) and for a flow range of 1 : 65.

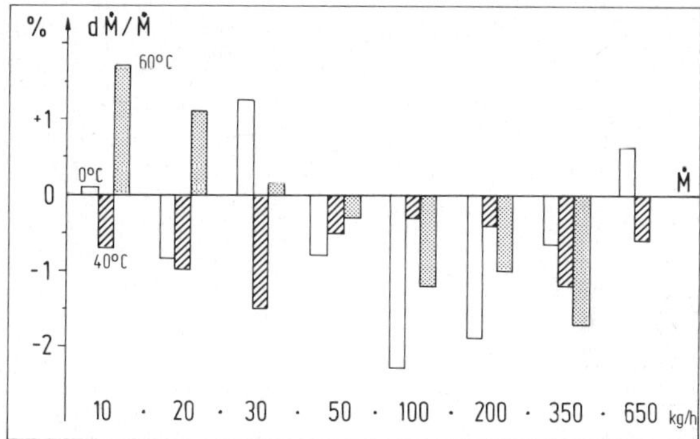

Figure 9-15.
Temperature compensation of a thin-film thermal mass-flow meter of the boundary-layer type from 0 to 60 °C: deviations relative to a measurement at 20 °C vs. mass-flow rate (after [19]).

As a measuring signal any voltage inside the active bridge circuit can be used or equivalently the current through the bridge. This ensures independence of the cable length and reduces the pick-up of noise.

Figure 9-16 illustrates how interchangeable instruments can be obtained despite unavoidable variations in the sensor characteristic of individual probes [18, 19]. In this digital signal-forming circuit each value of the bridge current I_B is associated with a digital correction value by A/D (analog to digital) conversion and direct memory access. After D/A (digital to analog) conversion this is transformed into a correction current I_C which is added to I_B, thereby yielding for all probes a signal equivalent to that of a reference. The correction current I_C represents only a small fraction of the output signal I_{out}. Variations of the electronic components are therefore negligible. Moreover, the correction can be performed with high resolution not only concerning its absolute value (I_C) but also the range of the sensor characteristic ($\dot{M}$) where it takes effect. With 8-bit converters the measured deviations are of the same order of magnitude as the reproducibility ($\lesssim 1\%$ of the reading).

The non-linearity of the sensor characteristic requires linearization. This can be performed by analog circuitry, which is rather complicated and inaccurate. With modern microprocessors

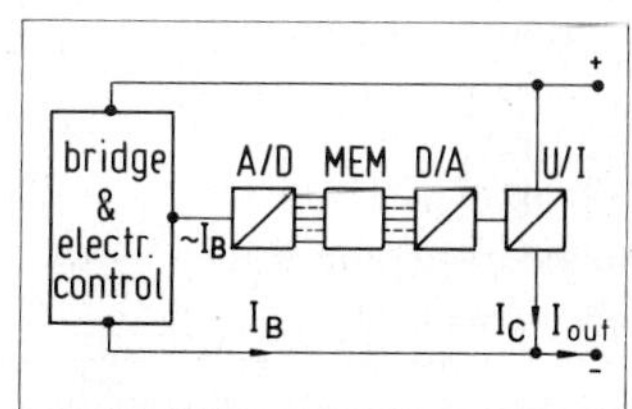

Figure 9-16.
Schematic diagram of a digital signal-forming circuit which provides for interchangable thin-film thermal mass-flow meters. A/D, analog-digital converter; MEM, memory; D/A, digital-analog converter; U/I, voltage-current converter (after [19]).

it can be executed much more precisely and with adequate speed. The sensor characteristic is thereby tabulated with high resolution in the memory. Moreover, the microprocessor can carry out additional jobs such as time integration, control of cirtical values, and digital display.

9.4 Technical Data, Examples of Application

Owing to the large number of different thermal mass-flow meters on the market, it is almost impossible to give a comprehensive overview of technical data. In the following sections a few typical examples of applications are covered and a list of technical data for commercial instruments designed for specific fields is given.

9.4.1 Microelectronics Processing

In semiconductor processing, the composition and concentration of process gases have to be precisely regulated [27]. This is accomplished by controlling the flow of each component or, more often, by maintaining accurate ratios among the gases involved. Generally the absolute amount of gas flow is small and has to be kept almost constant during the process. Therefore, calorimetric flow meters are very suitable. Normally the gases are clean, which is important in view of the flow restrictors used in this kind of instrument.

These flow meters are used for a variety of gases the composition of which changes frequently. It is therefore of advantage that the sensor characteristic can be converted fairly simply. The conversion factor basically includes the ratio of the specific heats of the gas under consideration and the gas used in the calibration (Equation 9-11). Some manufacturers also specify empirical conversion factors. However, maximum accuracy can only be achieved by calibration under the actual conditions of use.

In [27] a detailed list of technical data for a number of instruments is given. A summary of the most important features is presented in Table 9-1.

Table 9-1. Technical data for calorimetric flow meters for microelectronics processing. Summary of typical features in Ref. 27.

	Minimum	Typical	Maximum	Units
Flow range	0–0.01	0–100	0–20000	L_n^* min^{-1}
Accuracy		1–1.5		% of the full-scale reading
Maximum inlet pressure	0.2	1.4	17	MPa
Pressure coefficient		0.1		%/100 kPa of the full-scale reading
Operating temperature	−10	0–50	100	°C
Temperature coefficient:				
At zero	0	0.1	0.1	%/K of the full-scale reading
At full scale	0.001	0.1	1	%/K of the full-scale reading
Response time to within 2% of set-point	0.06	1–2	<60	s

* Subscript n: normal conditions, ie, 0 °C, 101 kPa.

9.4.2 Engine Test Rigs

Increasing emission restrictions and the reduction of fuel consumption for automobiles imply intensive testing of the engines. The measurement of the intake air is therefore of major importance [17]. The requirements for the flow meters are direct mass-flow measurement, simple handling, robust construction and reliable accuracy from idling to full-power operation for static measurements. Equally important are dynamic measurements where a rapid response is needed. Hot-film anemometers comply almost ideally with these requirements.

As an example, Figure 9-17 shows the response of a thermal mass-flow meter, virtually without delay, to the opening of a combustion-engine throttle valve. It is compared with the slow rise of the signal of a conventional vane. The time resolution of the measurement is about 6 ms.

Table 9-2 summarizes the most important features of this class of instruments.

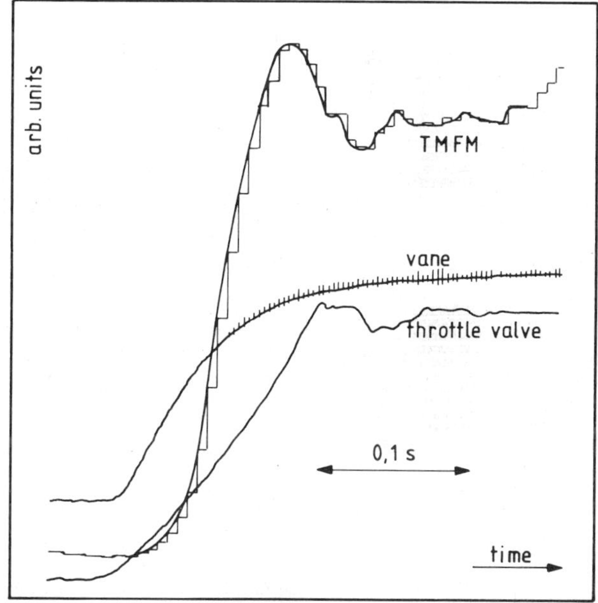

Figure 9-17. Dynamic behavior of a thermal mass-flow meter (TMFM) of the boundary-layer type on an engine test rig after opening of a throttle valve compared with a conventional vane (after [19]).

9.4.3 Application in Injection Control Systems

Electronic fuel-injection systems in automobiles allow the accurate control of the air-to-fuel ratio of internal combustion engines and thus optimize economy and performance and reduce emissions. One of the important inputs of the control system is the amount of intake air in mass units.

Table 9-2. Technical data for thermal mass-flow meters of the boundary-layer type for applications in engine test rigs.

	Minimum	Typical	Maximum	Units
Standard pipe diameters*	DN 25		DN 200	
Flow range	2–60		200–6000	kg/h
Accuracy		2		% of the reading
Reproducibility		1		% of the reading
Operating pressure	0		600	kPa
Pressure coefficient		0.2		%/100 kPa of the reading
Operating temperature	−25		100	°C
Temperature coefficient		0.03		%/K of the reading
Response time to within 63% of set-point		12		ms
Pressure drop at maximum flow		2	3	kPa

* DIN 2448.

Thermal mass-flow meters of the hot-wire or hot-film type have proved to have the most adequate overall capability [12, 14–16, 28]. They are now mass produced and have been installed in automobiles since 1980. Requirements are again a high accuracy over a large turn-down ratio, fast response, and robust construction. In addition, adaption to engine environment, reliability, and a low price are demanded. – Especially the last requirement favours the hot-film type mass-flow meter.

Sasayama et al. [28] gave a few technical data explicitly for a mass-flow meter where a wire-wound sensor is installed in a by-pass of the main passage. They are:

- dynamic range 40 : 1
- overall accuracy 4% of reading
- response time to within 63% of set-point 10 ms
- temperature range −40 to + 125 °C
- temperature coefficient ca. 0.1%/K of reading
- pressure drop <2 kPa

9.4.4 Process Control

In process control the orifice assembly is still the most frequently used flow meter. To measure mass flow a differential pressure (Δp), a pressure and a temperature sensor are needed in addition to an analysing unit which combines the measurements of the three sensors. Direct mass-flow measurement therefore reduces the expenditure of installation. Moreover, all the piping (clogging!) necessary for the Δp-method can be avoided.

Thermal mass-flow meters for applications in process control first of all have to withstand harsh environmental conditions as there are high temperatures and/or high pressures, high loads of dirt particles in the fluid, and explosion hazards, as in the following applications:

- The measurement of steam, which is characterized by extreme temperature and pressure conditions. It represents an important part of effective energy management in a process plant [29].
- The measurement of flue gas in power plants, which is required in order to optimize the performance of sulfur dioxide scrubbers [30]. The large amount of fly ash causes severe problems with differential pressure devices which can be solved by thermal mass-flow meters.
- In the chemical industry instruments often have to be installed in hazardous areas. Instruments must meet special specifications checked by competent authorities. Thermal mass-flow meters have been approved to meet the requirements of intrinsic safety. This excludes the ignition of explosions even under the assumption of various errors [31].

Calibration is normally possible only at moderate temperatures and pressures and with harmless gases, ie, none-explosive and none-toxic. This especially holds for high flow rates. High accuracy (0.5% or better) for up to several thousand kilograms per hour can be achieved with calibration rigs that include critical flow nozzles as standard elements [32]. In all other cases a theoretical conversion of the calibration with air to the specific gas has to be performed. King's law (Equation (9-14)) can be used as a basis where A and B include the thermal properties of the gas. With an additional empirical scaling function in λ, the thermal conductivity of the gas, excellent agreement between numerical conversion and test measurements can be achieved. Figure 9-18 shows the deviations between measurements and theoretical conversion calculations, which use a measurement with nitrogen as a starting point. The measurements were performed with different gases, the thermal conductivities of which cover a broad range, nz., Ar, Co_2, He and a mixture of N_2 (60%) and He (40%). The absolute accuracy of the measurement with the different gases was of the order of 2% of the reading [33].

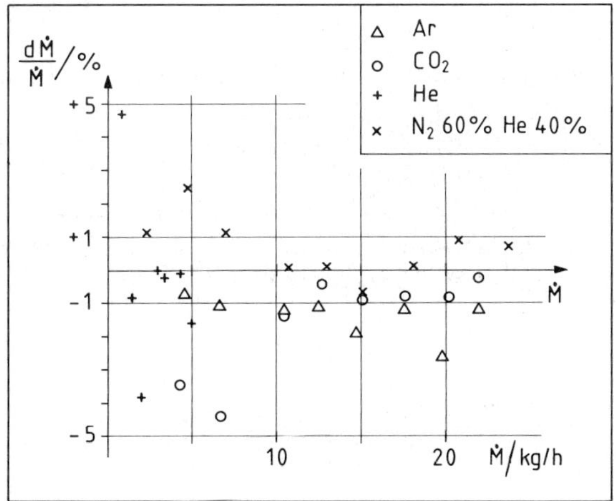

Figure 9-18.
Comparison between measurement (Ar, CO_2, He, N_2 – He (60:40) and theoretical conversion calculation on the basis of a calibration with N_2; relative deviations in mass flow vs. mass flow.

Table 9-3 gives an exemplary summary of important technical data for instruments of the hot-film type for this field of applications. The specification of accuracy relates to calibration with air and a straight pipe section upstream and downstream (15 × pipe diameter and

Table 9-3. Technical data for thermal mass-flow meters of the boundary-layer type for applications in process control.

	Minimum	Typical	Maximum	Units
Standard pipe diameters* **	DN 25		DN 300	
Flow range** ***	5–150		800–25000	kg/h
Accuracy	1			% of the full scale
			2	% of the reading
Reproducibility		1		% of the reading
Pressure range	0–4		0–10.4	MPa
Pressure coefficient		0.2		%/100 kPa of the reading
Temperature range	−25 to +150		−55 to +500	°C
Temperature coefficient		0.1		%/K of the reading
Response time to within 63% of set-point		200		ms

* DIN 2448. ** For probe/pipe combinations. *** For air at atmospheric conditions.

5 × pipe diameter in length, respectively). This ensures independence of flow conditions further upstream and downstream [18]. In process plants these straight pipe sections often cannot be guaranteed. In most cases this does not reduce the reproducibility. For optimum absolute accuracy a special environment can be employed in the calibration process.

9.5 Summary

A variety of different types of thermal sensors are employed in thermal mass-flow meters. Depending on the application, the sensors are adapted to specific measurement problems ranging from delicate sensor assemblies for optimum dynamic behavior to robust set-ups for use in process control. Additional to the capability of measuring mass flow directly, high accuracy, fast response, and simple handling are the important features that promote the application of these instruments in an increasing number of fields. The use of modern microelectronics provides for compatibility with process-control systems.

9.6 References

[1] Mettlen, D. in: *Discharge and Velocity Measurement,* Müller, A. (ed.); Rotterdam: Balkema, 1988, pp. 61–63.

[2] Gast, Th., Hönl, R., Köhn, H., *Technisches Messen* tm **55** (1988) 261–266.

[3] Wilde, K., *Wärme- und Stoffübergang in Strömungen;* Darmstadt: Steinkopff Verlag, 1978.

[4] Huijsing, J. H., Dorp, van A. L. C., Loos, P. J. G., *J. Phys. E.: Sci. Instrum.* **21** (1988) 994–997, and references cited therein.

[5] Jouwsma, W., *Sensor 88, Nürnberg, FRG, May 1988, Congress Proceedings Vol. B;* Wunstorf 2, FRG: ACS Organisations, 1988, pp. 219-232.
[6] Jouwsma, W., personal communication.
[7] Komiya, K., Higuchi, F., Ohtani, K., *Rev. Sci. Instrum.* **59** (1988) 477-479.
[8] King, L. V., *Philos. Trans. R. Soc. London, Ser. A* **214** (1914) 373-433.
[9] Freymuth, P., *A Bibliography of Thermal Anemometry*, St. Paul, MN: TSI Inc. 1982 and Addendum 1983.
[10] Strickert, H., *Hitzdraht- und Hitzfilmanemometrie;* Berlin: VEB Verlag Technik, 1974.
[11] Uhl, G., *Neue Technik* No. 8 (1985) 12-13.
[12] Sauer, R., *Technical Paper Series 880560,* 1988, Society of Automotive Engineers, Warrendale, PA, USA.
[13] Kuntz, W., Bährle, K., Sondergeld, M., *Technisches Messen tm* **55** (1988) 44-47.
[14] Sauer, R., *Technical Paper Series 800468,* 1980, Society of Automotive Engineers, Warrendale, PA, USA.
[15] Sumal, J. S., Sauer, R., *Technical Paper Series 840137,* 1984, Society of Automotive Engineers, Warrendale, PA, USA.
[16] Schäfer, W., *Regelungstech. Prax. rtp* **25** (1983) 468-471.
[17] Hohenstatt, M., Horlebein, E., *Messen, Prüfen, Automatisieren* **20** (1985) 668-671.
[18] Hohenstatt, M., *10th World Congress on Automatic Control, München, FRG, July 1987, Preprints Vol. 3;* Oxford, UK: IFAC Publication, c/o Pergamon Books 1987, pp. 313-317.
[19] Hohenstatt, M. in: *Discharge and Velocity Measurement,* Müller, A. (ed.); Rotterdam: Balkema, 1988, pp. 65-73.
[20] Knutti, J. W., *Sensors 88, Nürnberg, FRG, May 1988, Congress Proceedings Vol. A;* Wunstorf 2, FRG: ACS Organisations, 1988, Chapter A 1.4.
[21] Kirshman, S., *Microelectron. Manuf. Test.* **11** (1988) 11-12.
[22] Lymann, J., *Electronics* **60** October (1987) 85-86.
[23] Ramey, T., *Mech. Eng.* May (1985) 29-33.
[24] Freymuth, P., *J. Phys. E: Sci. Instrum.* **11** (1978) 915-918.
[25] Freymuth, P., *J. Phys. E: Sci. Instrum.* **14** (1981) 1373.
[26] Drubka, R. E., Tan-atichat, J., Nagib. H. M., *DISA Information No. 22,* December 1977, DISA Elektronik A/S, Skovlunde, Denmark.
[27] *Microelectron. Manuf. Test.* **10** No. 8 (1987) 19-22.
[28] Sasayama, T., Nishimura, Y., Sakamoto, S., Hirayama, T., *Sens. Actuators* **4** (1983) 121-128.
[29] *Control Instrum.* June (1987) 28-29.
[30] *Power Mag.* **131** No. 6 (1987) .
[31] Hohenstatt, M., *VDI Berichte,* No. 768 (1989), 169-179.
[32] Horlebein, E., *Sensors 88, Nürnberg, FRG, May 1988, Congress Proceedings Vol. B;* Wunstorf 2, FRG: ACS Organisations 1988, pp. 233-243.
[33] Göbel, R., Hohenstatt, M., Weiler, M., unpublished.

10 Cryogenics

Franco Pavese, CNR, Istituto di Metrologia "G. Colonnetti",
Turin, Italy

Contents

10.1 The Requirements for Cryogenic Thermal Sensors

The cryogenic region of temperatures extends from as near absolute zero as is technically feasible up to an indefinite boundary that is shared with the concept of "refrigeration". From an industrial point of view, this region is traditionally divided in two ranges: the upper range ($T > -200\,°C$) is the realm of "liquid nitrogen technology" and the lower range ($T < 50$ K and in most cases > 1.5 K) the realm of "liquid helium technology". This subdivision is a real separation and applies also to most laboratory work, which has mainly been developed in the lower range. This caused, for example, many laboratories to find themselves insufficiently equipped when suddenly the new high-T_c superconductors required research to be moved into the upper range. From the point of view of the main applications of cryogenics, and consequently of the market share of thermal sensors, the upper range is where most industrial plants work for production of technical gases and for liquefaction of LNG; the lower range is where, so far, the applications of superconductivity are involved.

The subdivision into the two ranges was also supported by different refrigeration requirements. Whereas large industrial applications have necessarily always relied on refrigerators, most medium and small applications, until recently, have found a simpler or more economical solution in using liquefied gases as refrigerating media, liquid nitrogen and liquid helium (or rarely hydrogen) being used in the two ranges, respectively. This makes a real difference in the hardware required. However, the development of multi-stage closed-cycle refrigerators is extending the upper range to lower and lower temperatures (down to 12 K with two stages or 4 K with three stages) for these applications also, confining the need for liquid helium as a refrigerant to a specialized and much more limited range of applications.

However, although from a technical point of view the boundary between the upper and lower temperature ranges will certainly move in the future to lower temperatures, from a physical point of view the bulk change of many properties of condensed matter (and, consequently, of technical materials) takes place in the temperature range between about 30 and 200 K. This is relevant to the design and use of thermal sensors, especially those intended for use over the *whole* range from room down to liquid helium temperature. It must also be realized that these temperatures differ by almost two orders of magnitude from ambient temperature (a few kelvin compared with 300 K); no similar requirement for such a wide working range is required by any of the practical thermometers designed for use above room temperature. In that broad temperature range, the physical laws responsible for the temperature dependence of thermal sensors cannot by assumed to apply over the whole range, especially in the case considered here of electrical thermometers. As a consequence, the technical feasibility of a sensor and the achievement of good metrological performances may be adversely affected by such a broad working range. As a final consideration, thermal sensors are often used in the low-temperature range under special conditions which are seldom or never encountered above room temperature.

It is not easy to give a comprehensive review of all the types of thermometers used at low temperatures, because of their extreme variety (Table 10-1). This variety reflects the difficulty of finding a single type with a broad range of applications. Figure 10-1 shows the change in sensitivity with temperature for the different types.

The characteristics of the thermometers listed in Table 10-1 (with the exception of the nonelectrical types, being outside the scope of this book, although they are frequently used

Table 10-1. Types of cryogenic thermometers

Type	Usual temperature range (K)	Typical reproducibility (mK)
1. Thermocouples[a] (Figure 10-2a)		
1.1 Type T	77->300	10[b]
1.2 Au + 0.07 at % Fe/KP	<4-100	10[b]
1.3 Au + 2.10 at % Co/KP	<4-300	10[b]
2. Resistance		
2.1 Platinum (Figure 10-2b) (Chapter 3)		
Standard	14->300	<0.5
Industrial	20->300	50[c]
2.2 Rhodium-iron (Figure 10-2c)		
Standard	0.5->30	0.3
Industrial	<4-300	10
2.3 Platinum-cobalt (Figure 10-2d)		
Industrial	2-300	10
2.4 CLTS	4-300	100
2.5 Germanium (Figure 10-2e)	1-100	$\delta T/T = 0.0002$[d]
2.6 Carbon (Figure 10-2e)	0.5-100	200
2.7 Carbon-glass (Figure 10-2e)	0.5-100	$\delta T/T = 0.002$[d]
2.8 Thermistor	77->300	100[e]
3. Capacitance (Figure 10-2f)	2-300	100[f]
4. Diode		
4.1 Gallium arsenide (Figure 10-2g)	4->300	50
4.2 Silicon (Figure 10-2h)	4-300	50[g]
5. Gas		
5.1 Vapor pressure	0.5-100	1-100[h]
5.2 Gas	2->300	1-100
6. Mechanical (bimetal)	200->300	1000

[a] Other types available, suitable only for industrial use (see Chapter 4): type E and K down to 77 K; type S and R down to 220 K. KP = Chromel P alloy.

[b] Best precision for laboratory use in the differential mode and with offset for parasitic emfs (up to several microvolts) corrected. Typical precision in the absolute mode (and for reference junction at refrigerant temperature): 100 mK.

[c] Can improve to 10 mK for selected units above 50 K.

[d] The relative precision is preferred for this stable thermometer because of its R-log (T) characteristics. Each unit does not cover the whole temperature range; the same applies for the carbon type (2.6).

[e] Can improve to a few mK in narrow ranges above 200 K.

[f] Not usable between 40 and 70 K because of a slope inversion.

[g] Not easily usable between 20 and 50 K because of a sharp change in slope.

[h] Range covered in narrow sub-ranges by different gases. No such possibility between 5 and 10 K and between 30 and 55 K. Accuracy depends largely on construction and on that of the pressure measurement.

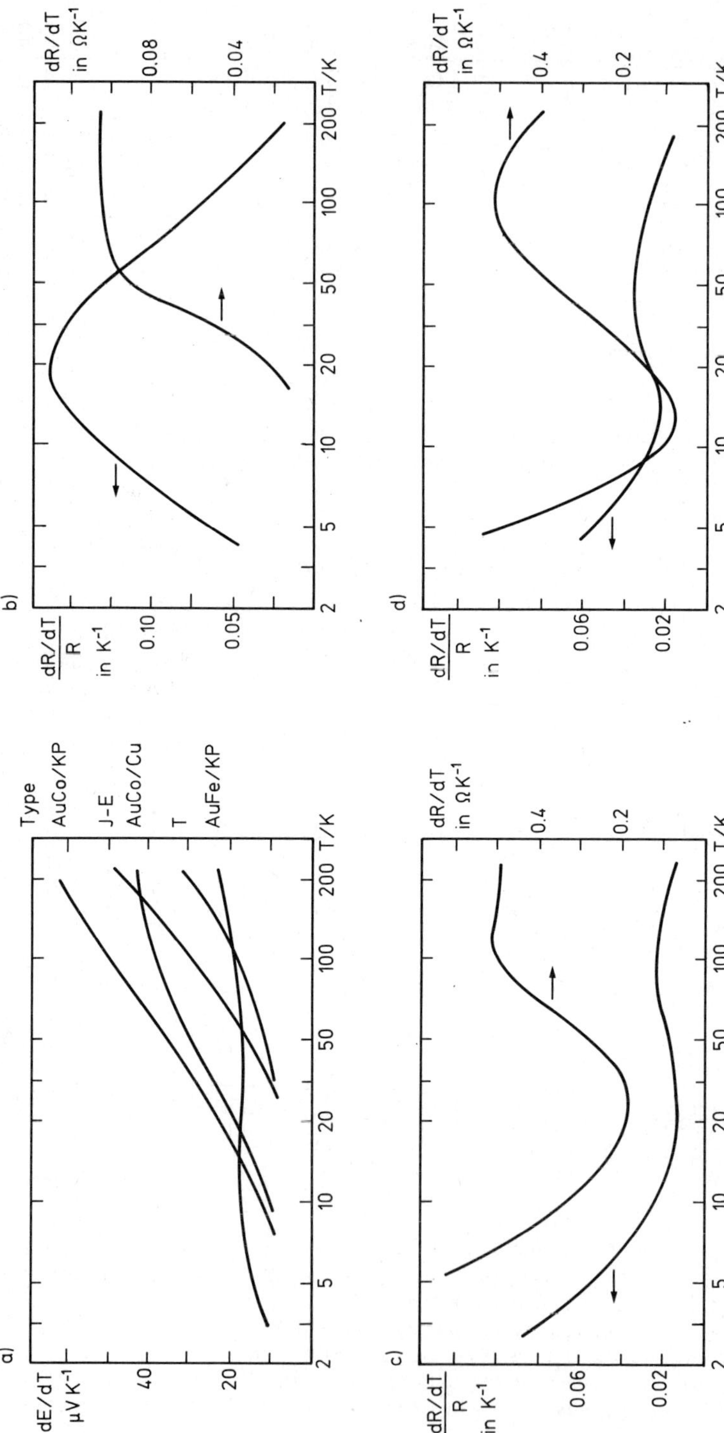
a)
dE/dT
μV K⁻¹
40
20
Type
AuCo/KP
J-E
AuCo/Cu
T
AuFe/KP
2
5
10
20
50
100
200
T/K
b)
dR/dT / R
in K⁻¹
0.10
0.05
dR/dT
in ΩK⁻¹
0.08
0.04
c)
0.06
0.02
0.4
0.2
d)
0.06
0.02
0.4
0.2

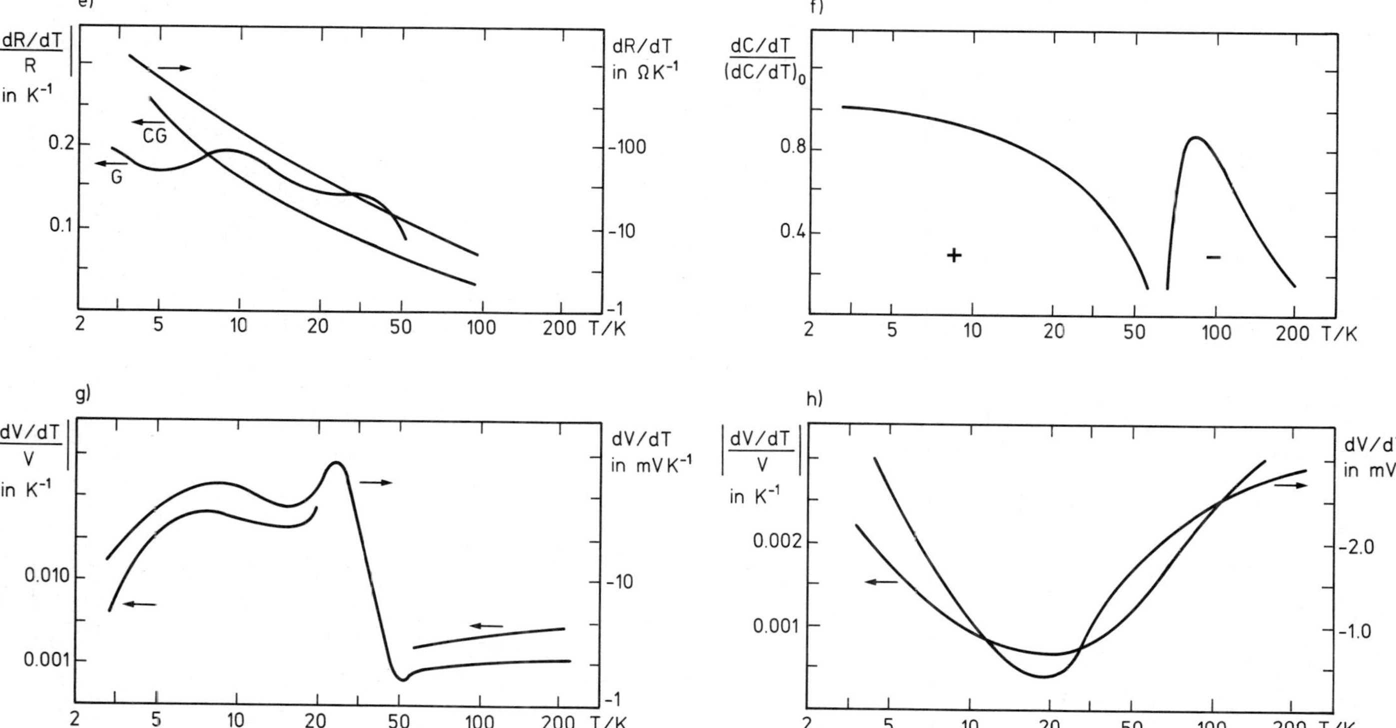

Figure 10-1. Typical characteristics of cryogenic thermometers. The behavior with temperature of the sensitivity and of the relative sensitivity is indicated. (a) Thermocouples: for the types refer to Table 10-1; KP = Chromel P alloy. (b) Platinum resistor. (c) Rhodium-iron resistor. (d) Platinum-cobalt resistor. (e) Germanium (G) and carbon glass (CG) resistors. Carbon resistors show a behavior similar to CG. (f) Capacitance thermometer; the sensitivity change relative to the value at 4.2 K is shown; note the change in sign. (g) Gallium arsenide diode. (h) Silicon diode.

in cryogenics in both laboratories and industry), will be reviewed in the following sections, with the aim of providing a guide to their selection for each particular application and for each accuracy level (Table 10-2). This will be done with account being taken of the parameters generally used for the characterization of sensors and with particular consideration of the typical problems in cryogenic applications.

A discussion of the calibration methods and of the accuracy that can be obtained with each type with respect to the International Temperature Scale can be found in [1].

Table 10-2. Accuracy required in different cryogenic applications

Better than 0.005 K	– Measurement of thermophysical properties
From 0.005 to 0.02 K	– Most laboratory measurements – Testing of industrial prototypes – Measurement of temperature differences – Monitoring on running apparatus (especially below 10 K) – High reliability requirements – Quality level for transducers permanently mounted in industrial equipments
From 0.02 to 0.1 K	– Most industrial monitoring requirements
Less than 0.1 K	– Cooldown and generic monitoring; alarms

10.2 The Choice of a Cryogenic Thermal Sensor

10.2.1 The Thermometric Parameter

The temperature-dependent electrical parameter that is measured is different for the different types, ie, resistance, capacitance, voltage, or emf. It is related to temperature by a functional relationship, which may itself affect the overall precision in the use of the sensor. In the worst case, a non monotonic behavior, as with some capacitance thermometers (Figure 10-1f), produces a gap in the useful temperature range, which must be split into two separate sub-ranges. A similar case is when a monotonic behavior shows a sudden change in slope, owing to a change in the underlying physical law, as with the silicon diode thermometers (Figure 10-1h); again, the characteristics must be split into two separate sub-ranges, and in the narrow region where the slope shows the sharp change the accuracy will be adversely affected. The change in the underlying physical law occurs very frequently for thermometers to be used in both the liquid nitrogen and the liquid helium temperature ranges. Therefore, the difficulty of fitting a single equation to the whole working range is common. It is well known that high-order polynomials are needed in order to represent accurately the characteristics of germanium thermometers (Figure 10-1e) [1]. With gallium arsenide diode thermometers (Figure 10-1g), the accuracy of fitting a single equation to the range 4–300 K is even lower than the reproducibility [2].

The usual temperature range of use of each thermometer, taking into account these difficulties, and the accuracy achievable, which will be discussed later, are shown in Table 10-1.

10.2.2 Purity of the Thermometric Substance, Interchangeability

The functional relationship relating the thermometric parameter to temperature is valid only for each well defined thermometric substance. Therefore, the purity in the case of pure elements (eg, platinum; see also Chapter 3), or the composition in the case of alloys (eg, rhodium-iron, gold-cobalt), must be precisely specified for the interchangeability of sensors of nominally the same type to be preserved. In addition, the adjustment of the nominal resistance value is of great importance. This is an important parameter in industrial applications, and is a weak point for most cryogenic thermometers, including the platinum type, which is one of the best in this respect above room temperature (see Chapter 3). For solid-state thermometers, interchangeability is even more difficult to achieve, as it depends on the uniformity and reproducibility of the doping. An insufficient degree of interchangeability can indirectly limit the overall accuracy of the measurements owing to insufficient flexibility of the measuring instrument in accounting for the spread of the sensor characteristics.

Table 10-3. Accuracy requirements for the measuring instrument

Sensor	Thermometric parameter (max.)	Sensitivity**	Max. accuracy (absolute and relative) required for: 1 K	0.1 K	0.01 K	0.001 K
Thermocouple	emf (mV)	(μV/K)	(μV and (% or ppm))			
Type T	6	10–40	10(0.6)	1(0.06)	0.1(60 ppm)	–
Au + 0.07 at % Fe/KP	5	12–22	12(0.4)	1(0.04)	0.1(40 ppm)	–
Au + 2.1 at % Co/KP	14	5–70	5(0.5)	0.5(0.05)	50 nV(50 ppm)	–
Resistance	R_0 (Ω)	(Ω/K)	(mΩ and (% or ppm))			
Platinum	25	0.004–0.1	4(0.4)	0.4(0.04)	0.04(40 ppm)	4 μΩ (4 ppm)
Platinum (ind.)*	100	0.08–0.8	80(8)	8(0.8)	–	–
Rh-Fe	100	0.15–0.6	150(0.4)	15(0.04)	1.5(40 ppm)	0.15 (4 ppm)
CLTS	290	0.23	230(0.1)	23(0,01)	2.3(10 ppm)	–
	R (4.2 K) (Ω)					
Germanium	1000	0.1–1000	100(10)	10(1)	1(0.1)	0.1(0.01)
Carbon-glass	1000	0.1–1000	100(10)	10(1)	1(0.1)	–
Diode	V (4.2 K) (V)	(mV/K)	(mV and (% or ppm))			
GaAs	1.5	0.6–3	0.6(0.05)	0.06(50 ppm)	60 μV(5 ppm)	–
Silicon	2	2–60	2(0.2)	0.2(0.02)	0.02(20 ppm)	–
Capacitance	C (4.2 K) (pF)	(pF/K)	(pF and (%))			
	15.000	(0)–300	150(1)	15(0.1)	1.5(0.01)	–

* $T > 20$ K
** absolute value in the range indicated in Table 10-1

10.2.3 Measurability

Today a sensor cannot be simply evaluated for its use in manual measurements. It is important also to consider, when selecting a type, the requirements set with respect to the associated electronic instrumentation. In this respect, measuring a resistance or a voltage is simpler than measuring a capacitance, and electronic or resonant thermometers (noise, quartz, quadrupole) require specialized and costly equipment; for this reason they are not considered here (see Chapter 6 for their general treatment). In Table 10-3 the requirements for relative and absolute accuracy of the measuring instrument are reported for each type of sensor; some of the best accuracy levels cannot actually be obtained for other reasons, as indicated in Table 10-1.

10.2.4 Effect of Magnetic Fields

The relationship between the thermometric parameter and temperature may also be affected by other physical parameters, which are to be considered as spurious. The thermometers must show as little sensitivity to them as possible. However, some of these effects are intrinsic to the thermometric material and cannot be avoided or minimized, unless another material is used. This applies to the effect of magnetic fields, which is a situation very specific to cryogenics because most high-field magnets are of the superconducting type. In addition, the thermometers cannot generally be considered as isotropic in this respect. Therefore, the direction of the field lines with respect to the axis of the element must be known for the effect of the field to be specified.

In Table 10-4, the change relative to the calibration in zero field is reported as an equivalent temperature change; this method of indication is not entirely correct when the effect of the magnetic field on the thermometer is very large, because in this instance it is also nonlinear as the thermometer sensitivity is changed by both temperature and magnetic field. This has been observed, for example, with platinum-cobalt thermometers: the magnetic field affects the sensitivity so much that below 4 K and above 5 T it decreases to zero [3]. The change in the thermometer sensitivity, due to the effect of the magnetic field, is more or less present in all instances, but is generally not reported in the literature [4]. Some cryogenic thermometers have been specifically developed for minimizing the calibration change in high magnetic fields, including capacitance, carbon glass and compensated germanium resistance thermometers. Sometimes the shape of the resulting characteristics is complicated, lowering the possibility of achieving a highly accurate fit, limiting the overall accuracy obtainable from the thermometers.

10.2.5 Reproducibility

The thermometric material must be mounted in a device to form a sensor. This fabrication process usually determines most of the overall quality of the sensor, especially its *reproducibility* on thermal cycling. In cryogenic sensors, the main source of instability comes from the mechanical effects induced by the different coefficients of thermal expansion of different parts, which produce a change each time the sensor is cooled. Strain is introduced in the wire-wound or film types of resistance thermometers and in thermocouples; in the com-

Table 10-4. Magnetic field-dependent temperature errors

Thermometer type	T/K	Magnitude of relative temperature error $\lvert \delta T/T \rvert$ (%) for certain values of B: magnetic flux density B (in T)					Notes	Reference
		1	2.5	8	14	19		
Carbon radio resistor (typical)	0.5		2–4	5–13	7–20		(a)	[4]
	1.0		2–4	6–15	9–25			
	2.5		1–5	6–18	10–30			
	4.2		1–5	5–20	10–35			
KVM carbon composite resistor	2.4		3	8 (at 5 T)			(b)	[5]
	4.2		1.5	5 ”				
	10		0.4	1.4 ”				
	20		0.1	0.4 ”				
	80		<0.01	<0.01 ”				
Carbon glass resistor	2.2		0.1	1.5	3	4	(c)	[6]
	4.2		0.5	2	5	7		
	10		0.2	1.1	3	4		
	20		<0.01	0.02	0.03	0.13		
	45		0.07	0.5	1.3	2		
	190		0.04	0.3	1	1.7		
Germanium resistor	2.0		8–10	60			(d)	[4]
	4.2		5–20	30–55	60–70			
	10		4–15	25–60	60–75			
	20		3–20	15–35	50–80			
	70		3–10	15–30	25–50			
Specially doped germanium resistor	2.0		<0.2	0.5	1		(e)	
	4.2		<0.2	1.5	2			[5, 7]
	10		<0.2	1.5	2			[8] [15]
	20		<0.5	3				
	30		<0.5	5				
	80		0.15	0.5				
Platinum resistor (industrial)	10		100				(f)	[3, 6]
	20		2–8	25				[9]
	40		0.5	3	6	9		
	87		0.04	0.4	1	2		
	190		<0.01	0.06	0.2	0.3		

(a) Relative temperature error monotonic in B and T and always positive. Higher nominal resistance units have smaller error. Radio resistors.

(b) Resistor specifically developed as a thermometer.

(c) Behavior similar to radio resistors, with improved stability to thermal cycling. Negative error below about 20 K.

(d) Not recommended except in very low fields. Strong orientation dependence.

(e) Specially developed for high magnetic fields. Orientation dependence only above 20 K. Stability as type (2.5) in Table 10-1.

(f) Small orientation dependence. Useful only for $T > 30$ K.

Table 10-4. continued

Thermometer type	T/K	Magnitude of relative temperature error $\lvert \delta T/T \rvert$ (%) for certain values of B: magnetic flux density B (in T) 1	2.5	8	14	19	Notes	Reference
Platinum resistor (pure)	20		20	100			(f)	[3, 6]
	40		<1	5	10			[9]
	87		<0.5	1	2			
Rhodium-iron resistor	2.0		22				(g)	[3, 6]
	4.2	2	11	40 (at 6 T)				[10]
	20	0.8	4	10 (at 5 T)				
	40		1.5	12	30	40		
	87		0.2	1.5	4	6		
	190		0.03	0.3	0.9			
Platinum-cobalt resistor	2.0	25	30				(h)	[3, 11]
	4.2	8	3	40 (at 5 T)				
	10	1	<0.1	12 ”				
	20	0.2	1	3.5 ”				
	30	0.2	0.3	3.5 ”				
CLTS linear resistor	4.2	20					(i)	[12]
	10	17	100					
	30	5	30					
$SrTiO_3$ capacitor	2.2		<0.02	<0.02	0.02		(j)	[6]
	4.2		<0.01	<0.01	0.01			
	20		<0.05	<0.05	<0.05			
	50		<0.05	<0.05	<0.05			
	87		<0.01	<0.01	<0.01	<0.01		
	190		<0.01	<0.01	<0.01	<0.01		
Silicon diode	4.2		75				(k)	[4]
	10		20	30	50			
	20		4	7	10			
	30		3	4	5			
	77		0.2	0.5	0.5			
Gallium arsenide diode	4.2		2–3	30–50			(k)	[4]
	10		1.5–2	25–40	75–200			
	20		0.5–2	20–30	60–150			
	40		0.2–0.3	4 -6	15–30			
	80		0.1–0.2	0.5–1	2–5			

(g) Little orientation dependence. Useful only in low fields.
(h) Little orientation dependence. Error negative below 12 K and 2.5 T. Useful below 3 T.
(i) Extremely high error. Useful only in very low magnetic fields, or as a magnetometer.
(j) Not useful between about 60 and 80 K because of a nonmonotonic behavior.
(k) Orientation dependent. Smaller values for junction parallel to B.

Table 10-4. continued

Thermometer type	T/K	Magnitude of relative temperature error $\mid \delta T/T \mid$ (%) for certain values of B: magnetic flux density B (in T)					Notes	Reference
		1	2.5	8	14	19		
Au + 0.07% Fe/KP thermocouple	4.2		2	10	15		(l)	[13]
	10		3	20	30			
	20		2	15	20			
	45		1	5	7			
	100		0.1	0.8				
KP/Constantan thermocouple	10		1	3	7		(l)	[13]
	20		<1	2	4			
	45		<1	<1	2			
Cu + 0.01% Fe/Cu thermocouple	4.2		2	3.5 (at 5 T)			(l)	[5]
	10		0.8	2 ”				
	20		0.6	1.5 ”				
	50		0.3	0.6 ”				

(l) Gradients in the magnetic field crossed by the wires can cause higher errors. KP = chromel P alloy.

pound types (carbon or cermet thermometers), the grains of conductive material tend to move relative to each other; in the solid-state types, problems with the connecting leads are considered to be the major source of instability. The reproducibility levels for the different commercial types are given in Table 10-1; they are realistic figures, even if some better levels are often claimed by the manufacturers.

10.2.6 Sensitivity to Spurious Parameters

A good thermometer must minimize the effect of spurious parameters which influence the thermometric parameter. Some of them have already been discussed, such as the sensitivity to magnetic fields (Section 10.2.4) and to strain (Section 10.2.5). Among others, it is worth mentioning, for cryogenic sensors, the spurious thermal emfs which are easily induced in thermocouples (see footnote b) in Table 10-1). However, if care is not taken they also develop, owing to the Seebeck effect, at the connections between the sensing element and the leads, especially in semiconducting thermometers, and may be a limiting factor for the achievement of the highest accuracy. Another limiting factor for resistance thermometers, when using an AC bridge for its measurement, is the size of the imaginary part of the actual thermometer impedance, which depends on the thermometer itself and on its connecting wires to the measuring apparatus.

10.2.7 Overheating and Mounting Errors

In cryogenic sensors these errors can be very high, unless great care is taken, because of the increasing difficulty in keeping the thermal resistance small between the sensing element and the body whose temperature is to be measured, and because the sensor is often working in vacuo. In Figure 10-2, a thermometer is shown mounted in a body B. The sensing element, with characteristics $f(T)$, is connected via two or four leads, which can be thermally anchored at temperature T_A, to room temperature; the measured T is an estimate of T_0. In general, the thermometer is inserted in the body B through a fitting F at temperature T_2 via a mounting M. All precautions must be taken so that $T_2 = T_0$. The thermometer itself cannot always be screwed to its mounting M. Especially in laboratory use, it is most often inserted in a well drilled in the mounting and some conducting medium, generally vacuum grease, is used. With dissipative thermometers, T_3 tends to be higher than T_2 because of this thermal resistance. For the same reason $T > T_3$; this is the overheating error, which can be much higher than the former, because the sensing element is thermally connected to the sensor case mainly through some gas contained in the sensor, rather than through the connecting leads. For operation in vacuo, the sensor must therefore be sealed, which is a generally unnecessary and uncommon requirement for use above room temperature. The size of the overheating error and of the heat introduced in the experiment by the sensor limits the use of some types to temperatures above 1 K. When small, it can be detected and corrected for, only with resistance thermometers, by measurement at two different current levels.

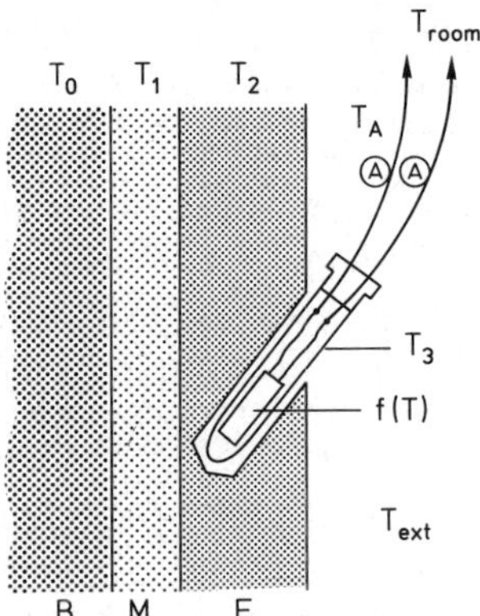

Figure 10-2.
Thermometer intended to measure the temperature T_0 of the body B. It requires, in general, a fitting F, which is connected to the body B through a mounting element M. The thermometer generally shows a thermal resistance to its fitting, so that T_3 can be different from the temperature T_0 to be measured. If the sensing element itself dissipates energy, its temperature T will be higher than T_3.

10.3 Future Trends

The goals for the improvement of cryogenic temperature sensors can be summarized as follows:

(a) availability of a single thermometer of good reproducibility over the whole cryogenic temperature range;

(b) availability of a reproducible thermometer which is also insensitive to magnetic fields up to at least 10 T;

(c) improvement of the interchangeability;

(d) lowering the cost of each unit;

(e) development of simple devices for in situ calibration or checking of the calibration stability.

Research efforts are concentrated on three different approaches. One is to develop solid-state devices matching a better interchangeability with higher sensitivity and good reproducibility, such as silicon diodes or germanium resistors additionally showing a lower sensitivity to high magnetic fields [7, 8]. Another is to develop film thermometers made of pure metals, eg, platinum, or of stable metallic alloys, eg, rhodium-iron, or of conductive oxides, eg, RuO_2 [14] for high magnetic fields. The third is to improve the technology of wire-wound thermometers, or to use it with new or recent (eg, platinum-cobalt [3, 11]) alloys.

10.4 References

[1] Bedford, R. E., Bonnier, G., Maas, H., Pavese, F., *Techniques for Approximating the International Temperature Scale* of 1990, Monograph **1**; Sèvres: BIPM 1990.

[2] Pavese, F., *Cryogenics* **14** (1974) 425–428.

[3] Pavese, F., Cresto, P., *Cryogenics* **24** (1984) 464–470.

[4] Sample, H. H., Rubin, L. G., *Cryogenics* **17** (1977) 597–606.

[5] Astrov, D. N., Abilov, G. S., Al'shin, B. I., *Meas. Tech. (USSR)* **20** (1977) 513–521.

[6] Rubin, L. G., Brandt, B. L., *Adv. Cryog. Eng.* **31** (1986) 1221–1230.

[7] Zinov'eva, K. N., Zarubin, L. I., Nemish, I. Yu., Vorobkalo, F. M., Boldarev, S. T., *Meas. Techn. (USSR)* **22** (1979) 845–847.

[8] Matacotta, F. C., Ferri, D., Giraudi, D., Pavese, F., Oddone, M., *Proc. II Nat. Cryog. Conf.;* Turin: Italian Cryogenic Society, 1984, pp. 120–125.

[9] Neuringer, L. J., Perlman, A. J., Rubin, L. G., Shapira, Y., *Rev. Sci. Instrum.* **42** (1971) 9–14.

[10] Rusby, R. L., in: *Temperature, Its Measurement and Control in Science and Industry,* Vol. 4, Plumb, H. H. (ed.); Pittsburg: ISA, 1972, pp. 865–869.

[11] Shiratori, T., Mitsui, K., Yanagisawa, K., Kobayashi, S., in: *Temperature, Its Measurement and Control in Science and Industry,* Vol. 5, Schooley, J. F. (ed.); New York: American Institute of Physics, 1982, pp. 839–844.

[12] McDonald, P. C., *Cryogenics* **13** (1973) 367–368.

[13] Sample, H. H., Neuringer, L. J., Rubin, L. G., *Rev. Sci. Instrum.* **45** (1974) 64–73.

[14] Bosch, W. A., Matho, F., Meijer, H. C., Willekers, R. W., *Cryogenics* **26** (1986) 3–8.

[15] Zarubin, L. T., Nemish, I. Y., Szmyrka-Grzebyk, A., *Cryogenics* **30** (1990), 533–537.

11 Automobiles

GERD KLEINERT and WOLFGANG PORTH, VDO Adolf Schindling AG, Schwalbach/Ts., FRG

Contents

11.1 Introduction

In automotive vehicles, various parameters and processes are monitored, eg, for direct indication or for further processing and evaluation. Sensors are used to detect the corresponding parameters and to convert them into electric signals. Exact indication and central monitoring of safety-relevant vehicle parameters and improved control behavior can be achieved by evaluation of the data obtained from the sensors. Consequently, thermal sensors play a significant role in the automotive world. Their main application is in temperature measurement in different places in the vehicle and in different media.

In addition, liquid-level sensors and volume flow meters and mass flow meters can also be realized as thermal sensors. One of the most important points is the special design of these sensors for automotive applications. The specifications for these sensors are very different to those used in laboratories or industrial installations. Depending on the type of installation, automotive sensors have to be resistant against:

vibrations
mechanical shocks
temperature shocks
salt water
sulfur dioxide
dead short-circuit
electric overload
electromagnetic interference
load dump
etc.

A load dump transient occurs in the event of a discharged battery disconnection while the alternator is generating charging current with other loads remaining on the alternator circuit at the moment of battery disconnection (eg, in the case of cable corrosion or an internal battery disconnection). In vehicles with a nominal 24 V supply voltage, peak voltages of 200 V, rise times of 10 ms and pulse durations up to 350 ms may occur.

11.2 Temperature Measurement

Temperature sensors for cars and trucks range from traditional sensors such as heavy-duty and compact switches and fast-response sensors to high-temperature-sensitive sensors. Traditional sensors are used to measure oil and water temperature in the vehicle. In combination with a warning contact, they can drive analog and digital indicators in combination with a warning light.

Table 11-1 shows a list of installation points, media and temperature ranges for different temperature sensors. It is important to note that the temperatures to be measured range from a minimum of about −40 °C up to a maximum of about 1200 °C (exhaust gas temperature). In the following we distinguish between temperature switches and temperature sensors. Temperature switches deliver only a switching signal (on/off) at a specified temperature point, whereas temperature sensors provide a continuous output signal depending on the changing temperature.

Table 11-1. Application of temperature sensors

Installation	Medium	Temperature range
Engine	Coolant	−40 to 150 °C
Oil basin	Oil	−40 to 180 °C
Gear box	Oil	−40 to 180 °C
Front bumper (outside temperature)	Air	−40 to 85 °C
Interior (inside temperature)	Air	−40 to 85 °C
Exhaust pipe	Exhaust gas	up to 1200 °C

11.2.1 Thermorelays

Bimetal strips are now the most commonly used sensors to indicate engine temperature. A simple bimetallic element consists of two inseparably joined metal layers with different coefficients of thermal expansion, so that a change in temperature causes the strip to bend. The

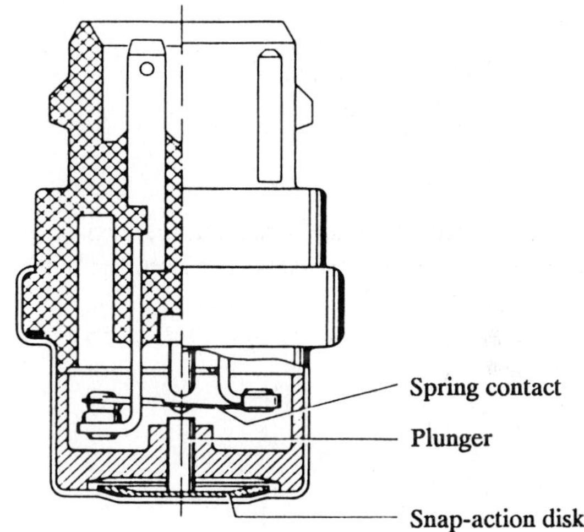

Typical values	
Operating temperature range	-40 °C to +130 °C
Measured media	coolant, engine oil
Switching points	acc. to the specification
Hysteresis	> 5 °C
Switching capacity	6 A non-inductive 5 A inductive load
Rated voltage	12 V

Figure 11-1. Thermorelay.

materials usually used for the strip with low thermal expansion (the so-called "passive" part) are alloys of Fe and about 36% by mass Ni (eg, Invar). The other component with high thermal expansion ("active" part) also consists of Fe alloys, the alloying constituents usually being Mn (about 6%) and Ni.

Figure 11-1 shows a snap-action temperature switch in which a bimetal disc spring undergoes an abrupt change of shape from concave to convex and vice versa when a defined temperature threshold is exceeded. This change in geometry is transferred to a spring contact by a plunger. This technical solution combines low-cost aspects with good reliability. One of the specialities in this area is the so-called temperature–timer switch shown in Figure 11-2. This switch is used to control the temperature-related timing of additional fuel injection during cold-start conditions in injection engines. The switch contains a bimetal leaf spring with a contact. The shape

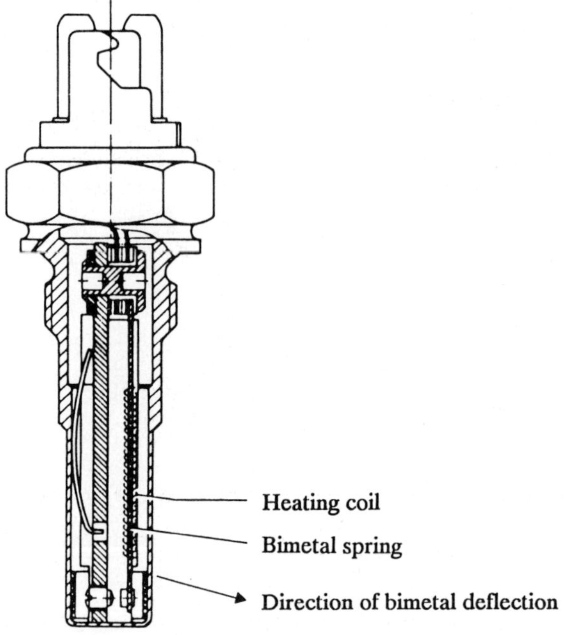

Typical values	
Operating temperature range	-25 °C to +125 °C
Switching temperature range	-10 °C to +80 °C (to be specified)
Switching delay	variable funktion of operating voltage and ambient temperature
Contact mode	break contact
Rated voltage	12 V
Switching capacity	typically 6 W (relay)
Measured media	coolant, oil

Figure 11-2. Temperature – timer switch.

of this bimetal spring is influenced by the temperature of the observed medium and the effect of an electric current flowing through the heating coil wound around the bimetal. Figure 11-3 shows the electrical block diagram of this sensor. If the temperature exceeds the threshold value, the switch will open and no additional fuel will be injected.

At low-medium temperatures and without electric heating, the bimetal contact is closed. When the heating coil is energized the bimetal will deflect and break contact. The delay time for the switch depends on the temperature of the medium and the heating current.

11.2.2 Wax Expansion Switches

Wax switches with a defined hysteresis are used for closed-loop control circuits with mechanical actuators. Coolant temperature, for example, can be controlled by a sensor of this type mounted on the radiator. If the coolant temperature is too high the sensor turns on the radiator fan until the temperature has decreased to a defined threshold.

Figure 11-4 shows the basic design of a wax expansion switch. The tank in the lower part of the housing contains a special wax which expands under the influence of the ambient temperature and, consequently, activates a snap-action switch via a plunger. Depending on the design of the snap-action switch, high switching capacities of up to 200 W are possible without the use of a separate relay.

11.2.3 Resistance Thermometers

11.2.3.1 Thermistors

NTC (negative temperature coefficient) sensors are nowadays the most commonly used temperature sensors for automotive applications. For example, the conversion of air temperatures, coolant temperatures, oil temperatures and gas temperatures into analog electrical signals for indication or computer-controlled regulation is mostly done by these sensors. These conductors change their resistance with respect to the ambient temperature. The typical thermistor characteristic has a very strong change in resistance with respect to temperature. The maximum accuracy in these applications is normally required in a narrow temperature range of about 20 K.

These sensors can provide precise temperature information at critical points. In other, less important temperature ranges, however, they are less accurate. Therefore, they can be produced at low cost.

Figure 11-5 shows one of the most common designs of these sensors. The thermistor is located at the bottom of the brass housing and is connected via a contact spring. An insulating tube is inserted to prevent short-circuits.

Another application of thermistors is shown in Figure 11-6. In this case, the sensor element is insulated from the housing and the sensor has a two-pin connector. The thermistor is placed at the top of the sensor and is embedded in a heat-conducting glue. This is a standard design for measuring the ambient temperature outside the vehicle's body.

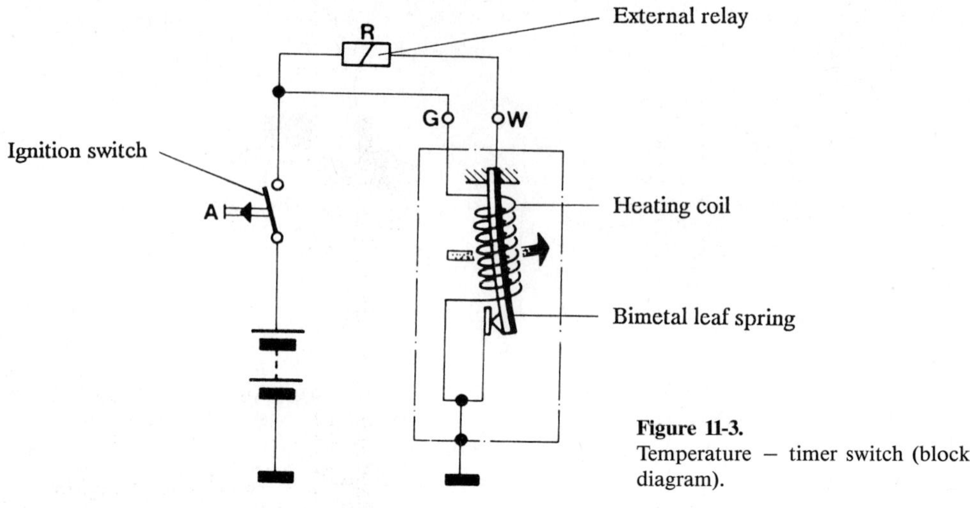

Figure 11-3.
Temperature – timer switch (block diagram).

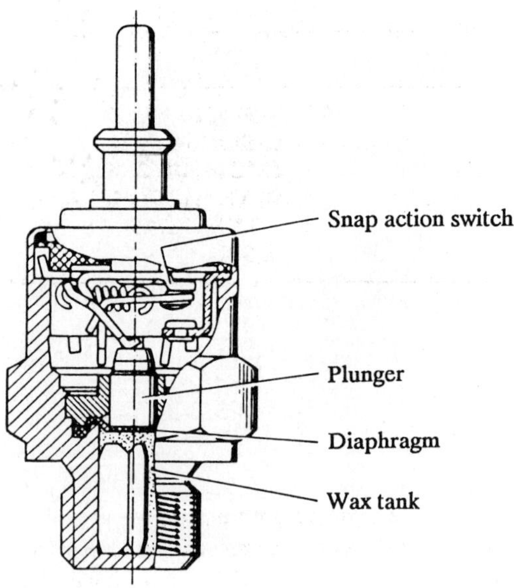

Typical values	
Operating temperature range	-40 °C to +130 °C
Measured medium	coolant, engine oil
Switching points	acc. to customer's specification in the range of 64 °C to 110 °C
Switching capacity	max. 200 W
Switching point tolerance	± 3 K

Figure 11-4. Wax expansion switch.

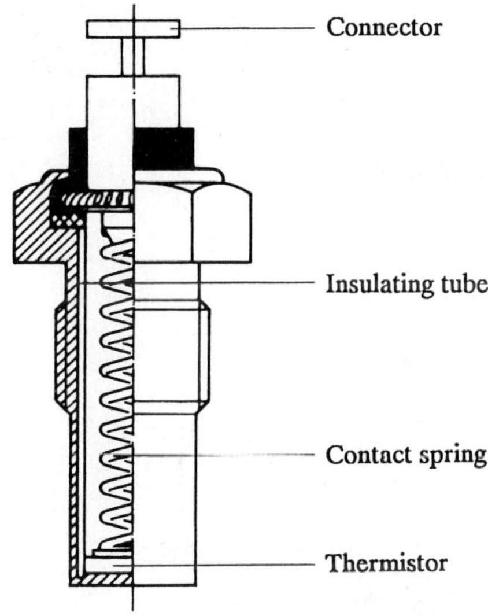

Typical values	
Operating temperature range	+50 °C to +150 °C
Measured medium	coolant, oil, air
Values of resistance	60 °C . . . 220 Ω
	90 °C . . . 83 Ω
	120 °C . . . 37 Ω
Accuracy	± 3 K

Figure 11-5. NTC sensor.

11.2.3.2 Metal Resistors

Metal resistors are used when high accuracy in a wide temperature range is required. The most common metal resistor type is the platinum resistor. Its best accuracy for automotive application is ±0.4 K at 0°C and ±0.8 K at the temperature limits of the sensor element. In addition, platinum resistors have an almost linear characteristic and excellent long-term stability. One disadvantage for automotive application is the low sensitivity because the relative change of the resistance with temperature is less than 0.4%/K.

Figure 11-7 shows a platinum thermometer for cylinder-head temperature measurement in trucks. The platinum resistor at the top of the housing is connected to the terminals via two thin wires. To insulate the two connecting wires, especially at high temperatures, the inner part of the housing is filled with alumina powder, which is compressed by ultrasonic vibration.

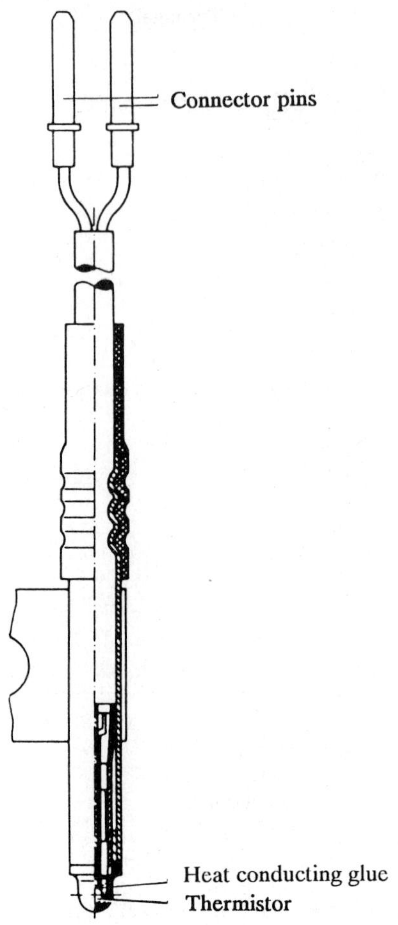

Typical values	
Operating temperature range	-40 °C to +85 °C
Storage temperature range	-40 °C to +90 °C (+100 °C max. 15 min)
Resistance at 0 °C	4076 Ω ± 26 Ω
Temperature adjustment time	3 min. maximum

Figure 11-6. Thermistor-type temperature sensor.

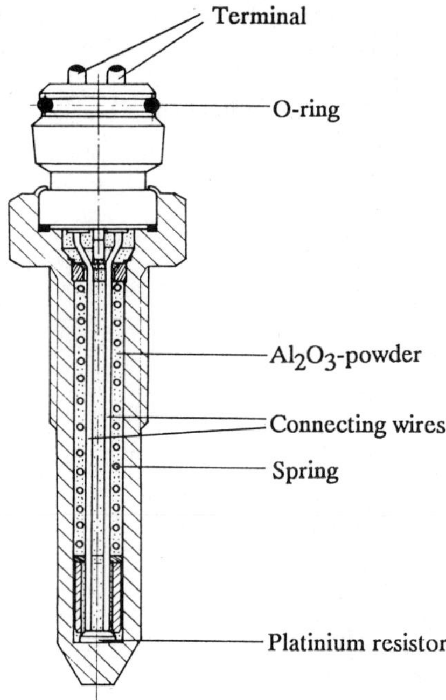

Typical values	
Operating temperature range	-40 °C to +300 °C
Application	cylinder head
Values of resistance	60°C . . . 616 Ω 150 °C . . . 785 Ω 210 °C . . . 896 Ω
Output	resistance vs. temperature almost linear
Accuracy	± 2 K

Figure 11-7. Platinum thermometer.

11.2.4 Thermocouples

Up to now, thermocouples have not played a significant role in the automotive world. Their great advantage is that they can be used in high-temperature areas which allows their application as exhaust-gas temperature sensors. Table 11-2 shows the advantages and disadvantages of thermocouples with respect to automotive applications.

With the use of an increasing number of catalysts in passenger cars, the necessity for exhaust gas measurement is growing. One main reason is the need to protect the catalyst against excess temperatures.

Figures 11-8 shows the design of an exhaust-gas NiCr-Ni thermocouple. The two thermoelements are covered with a metal tube and the sensor itself is fixed by a nut at the exhaust pipe.

Table 11-2. Thermocouples for automotive application (NiCr-Ni element)

Advantages	Disadvantages
High-temperature measurement up to 1200 °C	Low output signal t = 500 °C, E = 20.64 mV* t = 1000 °C, E = 41.269 mV* $dE/dt \approx 40\ \mu V/K$
Small size	Difficult handling
Fast response	For exact measurement a temperature reference is required
Good stability	Large tolerances ± 3 K (at 0 °C) ± 10 K (at limits)

* E = thermoelectric voltage relative to 0 °C reference
t = temperature

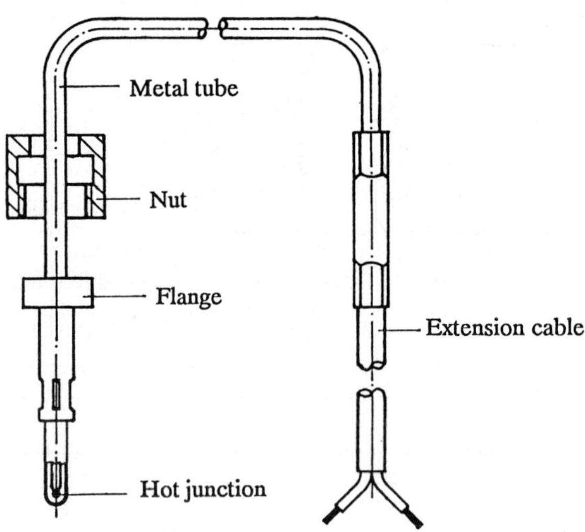

Typical values	
Operating temperature range	+300 °C to +1000 °C
Measured medium	exhaust gas
Output	typ. 4 mV / 100 K for NiCr - Ni

Figure 11-8. Thermocouple for exhaust- gas temperature.

11.2.5 Electronic Temperature Sensors

The electronic multi-level temperature sensor (Figure 11-9) consists of a thermistor and an electronic processing circuit. The combination of a thermistor with integrated electronics has the following advantages:

1. The temperature-resistance characteristics of the NTC can be converted into a linear output signal.
2. Different thresholds at different temperatures in one device are possible.
3. By laser trimming of the hybrid circuit, the thermistor tolerances are minimized and very good accuracy is provided.
4. The analog output signal can either be voltage or current.

The arrangement in Figure 11-9 shows the thermistor at the bottom of the sensor, connected via a contact spring with the PC-board. The PC-board carries a ceramic hybrid circuit which allows laser trimming. With this device, different thresholds in combination with different analog output signals, according to the customer's specification, are available. This offers the possibility of supplying different electronic modules in the vehicle by one sensor.

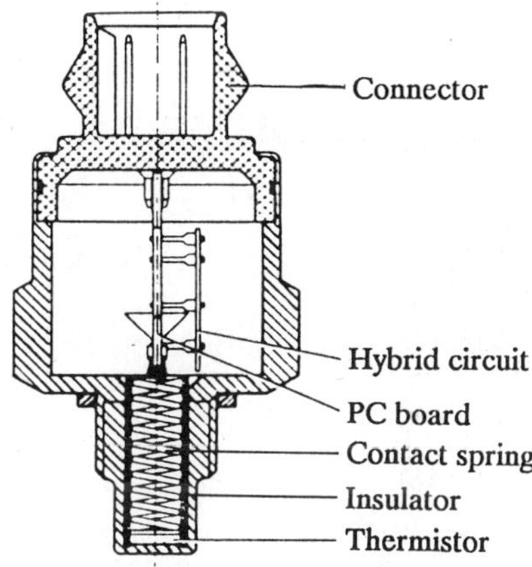

Typical values	
Operating temperature range	-40 °C to +125 °C
Supply voltage	10.8 V to 15 V
Output signal	according to customer's specification switching point tolerance ± 2.5 K switching current up to 2 A, depending on the relevant ambient temperature

Figure 11-9. Multi-level temperature switch.

11.2.6 Quartz Temperature Sensors

An accuracy of about ±1 K over wide temperature range is guaranteed by the quartz temperature sensor. The sensor is based on an oscillator quartz with a special temperature coefficient. Changes in the quartz frequency with the ambient temperature are converted by the integrated electronic into a temperature signal. The electronic SMD (surface-mounted device) parts are placed on a ceramic hybrid which can be adjusted by laser trimming.

According to the digital information of the sensor element itself, the signal calculation and the output signal are digital. This offers great advantages for the signal transportation in the vehicle. In fact, good protection against EMI (electromagnetic interferences) and RFI (radiofrequency interferences) and digital data links to different quartz sensors and/or electronic modules are possible.

Figure 11-10 shows the quartz in a standard packaging soldered to the hybrid. The lower part of the sensor housing is filled with heat-conducting glue to assure good heat transfer.

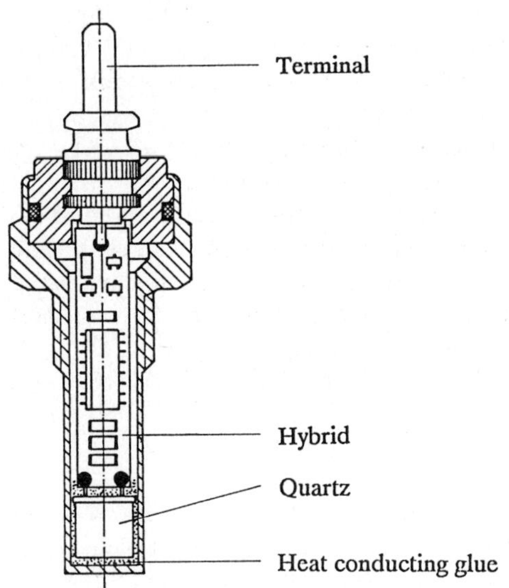

Typical values	
Operating temperature range	-40 °C to +140 °C
Measured medium	coolant, engine oil, air (ambient and outdoor)
Output signal	digital, acc. to customer's specification
Accuracy	± 1 K

Figure 11-10. Quartz temperature sensor.

11.3 Level Monitoring

Float devices are traditional measuring instruments for sensing fluid levels in automobiles. The position of the float changes the resistance of an electrical resistor. In the case of the so-called "float arm sensor", a lever arm connects the float to the sliding contact of a thick-film resistor.

The advantages of a tube-sheathed float sensor are good damping of sloshing liquid and of level variations due to lateral accelerations. The measured level is transferred by the float with a short-circuiting bar between two tense resistor wires.

In the following, two new level sensors are presented which have no movable parts because they operate on the electrothermal principle.

11.3.1 Heated Wire Principle

Metal wires normally have a positive temperature coefficient of resistivity. Temperature increases caused by Joule heating will therefore lead to a higher resistance. If the heating current is constant, the voltage drop is a direct measure of the resistance and hence of the mean temperature of the wire. The rise time of the temperature and the later thermal equilibrium value are functions of the heat transmission to the environment of the wire.

In the case of a gas or vapor, the temperature increase will be greater than that in a liquid. The oil level sensor shown in Figure 11-11 operates on this principle. A resistor filament is arranged in the probe capsule. When the ignition is switched on, a current pulse is supplied for approximately 0.5 s. The temperature and therefore resistance increase of the filament change with varying immersion depth. A subsequent electronic circuit processes the values measured to corresponding output signals. A limit comparator produces a warning signal if the oil level is too low. An analog reading of the level with the engine stationary is also possible.

11.3.2 Electrothermic Foil Sensor

The principle of the fuel level sensor shown in Figure 11-12 is similar to that of the oil-level sensor described in the previous section. The electrothermic foil sensor basically consists of thin-film resistors (preferably Ni or Cu) sputtered on a flexible polyimide foil and encapsulated in fluoroethene-propene. The thin-film elements are heated with constant current pulses and an ASIC (application-specific integrated circuit) performs the A/D conversion of the voltage drops every few milliseconds. One current pulse lasts about 500 ms. A microprocessor sums the digital values. In this way, single errors due to individual interferences become irrelevant on account of the large number of values and symmetric errors vanish completely. The measured values also include the information on the temperature of the medium being monitored. The processor computes the temperature correction from these values. The last step of the computation is to include the characteristic curves of the tank and instrument. The result is a digital value of the fuel level. Additionally, a warning signal of low levels is generated. A D/A converter finally creates the analog signal for the indicating instrument.

This electrothermal sensor has a number of advantages over conventional methods:

- long life due to the absence of mechanical movable parts or contacts;
- high media compatibility including ethanol, methanol, and water-containing fuels;
- high resolution in the "empty region";
- easy shaping of output characteristics with unusual tank shapes;
- electronics-compatible measuring system;
- universal application due to thc flexible design;
- damping improved by electronics;
- high accuracy.

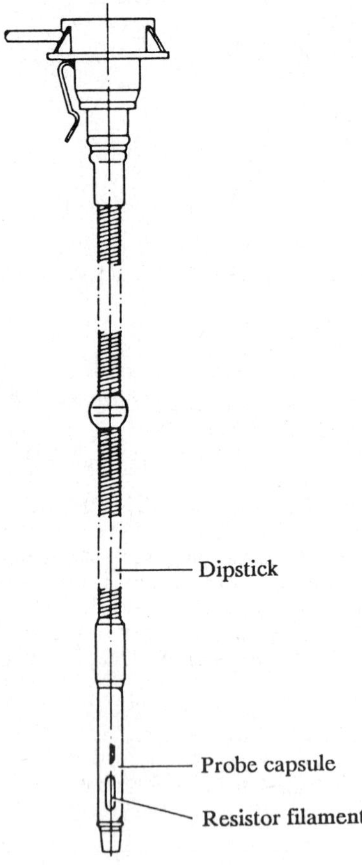

Typical values	
Operating temperature range	-30 °C to +150 °C
Measured medium	engine oil
Sensor design	acc. to customer's specification

Figure 11-11. Oil-level sensor.

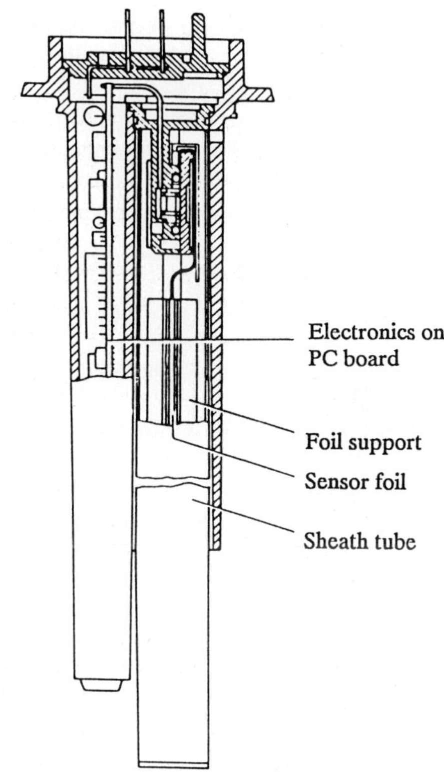

Typical values	
Operating temperature range	-40 °C to +85 °C
Supply voltage	10 V to 16 V
Output signal	current as function of filling height acc. to customer's specification
Measuring range	acc. to customer's specification

Figure 11-12. Electrothermal fuel-level sensor.

11.4 Flow Measurement

Thermal sensors for flow measurement have become a major alternative to mechanical and fluidic devices (eg, turbine meter moving vane, Prandtl's Pitot tube) in the last few years. Whereas the importance of metering of the intake air in injection engines is still increasing, the application of fuel flow meters in carburetor engines is decreasing.

Gasoline injection and Diesel engines do not need fuel flow meters for monitoring; fuel consumption can be found directly or indirectly from other quantities that are already available, eg, engine speed and valve-opening duration. As the number of cars fitted with carburetors is now decreasing, the production of fuel flow meters is stagnating.

In injection engines, metering of the intake air is necessary in order to determine the correct amount of fuel. Today these systems must be very accurate in order to meet the optimum of the following three quantities: pollutant emission, fuel consumption and engine performance. Low cost systems which use auxiliary quantities, such as manifold ambient pressure, engine speed, or angle of the throttle valve, do not fulfil this task sufficiently. Better results are obtained with direct-intake air metering systems. Today the most frequently used sensors are different versions of the moving vane sensor.

The deflection of the vane is transferred either mechanically by a lever directly to the fuel divider or electrically by a plastic potentiometer to the electronic unit. The main disadvantages of the moving vane system are fragile mechanical parts and inexact measurement of the air mass.

Thermal mass flow meters do not show these deficiencies, and they are therefore likely to acquire a larger share of the market. Because a comprehensive survey of thermal mass flow meters is given in Chapter 8, only a brief description of two examples is given in the following sections.

11.4.1 PTC Resistors for Fuel Flow Meters

PTC (positive temperature coefficient) thermistors made of a semiconducting oxide ceramic with barium titanate as the main component show a rapid increase in electrical resistivity at the so-called Curie temperature. In addition to their use as temperature sensors, their use as self-controlling constant-temperature heaters is another main application of PTC resistors. If supplied with a constant voltage, the PTC element heats up to a temperature near the Curie temperature (the Curie temperature range can be freely set from −20 to 330 °C by the raw material composition). This temperature is almost independent of the ambient temperature and the heat transmission to the environment. If the supply voltage is constant, the current is proportional to the electrical power and the heat dissipated. More heat is transferred to a streaming fluid than to a stationary fluid. Figure 11-13 shows a schematic view of an experimental fuel flow meter [1] which uses this effect. One PTC resistor, the so-called reference sensor, is situated in the dead water zone and the other is centered in the stream. It can be proved that the ratio of the two supply currents is a good temperature-compensated measurement of the throughput. The characteristic curve of the meter is non-linear.

11.4.2 Thermal Air Mass Flow Meters for Combustion Engines

The heat transmission resistance between a heated sensing element and a streaming fluid is a function of the local mass flow density. This is the principle of all heated wire and related instruments. A more detailed description is given in Chapter 9.

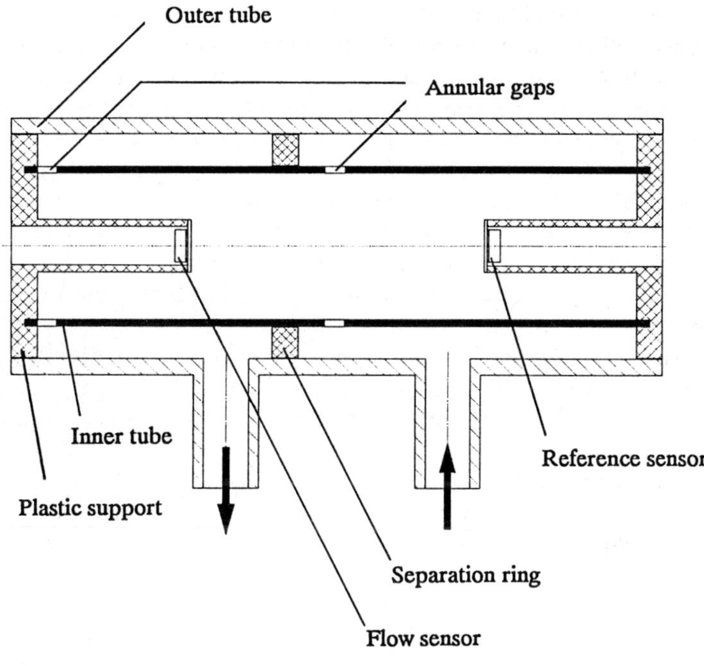

Target values	
Application	carburetor engines
Operating flow range	2 l/h to 60 l/h
Measured medium	motor gasoline
Operating temperature range	-35 °C to +85 °C
Max. fuel temperature	+65 °C
Accuracy	± 3 % of value
Pressure drop	max. 50 hPa for max. flow

Figure 11-13. Fuel flow meter with PTC resistors.

Heated wire sensors were the first thermal mass flow meters to be used in large numbers in the air intake of injection engines. The heated element is a thin, tense platinum wire. A second sensor element, measuring the air temperature for correction, consists of a Pt thin-film element on ceramic substrates. Later developments have heat foils or plates instead of the heated wire. Foil elements are made of polyimide films coated with a structured layer of Pt or Ni.

Figure 11-14 presents such a device, designed for a 2 l aspirating engine. Thin- or thick-film resistors on ceramic substrates are other possibilities for heated elements.

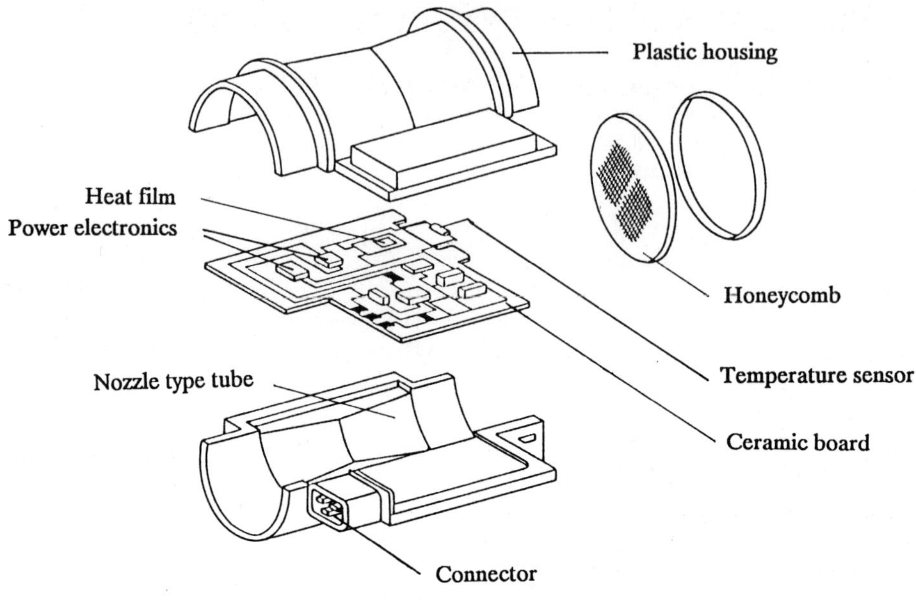

Typical values	
Operating mass flow range	10 kg / h to 430 kg / h for a 2 - liter - aspira - ting engine; range adaptable
Measured medium	atmospheric air
Output signal	acc. to customer's specification, the slope of the characteristic curve is degressive
Time constant	20 ms for a flow change of 63%
Pressure drop	less than 10 hPa at f. s.
Accuracy	± 3 p.c. of value, in a limited temperature range, otherwise correction by air temperature
Operating temperature range	-40 °C to +120 °C
Absolute pressure range	600 hPa to 1050 hPa

Figure 11-14. Thermal air mass flow meters.

11.5 References

[1] Uhl, G., *Temperaturkompensierte elektrokalorische Durchflußmessung mit selbstgeheizten Kaltleitern,* Ph. D. Thesis, Kaiserslautern, 1984.

12 Practical Problems in Process Control

Ingo Gessler, Degussa AG, Frankfurt a. M., FRG

Contents

12.1 Introduction

Temperature plays a major role in process control and figures in up to 50% of the number of sensors installed in a process unit [1]. The major fields of application are the chemical, pharmaceutical, and petrochemical industries, food and beverage domains, electrical and thermal energy-generating facilities, and thermal-conditioning installations for metal smelters as well as textile plants.

In most process control loops, accuracy and reproducibility are the key parameters responsible for the satisfactory performance of the thermal sensors, whether they are of the thermocouple (TC) or resistance thermometer (RTD) type. However, though these requirements seem easy to comply with, they are not in fact independent of each other.

Reproducibility demands that the reading at the start-up of a process stays consistant for the total service lifetime, or at least until the next shut down. Not only is a reliable reading of particular interest, but a reproducible time response may be of vital importance for the proper actuating of valves or, similarly, for safety loops. The time response depends on the convective heat transfer to the bulb, which may change due to deposits on the surface of the thermowell. It also depends on the conduction of internal heat to the actual sensing element. High vibration loads or thermal cycling may lead to a shift in the internal heat conduction resistance by mechanical wear or even abrasion. The in situ exchangeability must be maintained during the process being on-line. Thus thermal and mechanical wear and corrosion may lead to the consideration of additional parameters when the selection of a thermal sensor is made. It is evident that only a thorough technical evaluation of *all* parameters will finally lead to a satisfactory performance of the thermometer in process control.

12.2 Sensor Type

The choice between RTD or thermocouple is either based on international standards or on long-term experience, or even on the internal standards of trendsetting companies (e.g., in the petrochemical industry). The thermocouple is a simple device made from either precious or non-precious metal wires, one solid point of connection called the "hot junction" of the 2 different wire materials, and twin-bore insulators bearing the wires to the socket where either the extension or compensating cables lead to the cold junction box providing the reference temperature.

The development of *metal-sheathed mineral insulated thermocouples* (MITC) with high flexibility and ruggedness, where the sensor (the hot junction) itself is protected from the harsh surroundings, leads to *tube skin thermocouples* or wedge type thermocouples for monitoring surface temperatures or to *multistage TC assemblies* with an overall length of more than 30 m. At the same time, the outer diameters were reduced considerably, speeding up the response time as well as packing 24 TCs at different measurement levels into a 1/2 inch inox tube.

Another type of TC, which consists of a type S Pt-Rh/Pt wire combination in a high purity alumina tube fixed to a high pressure flange, is used to monitor a 50 bar hydrogen gas mixture at 1000 °C (in an ammonia-reactor).

The thermocouple could be the ideal type of sensor if high temperature, ruggedness, and/or fast response were the main parameters. Sensitive processes or technology may demand more accuracy and reproducibility. In process control the major domain for RTD's is −200 to 600 °C.

The environment of an industrial thermometer offers a variety of induced diffusion processes, e. g., diffusion in the "high purity" alumina insulator, diffusion related to the vapor pressure of "stabilizing" elements such as silicon or manganese in Chromel Alumel, or the hydrogen diffusion into the metal lattice which results in the chemical reduction of oxides. These processes lead to changes in the original complex composition of the thermocouple. [2]

In comparison, the RTD requires no special care for diffusion processes for up to 400 °C. During the recalibration procedure at 0 °C, the shift in the base resistance of 100 or 500 Ω allows first a judgment on the amount of decalibration, as well as the determination of the temperature coefficient between 0 °C and 100 °C [3]. Industrial RTDs are available which remain stable within 1 °C for several thousand hours of operational life at a fixed temperature of 650 °C or which will survive more than 2500 thermal cycles (8 °C/s) between 650 and 260 °C; a start-up shift within the first 1000 cycles of 0.07 Ω of the base resistance and 0.44% of the temperature coefficient α may occur [4].

In a narrow temperature range, one or two point calibration methods offer the possibility of a trim by balancing resistors to precise ohmic values for a large series of RTDs (an application in the textile industry) [5].

On the other hand the thermocouples can only be taken from one lot, or stretching of the wire or inhomogeneous annealing may already cause slight differences in the output without the chance of correction [6]. Recalibration may be a problem due to inhomogeneities either from diffusion, partial oxidation, or internal stress in the metal lattice.

The conclusion drawn in the past when choosing between TCs and RTDs was to evaluate each according to given priorities. This still is valid; however, today, head mounted transmitters with integrated cold junction compensation and a standard output of either 4–20 mA or 0–10 V offer the opportunity to select based on performance characteristics rather than on tradition.

12.3 Assembly

The conventional method of measuring the temperature of a medium is to establish a close thermal contact between the medium and the sensor in order to build up a steady state of heat transition. The standard assembly shows a thermowell (or protection tube) welded, flanged, or screwed into the wall of a pipe or other thermal component (pressure vessel, steam header, furnace) containing the process fluid or gas. Either the RTD or the TC is inserted into the cavity of the thermowell. Simple industrial installations still show this type of construction, especially for thermocouples: bare, solid wires combined to form a hot junction and insulated by two-bore ceramic capillaries and a terminal block at the interface leading into the connection head.

Obviously this is correct for measuring temperature – when the steady state has been achieved – but dynamic processes cannot be recorded adequately, e. g., because there is an air gap between the hot junction and the inner wall of the thermowell. In order to improve the internal heat transfer (response time), the hot junction could be welded to the bottom of the thermowell. Exchangeability of the thermocouple then could be a problem; the thermowell cannot be removed while in process; also, gas diffusion might occur and attack the bare TC-wires, humidity could corrode, e. g., the type J wires, or the insulators could decrease in mechanical stability under heavily vibrating loads.

The standard practice in process industry is to use metal-sheathed inserts at predetermined lengths and 3, 6, or 8 mm diameters which fit into the thermowell performing an optimum heat transition to the sensing elements, thus offering additional mechanical protection and nearly neutral ambient measuring conditions.

Also, for obtaining solid 2-, 3-, or 4-wire connections with a 3 × 6 mm thin film 500 Ω chip desirable for producing high sensitivity and accuracy, the measuring insert setup is the best solution.

12.4 Material Selection for Thermowells

The key to effective temperature measurement is the selection of the thermowell. However, depending on the complexity of the plant or process:

a) the temperatures may vary from near absolute zero (in helium liquefiers) to more than 1000 °C (in ethylene plants),
b) the low design pressure in bulk storage tanks or 250 hPa in ammonia plants or high pressure evaporators must be considered,
c) the media may be dry, inert gases, highly corrosive gases, or liquids with solid particles (oil ashes, abrasive products) leading to erosion, corrosion or carburization, or
d) the mass flow may be at high velocities (formation of Karman Vortex) or near zero in unstirred waste-water sinks or pools.

Primarily, it is the choice of the base material for the piping or the vessel which governs the selection of a component's material as well. For pressure vessel material, low alloy steels such as 15Mo3 or 13CrMo44 are predominant in applications up to 530 °C and 560 °C, respectively; for elevated temperatures, higher alloy Cr-Mo steels with higher yield strengths such as 10CrMo9 10 (T_{max} = 590 °C) or 12CrMo19 5 (T_{max} = 600 °C) can be used, provided that the mechanical values after heat treatment are guaranteed as specified by the regulations [7] [8] [9].

However, these regulations have become a critical point of discussion between the international engineering contractors and the national authorities. Whether it is the ASME Code or the VdTÜV (Association of Technical Surveillance Boards) which has been contracted for the material selection, without the ASME stamp of approval from a qualified material manufacturer a German material with the same analysis and strength is not accepted, likewise, the German Regulations for Pressure Vessels (AD-Merkblatt W10) do not provide the possibility of using an ASME material.

The following aspects, which are valid for the selection of a material for pressure vessels, should also be taken into consideration for the thermowells:

a) extremely low environmental temperature at the site of plant operation (Alaska, Siberia, e.g., compressor stations),
b) changes between high pressure/minimum temperature and low pressure/maximum temperature in regenerating processes,
c) varying temperatures across the piping system or vessel,
d) chemical or high temperature corrosion, and
e) malfunction in connection with pressure relief of liquid gases that have higher temperatures under pressure than in the non-pressurized state.

For example, owing to their usual inability to relieve local plastic deformation without cracking or to prevent already-existing cracks from extending, non-alloy, fine grain steels with only high strength characteristics may be preclusive, while TTStE36 or 20MnMoNi55 could be the better choice (T_{min} = − 60 °C).

Taking into account their thorough pre-production welding tests, the application of special cryogenic nickel steels is necessary for the production, transport, or storage of liquified cryogenic gases because of their high toughness characteristics, e.g., X8Ni9 (T_{min} = −196 °C, yield strength = 490 N/mm^2).

Austenitic chromium-nickel steels are widely used due to their high resistance to corrosion, their oxidation resistance, their high temperature strength and their toughness even down at extremely low temperatures (T_{min} = −253 °C for X5CrNi1810).

Above 350 °C a limited application is sometimes foreseen, provided that no intercrystalline corrosion can occur, but this holds valid only if the contact with the corrosive media is expected. This possibly can be excluded in many cases, e.g., in hot gases above 550 °C. Therefore, in the construction of process furnaces up to 800 °C, the X12CrNiTi18 9 type provides a satisfactory performance.

High-alloy ferritic chromium steels are used if stress corrosion cracking results from chlorine ions or considerable sulfur contents in reducing gases (18 Cr-2Mo, German.Code 1.4521) [10] [11]. Sea-water heat exchangers make use of CuNi10Fe or CuNi30Fe or even titanium as the piping material [12].

12.5 Design of the Thermowell

If the material selection has provided sufficient information, the next step is to define the outer dimensions of the thermowell depending on the location and piping schedules or the wall thickness of the vessel. The following aspects are to be considered:

a) The *stem conduction error* should be kept to a minimum. This means the transmission of a temperature deviation from the steady state condition to the sensing element with minimum dissipation via the thermowell's cross-section or, reverse, if in the environment of the tip a heat-sink develops, the element should not be "refuelled" from the pipe-wall's heat reservoir. Therefore, a large ratio of the heat pickup surface to the outlet cross-section, a large heat conduction resistance of the assembly (e.g., using air gaps or insulating materials for thermal insulation) versus the inner walls of the assembly (stainless steel has a lower thermal conductivity than the precious metal coated tip of the insert), a low mass for the sensor tip, and a long distance between the measuring point and ambient environment should be the design elements. Standard practice is a first-order calculation of the immersion length of the thermowell and afterwards a thorough check on the different parameters to achieve a safe mechanical performance of the thermometer.

Calculation of the Immersion Length

On the assumptions that heat transfer between the well and the measured medium is by convection only, and longitudinal heat transfer along the well is by conduction only, a fairly accurate prediction of a net immersion length L with acceptably low measurement errors can be calculated [13]:

$$L = 7/\sqrt{(\alpha \times U)/(k \times q)} \ [\text{m}] \tag{12-1}$$

where

α overall heat transfer coefficient [W/m^{2}s °C],
k thermal conductivity of well material [W/ms °C],
q cross section of the well $\pi/4 \times (d_0^2 - d_i^2)$ [m^2],
d_0, d_i external and bore diameters [m], and
U circumference of the thermowell $= \pi \times d_0$ [m].

Higher measuring errors can be expected in media with low convective heat transfer (gases) or short immersion lengths.

b) the *stress calculation* under static pressure must provide the wall thickness considering safety margins and adders for corrosion (decrease of the wall thickness per year of operational lifetime) [14].

c) the *immersion length* of the assembly may be limited at high media velocities by Karman Vortex-induced vibration of the stem.

In lines containing high-velocity fluids or gases, the vortex detachment from the tip of the thermowell will induce a frequency which may exceed the natural frequency of the pocket. If such a condition occurs the pocket will oscillate and is liable to snap off where the thermowell is fixed (e. g., welded into a flange). For standard probes the natural frequency may be derived from:

$$f_n = F_m \times 4.38 \times 104 \times (1/L^2) \times \sqrt{(E/d_s) \times (d_0^2 - d_i^2)} \ [\text{s}^{-1}] \tag{12-2}$$

where

d_s density of the pocket material [kg/m^3],
d_0 outside diameter [mm],

d_i inside bore diameter [mm],
E modulus of elasticity [kg/cm^2],
L length of the thermowell [m], and
F_m virtual mass factor, $\approx$ 0.9 for fluids, 1 for gases.

The shed frequency (f_s) of a cylinder is given by:

$$f_s = S \times V/d_0 \quad [s^{-1}] \tag{12-3}$$

where

V velocity of the fluid relative to the thermowell [m/s], and
S Strouhal number (which depends on the Reynolds number and can be assumed 0.21 between Re ≥ 103 and 105).

For safety reasons the induced vibration frequency should not exceed 80% of the natural frequency of the assembly [14] [15].

Time Constant and Measuring Error

Similar calculations are possible for the determination of the time-constant τ of an assembly or for the measuring error under the actual operating conditions [14]. However, for a special application under definite heat transfer conditions a test of the assembly is recommended under a standard flow of water, e.g., 0.2 or 1 m/s, to find the convective parameter and under 2 m/s flow of air to find the conduction parameter. Both define the thermometer constant, so that, for different operation conditions, the time constant can be calculated [16].

12.6 Lagging and Electrical Connection

Once a selection has been made for the thermowell, some aspects for the other parts of the assembly should be taken into consideration. In order to keep energy losses at a minimum, the piping or the pressure vessel is covered with a thick insulation material; in some cases (nuclear power plants) additional radiation shielding is required. As the thermowell must be fed and fixed through a bore via the pressure retaining wall, a cavity must be provided in the insulation layer to allow access to the assembly/flexible cable interface. If this interface (connection head or box, connector) is placed outside to obtain ambient conditions, the lagging extension of the thermowell may become quite long, therefore an extension tube containing a separable union may be mounted to the thermowell. Especially in the last case, this part of the assembly is sensitive to additional vibrations induced from pumps or shockwaves (e.g., from valve actuation).

Table 12-1 gives the resonance frequencies of some standard assemblies [17] in Hz.

Table 12-1. The extension resonance frequencies [Hz] of the C-Ir > cast iron head, Al > alumina head, Pla > plastic head. Dimensions: diameter × wall thickness [mm], l_2 length [mm] of the lagging extension.

l_2	extension 11 × 1			extension 11 × 2			extension 14 × 2.5			remarks
	C-Ir	Al	Pla	C-Ir	Al	Pla	C-Ir	Al	Pla	
145	42	79	90	44	89	102	59	107	130	w.t. block
165	35	68	75	36	71	85	46	98	111	dto.
170	34	66	74	34	70	82	45	95	110	dto.

For heavy duty assemblies, sometimes the wall thickness of the extensions are increased considerably thus avoiding any bend and possible damage during installation; however, it must be taken into account that the heat conduction cross section is enlarged. In some TC's installations, connectors of the same material as the legs of the TC are used in order to prevent parasitic em fs at the terminal blocks, generated by a non-uniform temperature distribution in the connectors (e.g., one side subjected to additional thermal radiation).

12.7 Thermowell and Assembly in Hazardous Locations

When temperature sensors are installed in areas with possible hazards due to a combination of an explosive gas or vapor and air (a process vessel, storage tank, or pipe), adequate means for protection must be provided. Figure 12-1 shows a possible area classification: a thermowell being immersed into a liquid, while part of it is also subjected to the vapor dome inside the

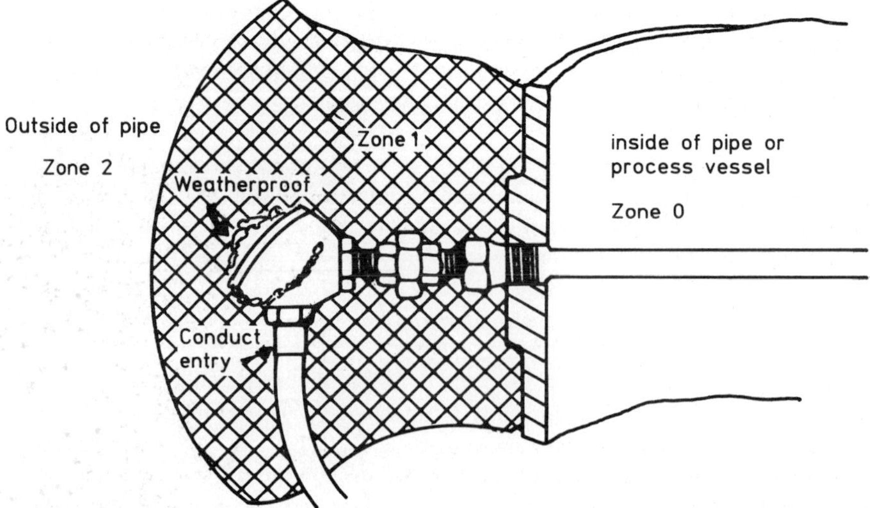

Figure 12-1. Typical thermocouple installation in a hazardous area.

closed storage tank. According to IEC 79-10, the classification is "Zone 0", in which "an explosive gas-air mixture is continously present or present for long periods." Intrinsic safety (i) Ia will be the mandatory form of protection.

The area classification outside can be evaluated differently: the thermowell gland be welded up and made leak-proof; if this has been substantiated, then direct transition to "Zone 2" and certification as type (n) of the weatherproof head (IP 54 or higher) with anti-vibration type terminals is possible. However, if the risk of a light leak during the whole life of the installation cannot be excluded or there are other possible leaks as such from adjacent control valves or sampling points in the closer surrounding, the area classification will be more on the "Zone 1" side: "the presence of an explosive atmosphere is likely to occur under normal operating conditions". The different possibilities are listed in Table 12-2.

Table 12-2. Possible Area Classifications.

ZONE	0	1		2	SAFE AREA
Explosive Atmospheres					
Presence of Explos. Atm	Continuously	Likely to occur in normal operation		Not likely	Practically never
Sources of Ignition					
VDE εX	i, s	d, f, o,	e, i, s	VDE 0165 § 22	
IEC Ex	i_a, s	d, p, o, q	e, i_b	(n)	
CENELEC EEX	i_a	d, p, o, q,	e, i_b	(n)	
IEC Marking	0 Ex ...	1 Ex d IIB T3		2 Ex n ...	–

i: intrinsic safety, d: flameproof enclosure, e: increased safety, p: pressurized enclosure, s: special protection, o: oil immersion, f: dust ignition enclosure, q: sand encapsulation

The extent of applicable protection for thermocouples is still under discussion, despite the fact that no thermocouple equipment operates under high voltages. There may be problems determining the temperature classification owing to the possibility of circulating earth currents, but it should be sound practice that a good installation would have anti-vibration terminals, weatherproof fittings, and no large circulating currents under normal operation (fault conditions are not considered in type "n" equipment in the Zone 2 environment) [18].

In "Zone 1" the installation of intrinsically safe head-mounted transmitters (2 wire) provides adequate alternatives for RTDs and thermocouples [19].

12.8 Assemblies for Special Applications

The variety of processes makes it necessary in some cases to look out for special technical solutions. For example, the protection of steam turbines from water induction may be a problem in older central stations if cool vapor, water, or condensate enters a rotating turbine; significant damages such as failures of blades, nozzles, or thrust bearings may occur. Until now the detection of lower metal temperature has been performed by standard spring loaded thermocouple assemblies, which may shorten the time for corrective action considerably if the response time of the TC is too slow. One of the methods, tested in a water induction prevention program by EPRI (Electric Power Research Institute), was the use of a *radially activated thermowell* (Figure 12-2). The thermowell is made of 2 thermoelectric materials; a rod of material "A" forms a thermoelectric junction in the outer wall of the assembly over a 360 ° arc with the well made of material "B". Thus an optimum use of all radial heat transfer to the sensor is obtained, a situation where response time is more important than accuracy. The fast thermowell is used as a primary sensor and the monitoring system is triggered by a crossover indication. It is obtained by comparing two temperature differences between a hot and a cold reference thermowell and the primary sensor in the extraction pipe [20].

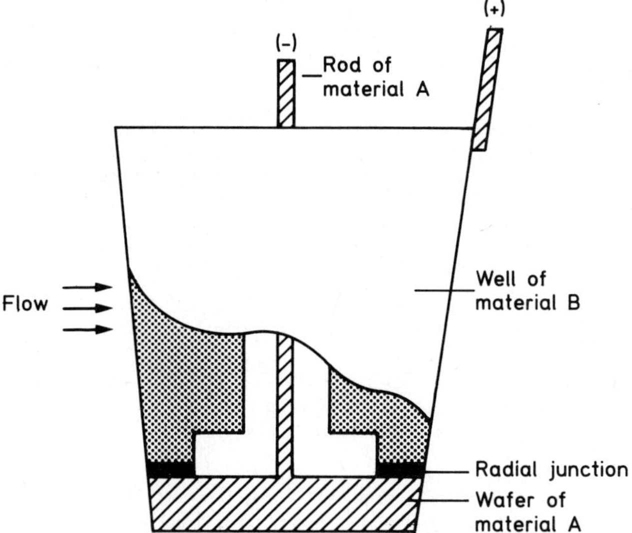

Figure 12-2. Cutaway view of new radial-junction thermowell.

Another interesting field of application is the use of *tube skin thermocouples* in primary ammonia or steam reformers or higher temperature pyrolysis furnaces. Given the alloy and the wall thickness of the tube, there is a maximum permissible operating temperature in order to warn against overheating of the feedstock at a given pressure. Most thermocouple assemblies produce temperatures higher than the actual metal and problems relate to longevity and placement. It is understood that the difficulties derive essentially from lead problems, because the leads must be taken out of the furnace with the least possible exposure to radiation and flame. The conventional TCs, a mineral insulated cable welded by means of a Hastalloy T/C pad to the tubing, still showed temperatures 15–70 °C higher than the actual temperature. Using the same weldpad type and adding a 310 SS shield with refractory filling showed satisfactory results; temperatures within 2–4 °C of the actual temperature. Of course placement of the probes at the proper side of the tube, which faces the fire, is necessary (Figure 12-3) [21].

In order to avoid corrosive effects in the cooler zone of the furnace, e. g., where the dew point is below the limit or during shut down, a second sleeve of Hastalloy X in these sensitive areas offers better performance. There is strong evidence that using Nicrosil/Nisil thermocouples instead of NiCr or type K thermocouples produces a good staying power, especially in the temperature range between 800 and 1000 °C.

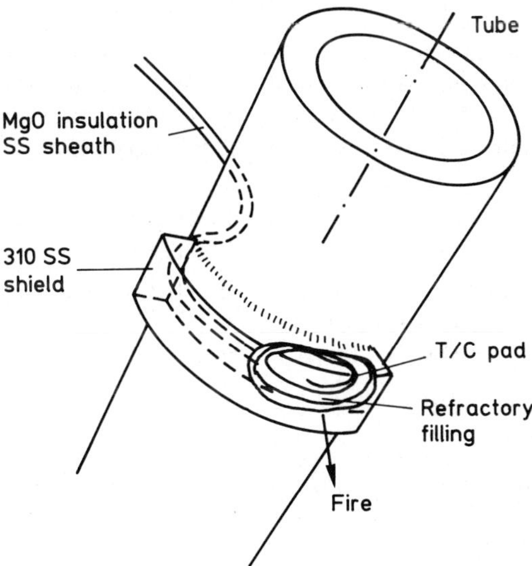

Figure 12-3.
Weldpad thermocouple.

12.9 Conclusions

Temperature sensors in the process industry provide important information of the process itself. It is clearly understood that, besides the in situ placement, a considerable number of temperature sensors are installed in the auxiliary systems, especially for surveillance and antipollution control: long mineral-insulated thermocouples for flame detection in exhaust stacks, lead- or Monel-cladded RTDs in waste or cooling water outlets, slot type RTDs in electrical generators, or special types of sensors in bearings of pumps or Diesel engines.

In many cases, some sort of type approval is requested, but it is not very often that extensive type-testing reflects the actual operating conditions. This short review encompassing only part of the theoretical background of thermometry already gives some idea of the complexity of the problem. Certainly step by step improvements are always possible; however, as somebody put it, improvements would gain wider acceptance if the heat transfer experts could explain more exactly why what ought to happen happens.

12.10 References

[1] Barthosiak, G., *Instrument & Control Systems* **11** (1976) 53.
[2] Kinzie, P. A., *Thermocouple Temperature Measurement;* New York, 1973.
[3] Carr, K. R., *An Evaluation of Industrial Platinum Resistance Thermometers* (NBS Symposium on Temperature Measurement); Washington: ISA, Pittsburgh, 1974, p. 971.
[4] Klappe, H. J. A., *Technische Messen TM* **54** (1987) 130–140.
[5] Harbert, F. C., "Automatische Regelung von Trocknungsvorgängen", *Messtechnik* **2** (1974) 52.
[6] Pollock and Finch, D. D., *The Effect of Cold Working upon Thermoelements* (NBS Symposium on Temperature Measurement); Washington: ISA, Pittsburgh, 1974, p. 237–241.
[7] *Werkstoffblätter des Verbandes der Technischen Überwachungsvereine;* Köln: Verlag TÜV Rheinland.
[8] ASME CODE, *American Society of Mechanical Engineers* (1986).
[9] Allgemeine Druckbehälter Verordnung, *AD Merkblatt* **W 10.**
[10] Fässler, K., "On the Corrosion Behaviour of a Ferritic 18Cr–2Mo Steel", *Z. Werkstofftechnik* **13** (1982) 8.
[11] Huppertz, R. H., "Selection of Materials for Pressure Vessels and Chemical Plants", *Z. Werkstofftechnik* **11** (1980) 124.
[12] Fässler, K., "Werkstoffverhalten unter dem Gesichtspunkt der kühlwasserseitigen Korrosionsbeanspruchung", *Z. Werkstofftechnik* **11** (1980) 227.
[13] Richmond, D. W., "Thermowell Selection", *ISA-87* (Preprint C. I. – 684) 445.
[14] Murdock, J. W., *Trans. ASME A* (1959) 403.
[15] ASME, *Performance Test Code* **19-3** (1974).
[16] Lieneweg, F., *Handbuch der Technischen Temperaturmessung;* Braunschweig: Vieweg, p. 200–213.
[17] Haberstroh, A., "Probleme bei der Anwendung von Berührungsthermometern", *VDI Bildungswerk BW* **5511** (1980).
[18] Towle, L. C., "The Use of Thermocouples in Hazardous Locations" (TEMPCON Conference Papers), *TP* **1008** (3/1973).
[19] Theede, W., "Temperatur Transmitter im Anschlußkopf von Thermometern", *Technische Messen* **47** (1980) 336.
[20] Benedict, R. P. and Beckmann, P., "A New Fast Response Thermowell", *ASME 83-JGPC-PTC-1* (1986) 1.
[21] Seebold, J. G., "Tube Skin Thermocouples", *Chemical Engineering Progress* **81** (1985) 57.

Index

List of Symbols and Abbreviations

Symbol	Designation	Chapter
a	absorption coefficient	5
a_m	nuclear spin substate population	2
a_v	function of molecular constants	7
A	area	
	power spectrum of sensing or reference resistor	6
	coefficient of King's law	9
$A(f)$	frequency dependent gain	6
$A_1(T)$	first acoustic virial coefficient	2
A_ε	molar polarizability	2
A_i	reciprocal of moment of inertia of ith mode of lattice vibration	2
A_{ki}	transition probability	2
A_λ	angular correlation coefficient	2
b	function of molecular constants	7
B	depth as fraction of effective depth	5
	magnetic flux density	10
	power spectrum of internal noise of amplifier chain	6
	coefficient of King's law	9
	activation energy of semiconductor electrical conductance	3
$B(T)$	second virial coefficient	2
B_F	frequency band-width	3
$B_\lambda(T)$	statistical tensor	2
c	velocity of sound	
c'_τ	elastic stiffness depending on cut of quartz plate	7
c_2	radiation constant	2, 5
c_{ijkl}	components of elastic stiffness tensor	7
c_p	specific heat capacity at constant pressure	
c_P	specific heat of the sample material	8
C	factor of reference temperature influence	4
	capacitance	4
	power spectrum of internal noise of amplifier chain	6
	Curie constant	2
	calibration constant	8
	elastic stiffness tensor	7
$C(P)$	capacitance at gas pressure P	2
$C(T)$	third virial coefficient	2

Symbol	Designation	Chapter
C_{inf}, C_{Bar}	velocity of sound in infinite solids, – bars	7
C_p, C_v	specific heat (constant pressure, constant volume)	7
d	acoustic path length	7
d_0	external diameter	12
d_{11}, d_{14}	piezoelectric strain constants	7
d_i	internal bore diameter	12
d_s	density of pocket material	12
D	coefficient	2
	diameter of the disc	8
D^*	detectivity	5
D_{eff}	effective depth	5
D_S	diameter of the sensitive area	8
e	dielectric permittivity	7
	elementary charge	
e_{11}, e_{14}	piezoelectric stress constants	7
E	thermoelectric voltage relative to 0 °C reference	11
	output of the thermopile	8
	electromotive force	4
	irradiance	5
	Young modulus	7
	energy	2
	modulus of elasticity	12
E_{AB}	Seebeck emf between materials A and B	2
E_d	displayed thermal emf	4
E_F	Fermi energy	2
E_g	energy of gas between conduction and valence band	3
E_g	electrochemical interference potential	4
E_k	energy of level k	2
E_m	nuclear substate energy	2
E_r	energy of formation of lattice vacancies	3
	thermal emf at t_r	4
E_s	secondary thermal emf	4
E_t	thermal emf for t with $t_r = 0$ °C	4
f	resonant frequency	7
	frequency	6
	function	2
Δf	electric bandwidth	5
	frequency band width	6
$f\ (T)$	sensing element characteristics	10
F	fraction of radiation	5
F	fitting	10
	freezing point	2
F_e	emissivity function	2
F_m	virtual mass factor	12

Symbol	Designation	Chapter
g	geometric factor	2
	Landé g-factor	2
G	shear modulus	2
G	gas thermometer point	2
h	thickness of quartz plate	7
	Planck constant	
	coefficient of heat transfer	2
H	thermal conductance per unit area	2
i	measuring current	2
i_0	chemical constant	2
I	current	
	radiant intensity	5
	nuclear spin	2
$\overline{I^2}$	mean square noise current	6
I_0	radiant intensity in direction normal to source	5
I_ϑ	radiant intensity in direction of ϑ	5
I_{ki}	spectral line intensity between levels k and i	2
J	total angular momentum quantum number	2
k	thermal conductivity of well material	12
k_B	Boltzmann constant	
K	function of a_v, b, T	7
	combined heat transfer coefficient	2
	capacitor compressibility	2
	differential volume coefficient of thermal expansion	2
K_0	factor of uncertainty of self-heating correction	3
K_L	thermal conductance	3
K_S	adiabatic bulk modulus	7
l	distance between transducers	7
	length	7
L	inductivity, inductance	
	radiance	5
	thickness	8
	linear expansion	7
	molar heat of vaporization	2
	length of thermowell	12
	strain tensor	7
$L(T)$	total radiance of a blackbody at temperature T	2
LO	load resonance frequency offset	7
m	mass	
	electron rest mass	3
	atomic mass	2
	exponential constant for Pr	9

Symbol	Designation	Chapter
M	mass of sample material	8
	radiant exitance	5
	mounting	10
	molecular weight of gasses	7
	mass of a portion of fluid	9
	Mach number	3
	molar mass	2
	molecular weight	2
	melting point	2
	mass flow	9
$M(T)$	total radiant exitance of a blackbody at temperature T	2
n	refractive index	5
	overtone order	7
	exponential constant for Re	9
	number of electrons per unit volume	3
	number of particles	2
	refractive index	2
	number of degrees of the emergent column of a liquid-in-glass thermometer	2
	number of differential thermocouples	8
N	number of atoms or molecules in a portion of fluid	9
	number of moles of a gas	2
N_a	Avogadro number	2
	acceptor atoms per unit volume	3
N_d	donor atoms per unit volume	3
p	fraction of specular reflectance	5
	pressure	7
	holes per unit volume	3
Δp	differential pressure	9
$p(f)$	correction factor (Planck factor)	6
P	pressure	2
	proportionality factor, Peltier coefficient	4
	fraction of specular reflectance	5
	thermal noise power	6
	holes per unit volume	3
	correction factor (Planck factor)	6
P_1, P_2	electric power in the heater elements of flow meters with two heated sensors	9
P_d	power of a heated body dissipated to moving fluid	9
P_{el}	electric power in heater element of calorimetric FM	9
$P_\lambda(\vartheta)$	Legendre polynomials	2
Pr	Prandtl number of a fluid	9
q	heat flow, heat flux	
	cross section of well	12
	heating power per unit length	9

Symbol	Designation	Chapter
q_{zz}	component of electric field gradient tensor	2
Q	radiant energy	5
	quality factor of oscillators	7
	heating energy	9
	quantity of heat	2
	nuclear electric quadrupole moment	2
Q'	power density	2
$\dot{Q}$	heating power	9
Q_λ	solid angle correction factor	2
Q_π	Peltier heat per second	2
Q_σ	Thomson heat per second	2
r	distance from the center	8
R	resistance, resistor	
	signal ratio	5
	universal gas constant	7
	gas constant	2
	optical pyrometer response	2
R_{100}, R_0	resistance at 100 °C, 0 °C	4
$R_{273.15\text{K}}$	resistance at 273.15 K, 0 °C	3
$R_{4.2\text{K}}$, R_{100}	resistance at 4.2 K, 100 °C	3
R_e	equivalent noise resistance	6
R_e, R_i	external (internal) radius of bulb mercury-in-glas thermometer	2
R_{HFS}	thermal resistance of the heat flux sensor (including the contact resistance)	8
Re	Reynolds number	9
s	thermometer responsivity	5
	distance of the wavefront from coordinate origin	7
$s(\lambda)$	spectral sensitivity of detector	2
S	thermometer output signal	5
	entropy	7
	pulling sensitivity	7
	spectral power density or power spectrum	6
	molar entropy	2
	Strouhal number	12
	thermal sensitivity	3
S_{14}	elastic compliance	7
S_A	absolute thermoelectric power of material A	2
S_{AB}	Seebeck coefficient of thermoelectric power of material A relative to material B	2
t	temperature	10, 11
	temperature in degrees Celsius	3
	time	
t_0	delay produced by all components except travel time	7
t_{90}	temperature (Celsius) on the ITS-90	2, 3
t_e	time error	7

Symbol	Designation	Chapter
t	thickness	7
T	temperature	8
	thermodynamic temperature in Kelvin	3
	triple point (temperature at which solid, liquid, and vapor phases are in equilibrium)	2
T_{68}	temperature (Kelvin) on the IPTS-68	2
T_{90}	temperature (Kelvin) on the ITS-90	2, 3
$T(x)$	spatial temperature distribution	9
T_t^n	temperature coefficient of order n	7
U	voltage	9
	U value	8
	circumference of the thermowell	12
U	general physical unit	2
U_λ	angular momentum deorientation coefficient	2
v	velocity	
V	voltage	
	volume	7
	velocity of fluid relative to thermowell	12
	vapor pressure point	2
$\overline{V^2}$	mean square noise voltage	2, 6
w	width	7
W	resistance ratio	2
	ratio $R(t)/R_{0.01\,°C}$	3
$W(\theta)$	γ-ray directional distribution	2
W_D	dissipated power over self-heating temperature	3
W_E	thermal conductance	3
W_J	power of Joule heating	3
x	cartesian coordinate	
X	force	7
y	cartesian coordinate	7
$y(T)$	quartic polynomial in T	2
Y	force	7
z	cartesian coordinate	7
Z	complex impedance	6
$Z(v)$	resistor impedance	2
$Z(T)$	Cragoe function	2
Z	force	7
α	temperature coefficient	
	absorptance, absorptivity	5
	index of impurity	3
	overall heat transfer coefficient	12

Symbol	Designation	Chapter
α_i	cosines of the direction of propagation	7
β	temperature coefficient	7
	activation energy of semiconductor electrical conductance	3
	pressure correction to mercury-in-glass thermometer reading	2, 7
γ	specific heat ratio	2, 7
	temperature coefficient	7
Γ	Christoffel elastic stiffness	7
	activation energy of semiconductor electrical conductivity	3
δ	salt property	2
Δ	salt property (shape dependent)	2
ε	emittance, emissivity	2, 5
	dielectric constant	2
	demagnetization coefficient	2
	modulus of elasticity	2
η	dynamic viscosity of a fluid	9
ϑ	elevation angle	5
ϑ_D	Debye temperature	2
Θ	AT-cut angle	7
Θ_R	Debye temperature	3
Θ_{SH}	self-heating temperature or error	3
λ	wavelength	7
	thermal conductivity	8, 9
$\Delta\lambda_D$	Doppler half-width	2
μ	Poisson number for metals	7
	heat-loss coefficient per unit length of capillary	9
	Fermi level	3
	expansion parameter involving C(P)	2
μ_B	Bohr magneton	2
ν	numerical value	2
	frequency	2
ξ_i	component of particle elongation	7
π_{AB}	Peltier emf of material A relative to material B	2
ρ	reflectance, reflectivity	5
	density	
	electrical resistivity	3
	specular reflectance	5
σ	Thomson coefficient	2, 7
	Stefan-Boltzmann constant	5
	electrical conductivity	3
	Stefan-Boltzmann constant	2
	stress tensor	7
τ	transmittance	5
	index for modes A, B, C	7
	measurement time (averaging time)	6
	relaxation time	3
	temperature on arbitrary scale	2
	time constant	2, 12

Symbol	Designation	Chapter
$\tau(\lambda)$	spectral transmittance	2
	individual relaxation times	3
υ	general physical quantity	2
φ	azimuthal angle	5
Φ	heat flow/flux	4, 9
	radiant flux	5
	radian phase shift	7
χ	magnetic susceptibility	2
ω	circular frequency	4
	solid angle	5
	circular frequency	7
ω_i	angular frequency of *i*th mode of lattice vibration	2

Abbreviation	Explanation
A-mode	bending mode
A/D	analog to digital
AC	alternating current
AC	chopped radiation
AD	Arbeitsgemeinschaft Druckbehälter
ADC	analog to digital converter
ASIC	application-specific integrated circuit
ASME	American Society of Mechanical Engineers
ASTM	American Society for Testing Materials
ATP	absolute thermoelectric power
B-mode	shear mode
C-mode	shear mode
CAL	continuous annealing line
CARS	coherent anti-Stokes Raman spectroscopy
CENELEC	Comité Europeen de Normalisation Electrotechnique
CG	carbon glass
CGPM	General Conference of Weights and Measures
CIPM	International Committee of Weights and Measures
CLTS	cryogenic linear temperature sensor
CMA	chromic methylammonium alum
CMN	cerium magnesium nitrate
CSIRO	Commonwealth Scientific and Industrial Research Organization
DC	direct current
DC	continuous radiation
DCC	digital cross correlat
DFP	disappearing-filament pyrometer
DIN	FRG standard
DSC	differential scanning calorimetry
DTA	differential thermal analysis

Abbreviation	Explanation
emf	electromotive force
EMI	electromagnetic interferences
EPT-76	Provisional 0.5–30 K Temperature Scale of 1976
FET	field-effect transistor
GOST	USSR standards
HFS	heat flow sensor
HIP	hot isostatic pressure
HTPRT	high temperature platinum resistance thermometer
IC	integrated circuit
ICT	CNR-Istituto di Metrologia „G. Colonnetti“
IEC	International Electrotechnical Commission
IEEE	Institute of Electrical and Electronics Engineers
IMGC	Istituto di Metrologia „G. Colonnetti“
IP	The Institute of Petroleum
IPRT	industrial platinum resistance thermometer
IPTS	International Practical Temperature Scale
IPTS 68	International Practical Temperature Scale of 1968
IR	infrared
IRAS	infrared astronomical sattelite
IRE	standards
ITS	International Temperature Scale
ITS-90	International Temperature Scale of 1990
IVDP	inside vapor deposition process
JNPT	Johnson noise power thermometer
K	designation of NiCr/Ni based thermocouples
KP	chromel P alloy
L	left quartz
LCSR	loop current step response
LH	left-handed
LNG	liquified natural gas
LTE	local thermodynamic equilibrium
MAS	manganous ammonium sulfate
MCVD	modified chemical vapor deposition
ME	monocrystalline elements
MI cable	mineral-insulated sheathed cable
MITC	metal-sheathed mineral insulated thermocouple
NBS	National Bureau of Standards, now NIST
NEP	noise equivalent power
NIST	National Institute of Standards and Technology
NPL	National Physical Laboratory

Abbreviation	Explanation
NT	noise thermometer
NTC	negative temperature coefficient
OIML	International Organization of Legal Metrology
OM	oxide mixture semiconductors
OVDP	outside vapor deposition process
PC	printed circuit
PRT	platinum resistance thermometer
PTB	Physikalisch Technische Bundesanstalt
PTC	positive temperature coefficient
R	right quartz
RFI	radio frequency interferences
RH	right-handed
rms	root mean square
RPA	remote preamplifier
RT	resistance thermometer
RTD	resistive thermal detector
S	synchronization signal
SI	Système International d'Unité
SMD	surface-mounted device
SPRT	standard platinum resistance thermometer
SQUID	superconducting quantum interference device
T	transducer
TC	thermocouple
TC-NT	combined thermocouple-noise thermometer
TERM	temperature and emissivity measurement by reflection method
TMFM	thermal mass flow meter
VDE	Verein Deutscher Elektrotechniker
VdTÜV	Vereinigung der technischen Überwachungsvereine